Atmospheric Turbulence and Air Pollution Modelling

A Course held in The Hague, 21-25 September, 1981

edited by

F. T. M. NIEUWSTADT

and

H. VAN DOP

Royal Netherlands Meteorological Institute, de Bilt

D. Reidel Publishing Company

A MEMBER OF THE KLUWER ACADEMIC PUBLISHERS GROUP

Dordrecht / Boston / Lancaster

Library of Congress Cataloging in Publication Data
Main entry under title:

Atmospheric turbulence and air pollution modelling.

(Atmospheric sciences library)
Bibliography: p.
Includes indexes.
1. Atmospheric turbulence–Mathematical models–Addresses, essays, lectures. 2. Boundary layer (Meteorology)–Mathematical models–Addresses, essays, lectures. 3. Atmospheric diffusion–Mathematical models–Addresses, essays, lectures. 4. Air–Pollution–Meteorological aspects–Mathematical models–Addresses, essays, lectures. I. Nieuwstadt, F. T. M. (Frans T. M.), 1946- . II. Dop, H. van (Han van), 1944- . III. Series.
QC880.4.T8A85 551.5'17 82-3758
ISBN 90-277-1365-6 (hbk.) AACR2
ISBN 90-277-1807-5 (pbk.)

CIP

Published by D. Reidel Publishing Company
P.O. Box 17, 3300 AA Dordrecht, Holland

Sold and distributed in the U.S.A. and Canada
by Kluwer Academic Publishers,
190 Old Derby Street, Hingham, MA 02043, U.S.A.

In all other countries, sold and distributed
by Kluwer Academic Publishers Group,
P.O. Box 322, 3300 AH Dordrecht, Holland

Reprinted with corrections.

Printed in The Netherlands

CONTENTS

PREFACE xi

SYMBOLS AND NOTATION xv

ACKNOWLEDGEMENTS xxi

1. EQUATIONS AND CONCEPTS - J.A. BUSINGER 1

1.1. INTRODUCTION 1

1.2. GOVERNING EQUATIONS 2

The Equations of Continuity 2
The Equation of State 3
Potential Temperature 4
The Equations of Motion 5
Conservation of Enthalpy 6
Conservation of Transferable Scalar Quantities 8

1.3. EQUATIONS OF THE MEAN FLOW 8

1.4. DISCUSSION OF THE BOUSSINESQ APPROXIMATIONS AND THE CONSERVATION OF ENTHALPY EQUATION 13

1.5. SUMMARY OF THE BOUSSINESQ SET OF EQUATIONS 15

1.6. THE CLOSURE PROBLEM, FIRST-ORDER CLOSURE 16

1.7. SECOND-ORDER VARIANCE AND COVARIANCE EQUATIONS 19

1.8. THE TURBULENT KINETIC ENERGY BALANCE; TEMPERATURE VARIANCE BALANCE 22

APPENDIX A. REQUIREMENTS FOR A DIVERGENCE-FREE VELOCITY FIELD 26

APPENDIX B. THE MAGNITUDE OF PRESSURE FLUCTUATIONS 30

APPENDIX C. THE ENTHALPY EQUATION FOR MOIST AIR 30

APPENDIX D. THE EKMAN SPIRAL 33

2. SIMILARITY RELATIONS, SCALING LAWS AND SPECTRAL DYNAMICS - H. TENNEKES 37

2.1. INTRODUCTION 37

2.2. ROSSBY-NUMBER SIMILARITY 39

The Surface Layer 42
The Process of Matching 43
The Constant-Stress Layer 46
The Von Karman Constant 47

2.3. DIABATIC EXTENSION OF ROSSBY-NUMBER SIMILARITY 47

2.4. MONIN-OBUKHOV SIMILARITY IN THE SURFACE LAYER 50

2.5. SCALING OF TURBULENCE QUANTITIES IN THE SURFACE LAYER 53

2.6. SCALING OF TURBULENCE OUTSIDE THE SURFACE LAYER 56

2.7. CORRELATION FUNCTIONS AND SPECTRA 60

2.8. INERTIAL SUBRANGES 64

3. BOUNDARY-LAYER MODELING - J.C. WYNGAARD 69

3.1. THE CALCULATION OF BOUNDARY-LAYER STRUCTURE 69

3.2. ENSEMBLE-AVERAGE MODELS 72

First-Order or Eddy-Diffusivity (K) Closure 72
Second-Order Closure 77

3.3. VOLUME-AVERAGE MODELS 97

Large-Eddy Models 98
Other Volume-Average Models 105

4. OBSERVED CHARACTERISTICS OF THE ATMOSPHERIC BOUNDARY LAYER - S.J. CAUGHEY 107

4.1. INTRODUCTION 107

4.2. CONVECTIVE BOUNDARY LAYER 110

Boundary-Layer Development 110
Spectra of the Velocity Components 114
Spectrum of Temperature 119
Cospectra of Heat Flux and Stress 121
Entrainment 127
Variances, Dissipation Rates and Structure Parameters 131
Turbulent Kinetic Energy Budget 137

4.3. STABLE BOUNDARY LAYER 139

General Characteristics of the SBL 139
Waves and Turbulence 143
Turbulence Spectra in the Stable Surface Layer 148
Turbulence Behavior through the SBL Depth 150
Depth of the SBL 156

4.4. CONCLUDING REMARKS 156

5. DIFFUSION IN THE CONVECTIVE BOUNDARY LAYER - R.G. LAMB 159

5.1. INTRODUCTION 159

5.2. FORMULATION OF A LAGRANGIAN DIFFUSION MODEL 166

5.3. NUMERICAL SIMULATIONS OF NON-BUOYANT MATERIAL DIFFUSION AND COMPARISONS WITH OBSERVATIONS 175

Results 176
Comparison of the Model Results with Observations 183

5.4. THE STRUCTURE OF TURBULENCE IN THE CONVECTIVE BOUNDARY LAYER 191

5.5. FORMULAS FOR APPLICATION 197

5.6. DISPERSION OF BUOYANT EMISSIONS IN A CONVECTIVE BOUNDARY LAYER 206

6. DIFFUSION IN THE STABLE BOUNDARY LAYER - J.C.R. HUNT 231

6.1. INTRODUCTION 231

6.2. BASIC IDEAS ABOUT MOLECULAR AND FLUID ELEMENT MOTION AND PROBABILITY DISTRIBUTIONS 232

6.3. TURBULENT DIFFUSION IN IDEALIZED FLOWS 236

Marked Fluid Elements in Unstratified Turbulence Away from Boundaries 236
Unidirectional shear flow 238
Straining flow 243
Flux gradient relations - when are they likely to go wrong? 244
Diffusion in Stably-Stratified Turbulence 247

6.4. TURBULENCE DIFFUSION IN THE STABLY-STRATIFIED ATMOSPHERIC BOUNDARY LAYER 253

Some Properties of the Stably Stratified Atmospheric Boundary Layer 253
Mean velocity profile 254
Mean temperature profile 254
Vertical turbulence and heat flux 254
Horizontal components of turbulence 255
Turbulence at heights of 50 to 300 m 256
Diffusivities and temperature fluctuations 256
Elevated Source above the Surface Layer 258
Sources in the Surface Layer 260
Elevated sources in the surface layer ($t < T_L$) 260
Elevated sources in the surface layer ($t > T_L$) 262
Comparison of vertical diffusion from ground-level and elevated sources 269

6.5. CONCLUDING REMARKS 271

Concentration Distributions 271
Complex Atmospheric Conditions 272
Topography 272

7. APPLICATIONS IN AIR POLLUTION MODELING - S.R. HANNA 275

7.1. INTRODUCTION 275

7.2. STATISTICAL MODELS OF DIFFUSION 277

Taylor's Statistical Theory 277
Monte Carlo Diffusion Models 280
Model description 280
Turbulent energy and Lagrangian time scales in the unstable PBL 282
Turbulent energy and Lagrangian time scales in the stable PBL 283
Turbulent energy and Lagrangian time scales in the neutral PBL 284
Results of application of the Monte Carlo model 285
The Langevin Horizontal Diffusion Model 286

7.3. IMPROVEMENTS TO THE GAUSSIAN MODEL 288

Wind Speed in the Gaussian Plume Model 288
Plume Rise Calculations 289
Estimation of σ_y and σ_z using σ_θ and σ_e 290
Determination of Stability Class 291
Revisions of Pasquill-Gifford (P-G) Sigma Curves 292

7.4. K-DIFFUSION MODELS 295

Analytical Solutions to the Diffusion Equation 295
Numerical Solutions to the Diffusion Equation 297

7.5. PROGRESS IN THE SIMILARITY THEORY OF DIFFUSION 300

7.6. RECENT SPECIAL APPLICATIONS 303

Skewness of Vertical Turbulent Velocity 304
Natural Variability of Pollutant Concentrations 306
Representativeness of Wind-Speed Observations 307

8. REPORT FROM THE PANEL DISCUSSION - L. KRISTENSEN 311

9. REFERENCES 323

AUTHORS INDEX 343

SUBJECT INDEX 350

PREFACE

The study of turbulence in the atmosphere has seen considerable progress in the last decade. To put it briefly: boundary-layer meteorology, the branch of atmospheric science that concentrates on turbulence in the lower atmosphere, has moved from the surface layer into the boundary layer itself. The progress has been made on all fronts: theoretical, numerical and observational.

On the other hand, air pollution modeling has not seen such a rapid evolution. It has not benefited as much as it should have from the increasing knowledge in the field of atmospheric turbulence. Air pollution modeling is still in many ways based on observations and theories of the surface layer only.

This book aims to bring the reader up to date on recent advances in boundary-layer meteorology and to pave the path for applications in air pollution dispersion problems.

The text originates from the material presented during a short course on Atmospheric Turbulence and Air Pollution Modeling held in The Hague during September 1981. This course was sponsored and organized by the Royal Netherlands Meteorological Institute, to which both editors are affiliated. The Netherlands Government Ministry of Health and Environmental Protection and the Council of Europe also gave support.

It is the responsibility of the Ministry of Health and Environmental Protection to carry out abatement policies concerning air pollution. To support the realization of such a policy the Ministry is well aware of the need for scientific research and development in the field of air pollution modeling, which must eventually lead to more understanding of the processes which govern the transport of contaminants. By supporting the course, the Ministry stimulates research from which the field of regulatory applications will certainly benefit.

The Council of Europe supports this course because it fits in excellently with the objectives of the Council, which are (i) the stimulation of cooperation among European nations in the field of higher education and research and (ii) the promotion of relations among European universities and institutions of higher education and research. The Council realizes that improvements and new developments are always brought about by bringing together a wealth of expertise - in this case on atmospheric turbulence and air pollution modeling - much greater than any institution in any country could muster for the purpose.

The editors are indebted to all people who have contributed to the completion of this book, a few of who we would like to mention separately: Dr. Z. Sorbjan, a participant to the course, provided us with cartoons of lecturers and editors. We feel that it would be a shame not to include them. (Dr. Lamb was unfortunately unable to give his lecture in person; his cartoon was modified accordingly.) Mrs. A. Haaken was responsible for so professionally typing the manuscript. Dr. A.G.M. Driedonks and Mr. A.A.M. Holtslag did a careful reading of the proofs.

Finally, a few words about the contents. Chapters 1-4 treat turbulence in the atmospheric boundary layer from various points of view. The emphasis is on recent advances in boundary-layer meteorology. Chapters 5-6 are focussed on dispersion by turbulence. Diffusion is formulated in terms of the relevant boundary-layer parameters, which are introduced in the first chapters. New and fundamental ideas are developed on diffusion during unstable and stable atmospheric conditions. Applications to dispersion problems encountered in practice are not neglected. These are discussed in Chapter 7.

Each chapter in the book corresponds with a lecture given during the course mentioned above. These lectures were prepared and presented by eminent specialists who have provided material of a very high standard. As editors, we have tried to combine and integrate these separate contributions to a coherent text. Our purpose was to compose an integrated and complete book on atmospheric turbulence and dispersion rather than a compilation of separate lectures. For that reason we have tried to avoid the repetition of material and have used cross-references where possible. We have also adopted a uniform notation for the whole text. For all errors and inconsistencies that have remained, we take all responsibility.

F.T.M. Nieuwstadt
H. van Dop

SYMBOLS AND NOTATION

The following symbols and notations are used throughout the book. Some symbols which are of local importance only will be defined when they are introduced and are not included in this list. First, a few general remarks concerning the notations should be made:

- A Cartesian tensor notation is adopted throughout the whole text. The notation x_i means the i-component of the vector (x_1,x_2,x_3). When used in the argument of a function, e.g., $f(x_i)$, x_i represents the whole vector, so that $f(x_i)$ stands for $f(x_1,x_2,x_3)$. Repeated indices indicate summation over all coordinate directions, except in the case of Greek indices. Integration will be indicated by one $\int$ sign, regardless of the dimension of the integration. The infinitesimal integration volume, $dx_1dx_2dx_3$ will be denoted as dx_i.

- Two kinds of averages are distinguished, ensemble averages and volume averages.

  Following the usual approach, it is assumed that a time average is a good representation of the ensemble average. Therefore, we maintain for both averages the notation with an overbar. (It should be noted here that in the original papers of R.G. Lamb, the ensemble averages were denoted by brackets, $\langle \dots \rangle$.) To avoid a frequent use of overbars we indicate, in the Eulerian description, the average values by upper case symbols and fluctuations by lower-case symbols. A tilde denotes the instantaneous value or one realization of the ensemble. An example is $\underset{\sim}{\theta} = \Theta + \theta$ etc. Exceptions to this rule are formed by $\underset{\sim}{q} = q + q'$, $\underset{\sim}{\rho} = \rho + \rho'$ and $\underset{\sim}{T} = T + T'$. In the Lagrangian description an ensemble average will be indicated by an overbar and a fluctuation by a prime, e.g., $V_x = \overline{V}_x + V_x'$.

  Volume averages will be denoted by upper-case symbols with braces, $\{ \dots \}$, and fluctuations by lower case symbols with primes. A tilde is used to indicate the value at a single point. An example is $\underset{\sim}{u}_i = \{U_i\} + u_i'$.

- Absolute signs are denoted by $\|\dots\|$.

a	speed of sound
$\underset{\sim}{c}$	(instantaneous) concentration $\underset{\sim}{c} = C + c$
$c_{p,v}$	specific heat at constant pressure, volume
c_T	temperature structure parameter defined by $c_T^2 = D_T(x_i,x_j)/[(x_i - x_j)^2]^{1/3}$, where $D_T(x_i,x_j)$ is the temperature structure function
c_V	velocity structure parameter defined by $c_V^2 = D_V(x_i,x_j)/[(x_i - x_j)^2]^{1/3}$, where $D_V(x_i,x_j)$ is the velocity structure function
$C_{w\theta}$, etc	cospectrum of $w\theta$, etc
C^y	cross-wind integrated concentration, given by $C^y = \int C(x,y,z,t)\, dy$
D	molecular diffusivity of matter
$D_T(x_i,x_j)$	temperature structure function, which is defined by $D_T(x_i,x_j) = \overline{[\theta(x_i) - \theta(x_j)]^2}$
$D_V(x_i,x_j)$	velocity structure function, which is defined by $D_V{}^k(x_i,x_j) = \overline{[u_k(x_i) - u_k(x_j)]^2}$
E_u, E_v, E_w, E_θ	one-dimensional Eulerian spectra of the velocity fluctuations u, v, w and of the temperature fluctuation θ, respectively, defined by $E_\alpha(k) = 2/\pi \int_0^\infty R_E^\alpha(x) \cos(kx)\, dx$
E_L	Lagrangian spectrum
f	Coriolis parameter, $f = 2\Omega \sin\phi$ (ϕ = latitude)
f, f_i	dimensionless frequencies nz/U and nh/U, respectively.
Fr	Froude number, $Fr = U^2/g\ell$
g_i	acceleration of gravity, $g_i = (0,0,-g)$
G	geostrophic wind speed, $G = (U_g^2 + V_g^2)^{\frac{1}{2}}$
h	boundary layer height, defined in unstable conditions as the height of the first

	inversion. In stable conditions it denotes the height above which turbulence is a negligible fraction, say 5%, of the turbulence level near the surface
$h_{m,v,d}$	enthalpy of moist air, water vapor and dry air, respectively
H	effective source height; $= z^s + \Delta z_e$
k	Von Karman constant
K	general notation for the scalar eddy transfer coefficient, e.g., K_h, K_m are eddy transfer coefficients for heat and momentum.
K_{ij}	eddy transfer coefficient in tensor form
ℓ	typical turbulent length scale
ℓ_m	mixing length
L	Obukhov length, $L = -\Theta u_*^3/(gk\overline{w\theta_o})$
$L_{x,y,z}^{u,v,w}$	Eulerian integral length scale for a specific component and direction, e.g., $L_x^w = {}_o\int^\infty \overline{w(x+x')w(x')}\, dx'/\sigma_w^2(x)$
M	Mach number, $M = U/a$
n	frequency
N	temperature fluctuation destruction rate $N = \kappa\overline{(\partial\theta/\partial x_j)^2}$
$O(..)$	order of magnitude sign
$p(x_i,t;x_i',t')$	probability density function for a particle released at x_i' and t'
$\underset{\sim}{p}$	(instantaneous) pressure $\underset{\sim}{p} = P + p$
q	velocity scale
$\underset{\sim}{q}$	(instantaneous) specific humidity $\underset{\sim}{q} = q + q'$
q_*	specific humidity scale, $q_* = -\overline{wq_o'}/u_*$

$\frac{1}{2}q^2$	turbulent kinetic energy ($= \frac{1}{2}(\overline{u^2}+\overline{v^2}+\overline{w^2})$)
Q	source strength
Q_o	surface heat flux ($= \rho c_p \overline{\theta w}_o$)
Ri	Richardson number ($= g/\Theta(\partial\Theta/\partial z)/(\partial U/\partial z)^2$)
R	Reynolds number
$R_{L,E}$	Lagrangian or Eulerian correlation function, $R_L^\alpha(\tau) = \overline{V'_\alpha(x_o,t)\ V'_\alpha(x_o,t+\tau)}$
Ro	Rossby number ($= G/fz_o$)
S	Source function, $S(x_i,t)$
t	(travel) time
$\underset{\sim}{T}$	(instantaneous) absolute temperature, $\underset{\sim}{T} = T + T'$
T_*	convective temperature scale ($= \overline{\theta w}_o/w_*$)
$T_{L,E}^{u,v,w}$	time scales for velocity components in a Eulerian or Lagrangian framework, e.g., $T_L^u = {}_o\int^\infty \rho_L^u(\tau)\ d\tau$
$\underset{\sim}{u}_i$	(instantaneous) Eulerian velocity with components $(\underset{\sim}{u},\underset{\sim}{v},\underset{\sim}{w})$ and $\underset{\sim}{u}_i = U_i + u_i$
u_*	friction velocity ($= (\tau_o/\rho_o)^{\frac{1}{2}}$)
U_g	x-component of the geostrophic wind speed
V_g	y-component of the geostrophic wind speed
V_i	instantaneous Lagrangian velocity defined by $V_i(t) = dX_i(x_{io},t)/dt$, with components $V_{x,y,z}$ or $V_{1,2,3}$ and $V_i = \overline{V}_i + V'_i$
w_*	convective velocity scale, $w_* = (h\overline{\theta w}_o g/\Theta)^{1/3}$
x_i^s	position of source or release point $x_i^s = (x^s,y^s,z^s)$
$X_i(x_{io},t)$	fluid element position with components X_1, X_2, X_3 or X, Y, Z, of a fluid element with initial position $x_{io} = (x_{1o},x_{2o},x_{3o})$.

$\overline{X_\alpha^n}$	n-th moment of the concentration distribution defined by: $\overline{X_\alpha^n} = \int x_\alpha^n\, C(x_i,t)\, dx_i / \int C\, dx_i$; the first moment $\overline{X_\alpha}$ is the mean fluid element position
z_o	surface rougness length
Γ	adiabatic lapse rate (= g/c_p)
δ_{ij}	Kronecker delta
$\delta(x)$	Dirac delta function, $\delta(x) = (1/2\pi)\int_{-\infty}^{\infty}\exp(ikx)dk$
Δz	plume rise
Δz_e	effective plume rise
ε	turbulent dissipation rate $\varepsilon = \nu\overline{(\partial u_i/\partial x_j)^2}$
ε_{ijk}	permutation tensor: $\varepsilon_{ijk} = 1$ when i, j, k are an even permutation of 1, 2, 3, and -1 for an odd permutation. Otherwise it is zero
ζ	dimensionless height (= z/L)
η	Kolmogorov length scale $(\nu^3/\varepsilon)^{\frac{1}{4}}$
$\underset{\sim}{\theta}$	(instantaneous) potential temperature $\underset{\sim}{\theta} = \Theta + \theta$
θ_*	temperature scale (= $-\overline{\theta w}_o/u_*$)
κ	thermal diffusivity (= $k/\rho c_p$, where k is the thermal conductivity)
λ	wave length, U/n
λ_m	peak wavelength in turbulent spectra
μ	Monin-Kazanski stability parameter (= h/L)
ν	kinematic viscosity
$\underset{\sim}{\rho}$	(intantaneous) density $\underset{\sim}{\rho} = \rho + \rho'$
ρ_o	representative boundary-layer density
ρ_L	dimensionless Lagrangian correlation function $\rho_L^\alpha(\tau) = R_L^\alpha(\tau)/\overline{V_\alpha^2}$

$\sigma^2_{u,v,w}$ variances of the Eulerian velocity fluctuations

$$\sigma^2_{u,v,w} = \overline{u^2},\ \overline{v^2},\ \overline{w^2} \text{ respectively}$$

σ_α rms fluid element displacement (in α-direction, $\alpha = 1,2,3$) with respect to the source position,

$$\sigma^2_\alpha = \overline{(X_\alpha - x^s_\alpha)^2} = \int (x_\alpha - x^s_\alpha)^2\, C(x_i,t)\, dx_i / \int C\, dx_i$$

σ^2_θ temperature variance ($= \overline{\theta^2}$)

Σ_α rms fluid element displacement (in α-direction), with respect to the mean position

$$\Sigma^2_\alpha = \overline{(X_\alpha - \overline{X}_\alpha)^2} = \int (x_\alpha - \overline{X}_\alpha)^2\, C(x_i, t)\, dx_i / \int C\, dx_i$$

τ_o absolute value of surface stress

$\phi_{m,h,\varepsilon,w,\theta}$ dimensionless functions in the surface layer

$$\phi_m = (kz/u_*)\, \partial U/\partial z,\quad \phi_h = (kz/\theta_*)\, \partial \Theta/\partial z,$$

$$\phi_\varepsilon = kz\varepsilon/u^3_*,\quad \phi_w = \sigma_w/u_*,\quad \text{and}\quad \phi_\theta = \sigma_\theta/\theta_*$$

ω_B Brunt-Vaissala frequency, $((g\, \partial\Theta/\partial z)/\Theta)^{\frac{1}{2}}$

Ω_i earth rotation vector

ACKNOWLEDGEMENTS

We are obliged to the following for permission to reproduce illustrations:

Figures 5.10, 5.11, 5.12, and 6.9 from Atmospheric Environment 12 (1978), 14 (1978) and 15 (1981), Pergamon Press, U.K.

Figures 4.15, 4.16, 4.26, 4.27, 4.28, 4.29, 4.31, 4.32, and 4.33, from Boundary Layer Meteorology 9 (1975), 11 (1977) and 16 (1979), D. Reidel Publishing Company.

Figures 4.1, 4.3, 4.24, and 4.25, from the Meteorological Magazine 104 (1975) and 106 (1977), Meteorological Office, U.K.

Figure 7.6, from the U.K. Meteorological Office, report TDN No. 86 (1977).

Figure 7.4, from the Los Alamos Scientific Laboratory, report LA-8667-MS.

Figure 6.5 from the Journal of Fluid Mechanics, 61 (1973), Cambridge University Press, U.K.

Figure 7.3, from the Proc. of 11th NATO-CCMS International Technical Meeting on Air Pollution and its Application, Amsterdam 1980, Plenum Press, New York.

Figures 2.2, 2.3, 2.4, 2.5, 4.2, 4.4, 4.5, 4.8, 4.9, 4.11, 4.13, 4.14, 4.22, 4.23, from the Journal of the Atmospheric Sciences 28 (1971), 31 (1974), 33 (1976) and 36 (1979), American Meteorological Society, U.S.A.

Figures 5.8, 5.9, 7.1, 7.2, 7.5, 7.7, from the Journal of Applied Meteorology 17 (1978), 18 (1979), 19 (1980) and 20 (1981), American Meteorological Society, U.S.A.

Figures 5.3, 5.4, 5.5, 5.6, 5.7, 5.17, 5.18, 5.21, from preprint volume of the 4th Symposium on Turbulence, Diffusion and Air Pollution, Reno, 1979, American Meteorological Society, U.S.A.

Figures 4.6, 4.10, 4.12, 4.17, 4.18, 4.19, 4.20, 4.21, 4.34, 4.35, 4.36, 4.37, from the Quarterly Journal of the Royal Meteorological Society 100 (1974) and 105 (1979), Royal Meteorological Society, U.K.

1. EQUATIONS AND CONCEPTS

J.A. Businger

University of Washington, Seattle, U.S.A.

1.1. INTRODUCTION

When we discuss atmospheric turbulence and air pollution then we place it automatically in the setting of the planetary boundary layer. The formation of the planetary boundary layer is a consequence of the interactions between the atmosphere and its underlying surface. The surface functions as a source or sink of energy and momentum to the atmosphere. The atmosphere provides the large scale environment which may modify the surface input over a wide range of values.

The boundary layer may be broadly defined as that portion of the atmosphere where the direct effect of the surface by turbulent transfer is noticeable. In most cases where the interaction is strong this definition does not present any serious problems; e.g., when there is a strong upward heat flux the boundary layer usually extends to an inversion layer above which the atmosphere is relatively undisturbed. When the interactions are weak, the direct effect of the surface is not always easily identified, the above definition becomes less useful and pollution problems become more severe.

The boundary layer plays a crucial role in more than one way: it is the layer in which energy is transferred from the surface to the atmosphere and vice versa in the form of water vapor, heat, and momentum; it is also the layer in which almost all human and biological activities (with their consequences) take place. A knowledge of the structure of the boundary layer is therefore basic to an understanding of the atmosphere's capability to dispose of pollutants. This seems ample justification to make thorough study of the boundary layer, both theoretical and experimental. Its structure, however, is discouragingly complex. The variability of the surface (roughness, change of terrain, albedo) and the atmosphere provide an infinite variety of boundary conditions. To this the rotation of the earth adds another variable. All this complexity could be handled with large enough computers, were it not for the fact that the flow within the boundary layer is predominantly turbulent.

Turbulence is the major source of confusion, but it is also the essential ingredient which makes the boundary layer livable. It facilitates the transfer of water vapor, heat, momentum, as well as the diffusion of pollutants and other trace elements. Without turbulent transfer, the lowest part of the boundary layer, the surface layer, would soon be polluted far beyond the danger point and life would suffocate or be poisoned.

1.2. GOVERNING EQUATIONS

The set of equations that form the foundation of the theory of the boundary layer is the set that describes the motions of a viscous, compressible, Newtonian fluid in a rotating system. (It should be remembered that turbulent flow has visco-elastic, non-linear properties (Rivlin, 1957).) The derivation of these equations can be found in Landau and Lifshitz (1959) or Hinze (1975).

The Equation of Continuity

This is an expression of conservation of mass which is easily derived in the form

$$\frac{\partial}{\partial t}\underset{\sim}{\rho} + \frac{\partial}{\partial x_i}\underset{\sim}{\rho}\,\underset{\sim}{u_i} = 0 \qquad (1.1)$$

where $\underset{\sim}{\rho}$ is the density and $\underset{\sim}{u_i}$ represents the components of velocity.

Equation (1.1) may also be written in the form

$$\frac{\partial u_i}{\partial x_i} = -\frac{1}{\rho}\frac{d\rho}{dt} \,. \tag{1.2}$$

The individual terms on the left hand side are of the order of 10^{-1} - 10^{-2} s^{-1} in moderate wind conditions. On the other hand the right hand side is typically of the order 10^{-4} s^{-1} or less. Consequently, for many practical purposes

$$\partial u_i/\partial x_i = 0 \tag{1.3}$$

is a good and useful approximation. However, some caution is needed in its application as we shall see later. The limitations of (1.3) are explored in Appendix A.

The Equation of State

For the atmospheric boundary layer one can simply use the equation of state for an ideal gas. For the oceanic boundary layer this equation is slightly more complex and salinity should be included. Its essential feature is that we can express variations in density in terms of variations in temperature,

$$\rho = \rho_r \{1 - \alpha(T - T_r)\} \tag{1.4}$$

where α is the coefficient of volume expansion and T_r is the temperature at which $\rho = \rho_r$. For an ideal gas $\alpha = 1/T_r$ However, we should keep in mind that the ideal gas law,

$$p = R\,\rho\,T \tag{1.5}$$

for small deviations from the reference state (r), leads to

$$\frac{p - p_r}{p_r} \simeq \frac{\rho - \rho_r}{\rho_r} + \frac{T - T_r}{T_r} \tag{1.6}$$

so that (1.4) is strictly valid only for constant pressure. Observations in the atmospheric boundary layer indicate that indeed the left-hand side of (1.6) is an order of magnitude smaller than each of the terms on the right hand side.

Equation (1.5) is strictly valid only for dry air. In case water vapor makes the air a variable gas mixture the gas constant, R, becomes also a variable because $R = R_*/m$, where R_* is the universal gas constant and m is the mean molecular mass which is made up of dry air (m_d = 28.9) and water vapor (m_w = 18). The ideal gas law is also a good approximation for water vapor, so instead of (1.5) we may write

$$\underset{\sim}{p} = \frac{R_*}{m} \underset{\sim}{\rho} \underset{\sim}{T} = \frac{R_*}{m_d} \underset{\sim}{\rho}_d \underset{\sim}{T} + \frac{R_*}{m_w} \underset{\sim}{\rho}_w \underset{\sim}{T} \ . \tag{1.7}$$

This equation may be written with $R = R_*/m_d$ in the form

$$\underset{\sim}{p} = R \underset{\sim}{\rho} \underset{\sim}{T} \{1 + (\frac{m_d}{m_w} - 1) \underset{\sim}{q}\} \tag{1.8}$$

where $\underset{\sim}{q} = \underset{\sim}{\rho}_w/(\underset{\sim}{\rho}_d - \underset{\sim}{\rho}_w)$ is the specific humidity. The term in brackets represents the correction coefficient of the gas constant for dry air when the specific humidity is q. However, rather than correcting the gas constant for dry air, the correction is usually applied to the temperature by introducing the virtual temperature

$$\underset{\sim}{T}_v = \underset{\sim}{T}\{1 + (\frac{m_d}{m_w} - 1) \underset{\sim}{q}\} = \underset{\sim}{T}(1 + 0.61\underset{\sim}{q}). \tag{1.9}$$

Substituting (1.9) into (1.8) gives the equation of state (1.5) for moist air in the form

$$\underset{\sim}{p} = R \underset{\sim}{\rho} \underset{\sim}{T}_v \tag{1.10}$$

and an equation similar to (1.6). This way we still have three variables rather than four.

Potential Temperature

Because the pressure decreases with height, a rising parcel of air expands and a sinking parcel compresses. If vertical mixing by turbulence takes place without exchange of heat, the atmosphere is neutrally stratified and the individual parcels of air behave adiabatically.

Even under strong turbulent conditions the hydrostatic equation

$$\frac{\partial P}{\partial z} = -\rho\ g \tag{1.11}$$

is a good approximation of the vertical pressure distribution. The averages are time averages. This equation combined with the First Law of Thermodynamics for adiabatic conditions ($d\underset{\sim}{p}/d\underset{\sim}{T} = \underset{\sim}{\rho} c_p$) and the equation of state (1.5) yields the adiabatic lapse rate,

$$\Gamma = -\left(\frac{\partial T}{\partial z}\right)_{ad.} = \frac{g}{c_p} . \quad (1.12)$$

It is convenient to introduce the potential temperature, $\tilde{\theta}$, which is defined as the temperature which would result if the air were brought adiabatically to the 1000 mbar level. Therefore,

$$\tilde{\theta} = \tilde{T}\left(\frac{1000}{\tilde{p}}\right)^{R/c_p} , \quad (1.13)$$

where $\tilde{p}$ is given in mbar. Using (1.11), (1.12), (1.13) and neglecting the contribution of $\partial p/\partial z$ we see that

$$\left(\frac{\partial \tilde{T}}{\partial z} + \Gamma\right) = \frac{\tilde{T}}{\tilde{\theta}} \frac{\partial \tilde{\theta}}{\partial z} . \quad (1.14)$$

In the following equations we will use this relation frequently, especially when the operator $\partial/\partial x_i$ is involved. The ratio $\tilde{T}/\tilde{\theta}$ is almost constant because the pressure variation across the boundary layer is negligible. For all practical purposes we have $\tilde{T}/\tilde{\theta} = T/\Theta$. Hereafter we shall take T/Θ equal to unity. This condition is acceptable when the boundary layer is near sea level. Corrections are necessary when the equations are used for boundary-layers at high altitudes, e.g., in Tibet or over Lake Titicaca

The Equations of Motion

An expression of conservation of momentum may be written to a fair degree of accuracy in the form (Batchelor, 1967)

$$\frac{\partial}{\partial t}\tilde{\rho}\, \tilde{u}_i + \frac{\partial}{\partial x_j}\tilde{\rho}\, \tilde{u}_j\tilde{u}_i = -\frac{\partial \tilde{p}}{\partial x_i} + \frac{\partial}{\partial x_j}\Big(\mu\Big(\frac{\partial \tilde{u}_i}{\partial x_j} + \frac{\partial \tilde{u}_j}{\partial x_i}\Big) \quad (1.15a)$$

$$- \frac{2}{3}\,\mu\,\frac{\partial \tilde{u}_k}{\partial x_k}\,\delta_{ij}\Big) + \tilde{\rho} g_i - 2\,\varepsilon_{ijk}\Omega_j\,\tilde{\rho}\,\tilde{u}_k$$

where μ is the dynamic viscosity, Ω_j is the angular vector of the earth's rotation and $g_i = (0,0,-g)$. The first term represents the local rate of change of momentum; the second term the advection of momentum; the first term on the right hand side is the pressure force, the second term the viscous force, the third term the force of gravity and the last term the Coriolis force. In turbulent flow the viscous force is usually very small in comparison to the other forces and is often omitted.

By applying (1.1) to the left-hand side of (1.15a) this may be written as

$$\frac{\partial \underset{\sim}{u}_i}{\partial t} + \underset{\sim}{u}_j \frac{\partial \underset{\sim}{u}_i}{\partial x_j} = - \frac{1}{\underset{\sim}{\rho}} \frac{\partial \underset{\sim}{p}}{\partial x_i} + \frac{1}{\underset{\sim}{\rho}} \frac{\partial}{\partial x_j} \left(\mu \left(\frac{\partial \underset{\sim}{u}_i}{\partial x_j} + \frac{\partial \underset{\sim}{u}_j}{\partial x_i} \right) - \frac{2}{3} \mu \frac{\partial \underset{\sim}{u}_k}{\partial x_k} \delta_{ij} \right) +$$

$$+ g_i - 2 \, \varepsilon_{ijk} \, \Omega_j \, \underset{\sim}{u}_k \, . \qquad (1.15b)$$

If we assume that (1.3) may be used and that μ is a constant, this equation simplifies to

$$\frac{\partial \underset{\sim}{u}_i}{\partial t} + \frac{\partial}{\partial x_j} (\underset{\sim}{u}_j \, \underset{\sim}{u}_i) = - \frac{1}{\underset{\sim}{\rho}} \frac{\partial \underset{\sim}{p}}{\partial x_i} + \nu \frac{\partial^2 \underset{\sim}{u}_i}{\partial x_j \partial x_j} + g_i - 2\varepsilon_{ijk} \, \Omega_j \, \underset{\sim}{u}_k \qquad (1.16)$$

where $\nu = \mu/\underset{\sim}{\rho}$. This equation is a good approximation for most applications in the boundary layer and will be used for determining mean, turbulent, variance and covariance equations.

Conservation of Enthalpy

The First Law of Thermodynamics which contains both the fluxes of sensible and latent heat, may be written

$$\frac{\partial \underset{\sim}{\rho} \, \underset{\sim}{h}_m}{\partial t} + \frac{\partial}{\partial x_i} \underset{\sim}{\rho} \underset{\sim}{u}_i \underset{\sim}{h}_m = \sum_{\ell} \underset{\sim}{C}_{h\ell} \qquad (1.17)$$

where $\underset{\sim}{h}_m$ is the specific enthalpy of moist air, and $\underset{\sim}{C}_{h\ell}$ represent sources and sinks of enthalpy.

The specific enthalpy, $\underset{\sim}{h}_m$, may be expressed as the weighted sum of the enthalpies of dry air, $\underset{\sim}{h}_d$, and water vapor, $\underset{\sim}{h}_v$

$$\underset{\sim}{h}_m = (1-\underset{\sim}{q}) \, \underset{\sim}{h}_d + \underset{\sim}{q} \, \underset{\sim}{h}_v . \qquad (1.18)$$

The specific enthalpies may be expressed by (see Iribarne and Godson, 1973)

$$\underset{\sim}{h}_d = c_p \underset{\sim}{T} + b_1 \quad \text{and} \quad \underset{\sim}{h}_v = c_{pv} \underset{\sim}{T} + b_2 \qquad (1.19)$$

where c_p is the specific heat at constant pressure for dry air,

c_{pv} the specific heat at constant pressure for water vapor, and b_1 and b_2 are constants reflecting energies present in the respective substances.

The right-hand side of (1.17) may be written

$$\sum_{\ell} C_{h\ell} = k\frac{\partial^2 T}{\partial x_i^2} - \frac{\partial R_{ni}}{\partial x_i} + \varepsilon + (c_{pv}T + b_2)D_q\rho\frac{\partial^2 q}{\partial x_i^2} \qquad (1.20)$$

where k is the thermal conductivity of air, R_{ni} is the i-th component of net irradiance, ε is the rate of dissipation of turbulent kinetic energy (see (1.56)) which enters as a source, and D_q is the diffusion coefficient for water vapor in air.

All terms on the right-hand side of (1.20) are usually small, except very close to the earth's surface. Especially ε may be neglected in comparison to the other terms, so we shall assume ε = 0.

Using (1.18) - (1.20) in (1.17) yields

$$\frac{\partial}{\partial t}\rho h_m + \frac{\partial}{\partial x_i}\rho u_i h_m = k\frac{\partial^2 T}{\partial x_i^2} - \frac{\partial R_{ni}}{\partial x_i} + (c_{pv}T + b_2)D_q\rho\frac{\partial^2 q}{\partial x_i^2},$$

$$h_m = T(1 - q)c_p + q\,Tc_{pv} + b_4 \quad \text{and} \quad b_4 = (1-q)b_1 + qb_2. \qquad (1.21)$$

The fact that c_p must be used instead of c_v, the specific heat at constant volume, has been demonstrated by Montgomery (1948); see also Fleagle and Businger (1980). It can be recognized intuitively when we realize that the pressure is essentially constant near the earth's surface over periods long enough to average the turbulent fluctuations (see also Appendix B).

However, it is easy to arrive at the wrong conclusion that c_v should be used instead of c_p by misapplying Equation (1.3) (Chandrasekhar, 1961). Starting with the first law of thermodynamics for dry air, to keep it simple in the form

$$\frac{dQ}{dt} = c_v\frac{dT}{dt} - \frac{p}{\rho^2}\frac{d\rho}{dt}$$

where dQ/dt is the rate at which heat is conducted into the system per unit mass. Thus ρ dQ/dt is equal to the right-hand side (r.h.s.) of (1.21) for $q = 0$, or

$$c_v\rho\frac{dT}{dt} - \frac{p}{\rho}\frac{d\rho}{dt} = \text{r.h.s. of (1.21)}.$$

Using (1.2) we find

$$c_v \rho \frac{dT}{dt} + \rho \frac{\partial u_i}{\partial x_i} = \text{r.h.s. of (1.21)}.$$

And with (1.3) we find

$$c_v \rho \frac{dT}{dt} = \text{r.h.s of (1.21)}$$

which is in contradiction with the left hand side of (1.21). (The reason why in this case (1.3) cannot be used is that it is multiplied by a large quantity which is ρ.)

Nevertheless, (1.21) may be simplified by the use of (1.1) and (1.3). If we furthermore assume that c_p and k are constant, (1.21) may be reduced to a form similar to (1.16).

Conservation of Transferable Scalar Quantities

Many scalar quantities are transported through the boundary layer, notably water vapor, ozone, carbon dioxide etc. Their transport is usually described in terms of (mass) mixing ratio c, or concentration $\chi(=\rho c)$. (for water vapor usually specific humidity, q, is used). In general we may write

$$\frac{\partial}{\partial t}\rho c + \frac{\partial}{\partial x_i}\rho u_i c = \rho D \frac{\partial^2}{\partial x_i^2} c + S \tag{1.22}$$

where S is the sum of sources and sinks of the quantity per unit mass of air, and D is the molecular diffusivity for the quantity c in air.

If c is replaced by q we have the equation of conservation of water vapor. The sources and sinks are then represented by evaporation and condensation respectively.

The Equations (1.16), (1.21) and (1.22) will be used for determining mean, turbulent, and variance equations.

1.3. EQUATIONS OF THE MEAN FLOW

Although the set of Equations (1.3), (1.10), (1.16), (1.21) and (1.22) is complete in the sense that, given adequate boundary conditions, the number of equations and unknowns is the same, the theory of turbulence and, consequently, of the planetary boundary layer is incomplete. The reason is that the equations of motion are non-linear. It is possible to find solutions of these equations which may correspond to laminar flows. In order for such a flow to be oberved in a real fluid it must be stable to small perturbations. If it is unstable, small perturbations will grow and eventually disrupt the original flow. Considerable effort has gone into the mathematical theory of hydrodynamic instability (e.g. Lin, 1945; Chandrasekhar, 1961; Monin and Yaglom, 1971) and some general understanding has been gained.

The complexity of a turbulent flow is so formidable that even if we were able to describe its detailed structure, it would be impossible to comprehend. Consequently the study of turbulent flow is directed towards describing its statistical characteristics. We assume therefore that the fluid motions can be separated into a slowly varying mean flow and a rapidly varying turbulent component. In order to determine the mean flow we must take an ensemble average, whereby the average is taken over a number of flow realizations. Such an average is rarely available to us; it is technically impractical and often impossible to obtain. We assume therefore that the flow is statistically stationary, which means the ensemble averages do not change with time and may be replaced with time averages at one point in the fluid (also called Eulerian averages). This point may be at rest, e.g., on a fixed platform such as a tower, or moving, e.g., on an airplane. The limitations that this assumption, which has been formulated as the 'ergodic hypothesis' (see Lumley and Panofsky, 1964, and Tennekes and Lumley, 1972), poses on the range of problems that can be studied, have not been fully explored. In practice one looks at averages over increasingly long time periods and determines whether or not the variance of these averages levels off with increasing sampling time. Such a leveling off will only exist if there is a well-developed 'spectral gap' in the overall variance spectrum. Over uniform terrain in the atmospheric surface layer this requirement is frequently fulfilled, but over non-uniform terrain and in the bulk of the boundary layer the spectral gap may often be missing. (For further information see Lumley, 1970; Wyngaard, 1973; Coantic, 1975; Favre, 1976.)

After these preliminaries we write the variables in a mean and fluctuating part following the convention which was initiated by Reynolds in 1895. We follow the notation for Eulerian averages outlined in the list of symbols and notations. Furthermore we shall apply the following simple rules

$$
\begin{aligned}
&\underset{\sim}{u}_i = U_i + u_i\,, \quad \underset{\sim}{\theta} = \Theta + \theta \quad \text{etc.} \\
&\overline{u}_i = \overline{\theta} = 0\,, \quad \overline{\underset{\sim}{\theta}\,\underset{\sim}{u}_i} = \Theta\, U_i + \overline{\theta u}_i\,, \quad \overline{\Theta\, \underset{\sim}{u}_i} = \Theta\, U_i\,, \\
&\overline{U_i(x_i) + U_i(x'_i)} = U_i(x_i) + U_i(x'_i)\,,
\end{aligned}
\qquad (1.23)
$$

$\overline{a\,\underset{\sim}{u}_i} = a\,U_i$ with a constant.

The governing equations (1.3), (1.10), (1.16), (1.21) and (1.22) may now be written for the average and turbulent components:

Continuity Equation

$$\frac{\partial U_i}{\partial x_i} = 0 \text{ and } \frac{\partial u_i}{\partial x_i} = 0 \,. \qquad (1.24)$$

Equation of State

$$P = R\,\rho\,T_v\left(1 + \frac{\overline{\rho' T_v'}}{\rho\,T_v}\right) \text{ and } \frac{p}{P} = \frac{\rho'}{\rho} + \frac{T_v'}{T_v}$$

However $\| p/P \| << \| \rho'/\rho \|$ and $\| T_v'/T_v \|$; and $\| \rho'/\rho \| << 1$ and $\| T_v'/T_v \| << 1$ in the planetary boundary layer, thus

$$P = R\,\rho\,T_v \quad \text{and} \quad \rho'/\rho \simeq -T_v'/T_v \qquad (1.25)$$

A discussion of the magnitude of the pressure fluctuations is given in Appendix B.

Equations of Motion

$$\frac{\partial U_i}{\partial t} + \frac{\partial}{\partial x_j}[U_j U_i + \overline{u_j u_i}] = -\frac{1}{\rho}\frac{\partial P}{\partial x_i} + \nu\frac{\partial^2 U_i}{\partial x_j^2} + g_i - 2\varepsilon_{ijk}\Omega_j U_k \qquad (1.26a)$$

$$\frac{\partial u_i}{\partial t} + \frac{\partial}{\partial x_j}[U_j u_i + u_j U_i + u_j u_i - \overline{u_j u_i}] =$$

$$-\frac{1}{\rho}\frac{\partial p}{\partial x_i} + \frac{\rho'}{\rho^2}\frac{\partial P}{\partial x_i} + \nu\frac{\partial^2 u_i}{\partial x_j^2} - 2\,\varepsilon_{ijk}\,\Omega_j\,u_k \,. \qquad (1.26b)$$

At this stage we only consider the equation for conservation of enthalpy in the case of dry air. The addition of humidity is discussed in Appendix C. We thus now substitute $\underset{\sim}{h}_d$ for $\underset{\sim}{h}_m$ in (1.17) and set $b_1 = 0$.

If we further use (1.1) and (1.3) in the right order, introduce the potential temperature (1.13) and apply $\underset{\sim}{T} \simeq \underset{\sim}{\theta}$ we obtain

$$\frac{\partial \underset{\sim}{\theta}}{\partial t} + \frac{\partial \underset{\sim}{\theta}\, \underset{\sim}{u}_i}{\partial x_i} = \kappa \frac{\partial^2 \underset{\sim}{\theta}}{\partial x_i^2} - \frac{1}{\rho \underset{\sim}{c}_p} \frac{\partial \underset{\sim}{R}_{ni}}{\partial x_i} \tag{1.27}$$

where κ the thermal diffusivity. In the following we shall neglect the radiative term $\underset{\sim}{R}_{ni}$ (see Section 2.2 on second-order closure), because its contribution to (1.27) is typically of the order 10^{-5} ^{o}C s^{-1}. The resulting equation can be decomposed in equations for the mean and fluctuating components

Conservation of Enthalpy

$$\frac{\partial \Theta}{\partial t} + \frac{\partial}{\partial x_i} [U_i \Theta + \overline{u_i \theta}\,] = \kappa \frac{\partial^2 \Theta}{\partial x_i^2} \tag{1.28}$$

$$\frac{\partial \theta}{\partial t} + \frac{\partial}{\partial x_i} [u_i \Theta + U_i \theta + u_i \theta - \overline{u_i \theta}\,] = \kappa \frac{\partial^2 \theta}{\partial x_i^2} . \tag{1.29}$$

Similarly we find for the transfer of a scalar quantity after using (1.1) and (1.3) on (1.22)

Conservation of a Scalar Quantity

$$\frac{\partial C}{\partial t} + \frac{\partial}{\partial x_i} [U_i C + \overline{u_i c}\,] = D \frac{\partial^2 C}{\partial x_i^2} + S \tag{1.30}$$

$$\frac{\partial c}{\partial t} + \frac{\partial}{\partial x_i} [u_i C + U_i c + u_i c - \overline{u_i c}] = D \frac{\partial^2 c}{\partial x_i^2} . \tag{1.31}$$

Of special interest is the case of steady state and horizontal homogeneity. In this case $\partial/\partial t = \partial/\partial x = \partial/\partial y = 0$. The equation of continuity becomes especially simple.

When we use (1.24), the result is

$$\partial W/\partial z = 0 \quad \text{or} \quad W = 0 . \tag{1.32}$$

The equations for the averaged quantities may now be written conveniently in component form. Thus we find for (1.26a)

$$\frac{\partial \overline{uw}}{\partial z} = f(V - V_g) + \nu \frac{\partial^2 U}{\partial z^2} , \tag{1.33a}$$

$$\frac{\partial \overline{vw}}{\partial z} = - f(U - U_g) + \nu \frac{\partial^2 V}{\partial z^2} , \tag{1.33b}$$

$$\frac{\partial \overline{w^2}}{\partial z} = -\frac{1}{\rho}\frac{\partial P}{\partial z} - g - 2(\Omega_1 V - \Omega_2 U) \tag{1.33c}$$

where the Coriolis parameter $f = 2\Omega \sin \phi$, and $\Omega = (\Omega_i^2)^{\frac{1}{2}}$.

The U_g and V_g are the components of the geostrophic wind given by

$$U_g = -\frac{1}{\rho f}\frac{\partial P}{\partial y}, \qquad V_g = +\frac{1}{\rho f}\frac{\partial P}{\partial x}. \tag{1.34}$$

They represent the large scale force acting on the fluid. The latitude, ϕ, will be considered a constant. In appendix D the Ekman spiral is discussed, which represents an exact solution of (1.33a) and (1.33b) for laminar flow. This solution is of diagnostic value. The first and last term in Equation (1.33c) are typically of the order of 10^{-3} ms^{-2} and therefore negligible in comparison with the two middle terms, so this equation reduces to (1.11)

Equation (1.28) simplifies to

$$\frac{\partial \overline{w\theta}}{\partial z} = \kappa \frac{\partial^2 \Theta}{\partial z^2}. \tag{1.35}$$

In the bulk of the boundary layer the term on the left-hand side is typically of the order of 10^{-3} ^oC s^{-1}, whereas the term on the right-hand side is of the order of 10^{-9} ^oC s^{-1}. Because the right-hand side is usually negligible with respect to the left-hand side $\overline{w\theta}$ = constant with height. In reality, steady state cannot be maintained in the presence of fluxes which are distributed over a layer of finite thickness, so we must write

$$\frac{\partial \Theta}{\partial t} = -\frac{\partial \overline{w\theta}}{\partial z}. \tag{1.36}$$

Similarly, we find for other scalar quantities in the absence of sinks and sources

$$\frac{\partial C}{\partial t} = -\frac{\partial \overline{wc}}{\partial z}. \tag{1.37}$$

1.4. DISCUSSION OF THE BOUSSINESQ APPROXIMATIONS AND THE CONSERVATION OF ENTHALPY EQUATION

The set of assumptions, which allow the use of (1.24) and (1.25) together with the assumptions of constancy of the molecular diffusivities, form the Boussinesq approximations. The flows are essentially treated as incompressible with a temperature-dependent density. By carefully ordering the steps from the more fundamental compressible equations to the Boussinesq equations, the fluctuations in density become only significant when multiplied by the acceleration due to gravity.

Especially the use of the incompressibility assumption (1.3) or (1.24) in combination with the conservation of enthalpy, equation (1.21), deserves special attention for the case of dry air, i.e., $\tilde{q} = 0$. In Section 1.2 on the conservation of enthalpy we have already seen an example where the use of (1.3) leads to an erroneous result. It is therefore useful to analyze what is involved in going from Equation (1.21) to (1.28) and (1.35). If we assume c_p = const, take the average of the left hand side of Equation (1.17) with $\tilde{q} = 0$, use (1.19) with $\tilde{\theta} = \tilde{T}$ and expand the triple product we obtain

$$c_p \left\{\frac{\partial}{\partial t} (\rho\ \Theta + \overline{\rho'\theta}) + \frac{\partial}{\partial x_i} (\rho\ U_i\Theta + \rho\ \overline{u_i\theta} + U_i\overline{\rho'\theta} + \right.$$

$$\left. + \Theta\ \overline{u_i\rho'} + \overline{\rho' u_i\theta})\right\} = \ldots \qquad (1.38)$$

Because the pressure may be considered constant, $\partial\tilde{\rho}\tilde{\theta}/\partial t \simeq 0$; hence the first term of (1.38) is negligible. Furthermore by using (1.25) and $\theta'/\Theta = T'/T$ we can show that $\overline{\rho' u_i\theta}$ and $U_i\overline{\rho'\theta} < \rho\ \overline{u_i\theta}$. We can also show that $\rho\ \overline{u_i\theta} = -\Theta\ \overline{u_i\rho'}$. Thus (1.38) reduces to $c_p\partial(\rho\ U_i\Theta)/\partial x_i = \ldots$. If we now assume horizontal uniformity and divide by $c_p\rho$ the result is

$$\frac{1}{\rho}\frac{\partial}{\partial z}\rho\ W\ \Theta\ = \text{r.h.s. of (1.35)} \qquad (1.39)$$

If (1.35) and (1.39) are both correct we recognize that

$$\overline{w\theta} = W\ \Theta \qquad (1.40)$$

and if we apply approximation (1.24), which led to (1.32) for horizontal uniformity, we find that $\overline{w\theta} = 0$, which means that no turbulent heat flux can exist. This last result is obviously erroneous and the approximation (1.24) may not be used in this case.

Using Equation (1.1) after averaging and applying horizontal uniformity and steady-state, we find instead for (1.32)

$$\partial\,(\overline{\underset{\sim}{\rho}\,\underset{\sim}{w}})/\partial z = 0 \tag{1.41}$$

which leads for dry air to $\overline{\underset{\sim}{\rho}\,\underset{\sim}{w}} = 0$, or

$$\rho\,W = -\,\overline{\rho' w}\;. \tag{1.42}$$

Combining this result with (1.25) we find (1.40) again. We thus see thus that Equation (1.40) is correct and consequently also (1.35) and (1.36). The Boussinesq approximations which led to (1.35) resulted in the juggling act to eliminate both $\rho\,W\,T$ and $T\,\overline{\rho' w}$, each of which is equal in magnitude to $\rho\,\overline{wT'}$, but they fortunately cancel each other because they have opposite signs.

Things do not work out equally well when we consider the horizontal heat flux $c_p\,\underset{\sim}{\rho}\,\underset{\sim}{U}\,\underset{\sim}{T}$. The result here is that the mean horizontal heat flux is $c_p\,\rho\,U\,T$. The term $c_p\,\rho\,\overline{uT'}$ is cancelled by $c_p\,T\,\overline{u\rho'}$. The Boussinesq approximations lead to $c_p\rho(UT + \overline{uT'})$, which is not far off because usually $\|\overline{uT'}\| < \|U\,T\|$, but misleadingly incorrect, because it suggests that $\overline{uT'}$ provides a correction that should be applied if one wants an accurate determination of the horizontal heat flux.

The fact that for constant pressure $\partial\,(\rho\,T)/\partial t = 0$, even under quasi steady-state conditions where $\partial\rho/\partial t$ and $\partial T/\partial t$ have to be taken into account, indicates that the divergence of the enthalpy flux, $\partial(c_p\rho\,U_i T)/\partial x_i$, is not relevant as a cause of local temperature change (Businger and Deardorff, 1968). To see what is relevant we expand the left-hand side of (1.21) using $T \simeq \Theta$. After averaging and setting $\underset{\sim}{q} = 0$ we find

$$c_p\rho\,\frac{\partial\Theta}{\partial t} + c_p\,\overline{\underset{\sim}{\rho}\,\underset{\sim}{u}_i}\,\frac{\partial\Theta}{\partial x_i} + c_p\,\frac{\partial}{\partial x_i}\,\overline{(\underset{\sim}{\rho}\,\underset{\sim}{u}_i)'\theta}$$

$$= c_p\,\rho\ (\text{r.h.s. of (1.28)}).$$

If we now consider the horizontally uniform case and neglect the right-hand side, we have

$$c_p\,\rho\,\frac{\partial\Theta}{\partial t} + c_p\,\overline{\underset{\sim}{\rho}\,\underset{\sim}{w}}\,\frac{\partial\Theta}{\partial z} + c_p\,\frac{\partial}{\partial z}\,\overline{(\underset{\sim}{\rho}\,\underset{\sim}{w})'\theta} \simeq 0\;. \tag{1.43}$$

The term

$$\overline{\underset{\sim}{\rho}\,\underset{\sim}{w}}\,\frac{\partial\Theta}{\partial z} = -\,\frac{\partial\Theta}{\partial z}\int_{o}^{z}\frac{\partial\rho}{\partial t}\,dz$$

may be approximated in the surface layer (below 20 m) by

$$\overline{\underset{\sim}{\rho}\,\underset{\sim}{w}}\frac{\partial\Theta}{\partial z} = \frac{z\,\rho}{T}\frac{\partial\Theta}{\partial z}\;\frac{\partial T}{\partial t}$$

Because $z/T\ \partial\Theta/\partial z$ is usually of a magnitude of 0.01 or less, the second term in (1.43) is negligible in comparison to the first, therefore,

$$c_p\rho\frac{\partial\Theta}{\partial t} = -\,c_p\frac{\partial}{\partial z}\overline{(\underset{\sim}{\rho}\,\underset{\sim}{w})'\theta}.$$

Furthermore because $(\underset{\sim}{\rho}\ \underset{\sim}{w})' = \rho\,w + W\rho' - \overline{\rho' w} + \rho' w \simeq \rho\,w'$ and ρ does not vary more than 10% in the boundary layer

$$c_p\,\rho\frac{\partial\Theta}{\partial t} \simeq -\,c_p\rho\frac{\partial}{\partial z}\overline{w\theta}$$

which is the same result as derived with the Boussinesq approximations (1.36).

It is clear from the preceding discussion that $\partial(\underset{\sim}{\rho}\,\underset{\sim}{u}_i\,\underset{\sim}{\theta})/\partial x_i = 0$ is a much better approximation in the bulk of the boundary layer than $\partial\underset{\sim}{u}_i/\partial x_i = 0$.

In the following sections we shall use the equations in the Boussinesq approximated form, i.e., equations (1.24), (1.25), (1.26a,b), (1.28), (1.29), (1.30), (1.31) and (1.33) - (1.37).

1.5. SUMMARY OF THE BOUSSINESQ SET OF EQUATIONS

Having been informed of the limitations and pitfalls of the Boussinesq approximations, the reader is encouraged to use the Boussinesq set of equations, which will be summarized.

The density will only appear as an average quantity for which we choose an appropriate value, ρ_0, representative of the entire boundary layer $\rho = \rho_0$ The temperature will usually be given in the notation of potential temperature.

Equation of Continuity

$$\frac{\partial U_i}{\partial x_i} = 0\ ,\qquad \frac{\partial u_i}{\partial x_i} = 0\ . \tag{1.24}$$

Equations of Motion

$$\frac{\partial U_i}{\partial t} + \frac{\partial}{\partial x_i}[\,U_jU_i + \overline{u_ju_i}\,] = -\frac{1}{\rho_o}\frac{\partial P}{\partial x_i} + \nu\frac{\partial^2 U_i}{\partial x_j^2} + g_i - 2\,\varepsilon_{ijk}\Omega_j U_k \qquad (1.26a)$$

$$\frac{\partial u_i}{\partial t} + \frac{\partial}{\partial x_j}[\,U_ju_i + u_jU_i + u_iu_j - \overline{u_iu_j}\,]$$

$$= -\frac{1}{\rho_o}\frac{\partial p}{\partial x_i} - \frac{g_i}{\Theta}\theta + \nu\frac{\partial^2 u_i}{\partial x_j^2} - 2\,\varepsilon_{ijk}\Omega_j u_k . \qquad (1.26b)$$

Equation of Heat Conduction (Enthalpy of Dry Air)

$$\frac{\partial \Theta}{\partial t} + \frac{\partial}{\partial x_i}[\,U_i\Theta + \overline{u_i\theta}\,] = \kappa\frac{\partial^2\Theta}{\partial x_i^2} \qquad (1.28)$$

$$\frac{\partial \theta}{\partial t} + \frac{\partial}{\partial x_i}[\,u_i\Theta + U_i\theta + u_i\theta - \overline{u_i\theta}\,] = \kappa\frac{\partial^2\theta}{\partial x_i^2} . \qquad (1.29)$$

Equation for Transfer of a Scalar Quantity

$$\frac{\partial C}{\partial t} + \frac{\partial}{\partial x_i}[\,U_iC + \overline{u_ic}\,] = D\frac{\partial^2 C}{\partial x_i^2} + S \qquad (1.30)$$

$$\frac{\partial c}{\partial t} + \frac{\partial}{\partial x_i}[\,u_iC + U_ic + u_ic - \overline{u_ic}\,] = D\frac{\partial^2 c}{\partial x_i^2} . \qquad (1.31)$$

Because in the boundary layer a representative density ρ_o has been chosen, C may denote both mixing ratio and concentration in (1.30) and (1.31). Only the source and sink term (S) should be modified correspondingly.

1.6. THE CLOSURE PROBLEM, FIRST ORDER CLOSURE

Although the mean flow is much simpler than the total turbulent flow, the equations for the mean flow are no longer a closed set of equations. New unknowns, such as $\overline{u_iu_j}$ and $\overline{u_i\theta}$, had to be introduced, making the number of unknowns larger than the number of equations. If we construct equations for these covariances and variances, which we will do later in this section, new unknowns in the form of triple correlations will appear. This procedure may be repeated, but every time a higher order correlation will appear as a new unknown. Thus, it is fundamentally impossible to close the set of equations in a mathematical sense. This problem is referred to as the closure problem of turbulence.

The covariances $\overline{uw}$, $\overline{vw}$, $\overline{w\theta}$ are of special importance in the boundary layer as can be seen from Equations (1.33 - 1.37). In fact, they often represent the largest terms in the equations. In order to make these terms tractable, assumptions have to be made in such a way that the set of equations is closed again. The simplest scheme which has been used with considerable success is to model turbulent transfer after its molecular counterpart. Boussinesq (1877) was the first to apply this to the momentum flux, by introducing an eddy viscosity, K_m, so that

$$\overline{uw} = -K_m \frac{\partial U}{\partial z} \; . \qquad (1.44)$$

This approach may be generalized to the turbulent transfer of any quantity $\underset{\sim}{c}$, using an eddy diffusivity, K

$$\overline{wc} = -K \frac{\partial C}{\partial z} \; . \qquad (1.45)$$

Unlike the molecular viscosity and diffusivities, these eddy coefficients or turbulence transfer coefficients may not be assumed constants in the flow, because they have to absorb the complexities of turbulence. The next step, therefore, is to make appropriate assumptions concerning these eddy coefficients so that the mean characteristics of the flow are adequately described.

The molecular coefficients are proportional to the average velocity of the molecules and the mean free path between collisions of the molecules. In analogy to this Prandtl (1932) assumed that in the surface layer the transfer is due to eddies which move vertically at velocity w over a distance ℓ_m, also called mixing length, before adjusting their momentum to the surrounding fluid. The u-fluctuation in this process is approximately $-\ell_m\, \partial U/\partial z$, where ℓ_m can be positive or negative. The average momentum transfer per unit mass through this exchange is

$$-\overline{uw} = \overline{\ell_m w} \frac{\partial U}{\partial z} \qquad (1.46)$$

so that $K_m = \overline{\ell_m w}$. Analogous arguments can be made for scalar transfer, but we need not distinguish the K's here. Thus it was recognized early that eddy diffusivity is of the order of $q\ell$, where q and ℓ are velocity and length scales of the energy-containing eddies - those responsible for the turbulent transfer.

From the result $K \sim q\ell$ we can infer two important features of turbulent transfer. First, the ratio of eddy and molecular diffusivities, being a turbulent Reynolds number, is typically large:

$$K/\nu \sim q\ell/\nu = R_\ell . \tag{1.47}$$

For a typical boundary layer ($q \sim 1$ m s^{-1}, $\ell \sim 300$ m) $K/\nu \sim 2\ 10^7$, so turbulent transfer overwhelms that due to molecular processes. Second, the result $K \sim q\ell$ means that K is a property of the flow, not a property of the fluid, so that we need to know something about the flow structure in order to specify K values.

The dependence of K on the flow structure and the difficulty of determining this dependence are the most serious problems with the K-approach.

In spite of these difficulties, and while, as is shown in Chapter 3, there are fundamental objections to the parameterization of turbulent stress purely in terms of the local mean shear, K-models remain the most common approach to turbulent flow calculation. More sophisticated approaches are available, as we will discuss later (see Chapter 3). However, they often cannot be used in larger-scale flows because of their excessive computer time requirements. Thus K-models remain very important, and the specification of proper K values is of considerable interest.

Both Reynolds stress and mean strain are second-order tensors, and a general linear relation between them is

$$\overline{u_i u_j} = \frac{1}{3}\delta_{ij}\,\overline{u_k u_k} - K_{ijk\ell}\left(\frac{\partial U_k}{\partial x_\ell} + \frac{\partial U_\ell}{\partial x_k}\right) . \tag{1.48}$$

In this formulation the eddy viscosity is a fourth order tensor quantity having $3^4 = 81$ components; even with symmetry constraints a large number of them are independent. Similarly, the general linear relation between temperature flux and mean temperature gradient, both first order tensors, is

$$\overline{\theta u_i} = - K_{ij}\frac{\partial \Theta}{\partial x_j} . \tag{1.49}$$

This eddy diffusivity is a second-order tensor with nine components. This allows for such effects as horizontal as well as vertical heat fluxes in the presence of a vertical mean temperature gradient, which is observed in the surface layer (see Section 3.2 on second-order closure):

$$\overline{\theta u} = -K_{13}\,\partial\Theta/\partial z\ ,$$
$$\overline{\theta w} = -K_{33}\,\partial\Theta/\partial z\ . \tag{1.50}$$

The tensorial nature of K for scalars has also received attention in diffusion applications (see Monin and Yaglom (1971) for a review).

Because a tensorial formulation of K drastically complicates its determination, it is not often used and only then in the scalar case. We will deal here only with the specification of scalar K values such that

$$\overline{uw} = -K_m \frac{\partial U}{\partial z}\ , \tag{1.44}$$

$$\overline{wc} = -K \frac{\partial C}{\partial z}\ , \tag{1.45}$$

$$\overline{w\theta} = -K_h \frac{\partial \Theta}{\partial z}\ . \tag{1.51}$$

1.7. SECOND-ORDER VARIANCE AND COVARIANCE EQUATIONS

Some of the fundamental problems with the first-order closure may be resolved with higher order closures, but not without introducing additional difficulties. Especially, the second order closure has been studied in considerable detail. The procedure is to derive the variance and covariance equations and make assumptions concerning the higher order terms and those terms that cannot yet be measured. The derivation of the variance and covariance equations is most easily demonstrated by giving an example. The shear stress budget equation is obtained by multiplying Equation (1.26b) for u_i by u_k and adding to this the equation for u_k multiplied by u_i. After some manipulation and averaging the result is

$$\frac{\partial \overline{u_i u_k}}{\partial t} = \underset{(1)}{- U_j \frac{\partial}{\partial x_j} \overline{u_i u_k}} - \underset{(2)}{\{\overline{u_k u_j}\frac{\partial U_i}{\partial x_j} + \overline{u_i u_j}\frac{\partial U_k}{\partial x_j}\}} - \underset{(3)}{\frac{\partial}{\partial x_j}\overline{u_j u_i u_k}} -$$

$$\overset{(4)}{-\frac{1}{\rho_0}\{\overline{u_k\frac{\partial p}{\partial x_i}}+\overline{u_i\frac{\partial p}{\partial x_k}}\}}-\overset{(5)}{\{\frac{g_i}{\Theta}\overline{u_k\theta}+\frac{g_k}{\Theta}\overline{u_i\theta}\}}+ \qquad (1.52)$$

$$\overset{(6)}{+\nu\{\overline{u_k\frac{\partial^2 u_i}{\partial x_j^2}}+\overline{u_i\frac{\partial^2 u_k}{\partial x_j^2}}\}}-\overset{(7)}{2\,\Omega_j\{\varepsilon_{ijl}\overline{u_k u_l}+\varepsilon_{kjl}\overline{u_l u_i}\}}\,.$$

In the derivation, Equations (1.11), (1.24) and (1.25) have been used and $\rho'/\rho_0^2\ \partial P/\partial x_1$ and $\rho'/\rho_0^2\ \partial P/\partial x_2$ have been neglected in Equation (1.26b).

The term on the left-hand side is the rate of change of the covariance $\overline{u_i u_k}$, which represents a momentum flux when $i \neq k$ and the turbulent kinetic energy when $i = k$. The terms (1) are advection terms which are only important when the terrain is non-uniform. When the turbulence is horizontally homogeneous, these terms are negligible. Terms (2) are production terms, due to the interaction of the mean flow with turbulence. Terms (3) are transport terms due to divergence of the variances or covariances. Terms (4) indicate the interaction of pressure and velocity components. Terms (5) represent buoyant production, which may be positive or negative. Terms (6) represent the viscous effects. Terms (7) represent the effect of the Coriolis force on the variances and covariances. These terms are usually negligible if we consider an averaging time < 1 hr (Busch, 1973).

The covariance equations involving the fluxes of scalar quantities may be derived in a similar way. Multiplying (1.26b) by θ and (1.29) by u_i, adding and averaging yields

$$\frac{\partial}{\partial t}\overline{u_i\theta}=\overset{(1)}{-U_j\frac{\partial}{\partial x_j}\overline{u_j\theta}}-\overset{(2)}{\{\overline{u_i u_j}\frac{\partial\Theta}{\partial x_j}+\overline{u_j\theta}\frac{\partial U_i}{\partial x_j}\}}-\overset{(3)}{\frac{\partial}{\partial x_j}\overline{u_i u_j\theta}}+ \qquad (1.53)$$

$$\overset{(4)}{-\frac{1}{\rho_0}\overline{\theta\frac{\partial p}{\partial x_i}}}\overset{(5)}{-\frac{\overline{\theta^2}}{\Theta}g_i}+\overset{(6)}{\{\nu\,\overline{\theta\frac{\partial^2 u_i}{\partial x_j^2}}+\kappa\,\overline{u_i\frac{\partial^2\theta}{\partial x_j^2}}\}}-\overset{(7)}{2\,\varepsilon_{ijk}\Omega_j\overline{u_k\theta}}\,.$$

The physical meaning of the terms in this equation is similar to the corresponding terms in Equation (1.52). Terms (1) again are advection terms which are negligible in horizontal uniform conditions. Terms (2) are production terms, (3) is a transport term, (4) a pressure temperature interaction term, (5) a buoyancy production term, (6) are molecular terms and finally (7) represent the effect of the Coriolis force on the covariance. The possible effects of condensation and evaporation have been omitted.

A similar covariance equation may be derived for a scalar quantity in general using Equations (1.26b) and (1.31).

$$\frac{\partial}{\partial t}\overline{u_i c} = \underset{(1)}{-\, U_j \frac{\partial}{\partial x_j}\overline{u_i c}} \underset{(2)}{-\,\{\overline{u_i u_j}\frac{\partial C}{\partial x_j} + \overline{u_j c}\frac{\partial U_i}{\partial x_j}\}} \underset{(3)}{-\,\frac{\partial}{\partial x_j}\overline{u_i u_j c}} +$$

$$\underset{(4)}{-\,\frac{1}{\rho_o}\overline{c\frac{\partial p}{\partial x_i}}} \underset{(5)}{-\,\frac{g_i}{\Theta}\overline{\theta c}} + \underset{(6)}{\{\nu\,\overline{c\frac{\partial^2 u_i}{\partial x_j^2}} + D\,\overline{u_i\frac{\partial^2 c}{\partial x_j^2}}\}} \underset{(7)}{-\,2\varepsilon_{ijk}\Omega_j\overline{u_k c}}\,. \qquad (1.54)$$

The variance equations for the velocity components form a subset of the Equations (1.52). The equation for the variance of temperature is obtained by multiplying Equation (1.29) by θ and averaging. The result is

$$\frac{1}{2}\frac{\partial\overline{\theta^2}}{\partial t} = \underset{(1)}{-\,\frac{1}{2}U_i\frac{\partial\overline{\theta^2}}{\partial x_i}} \underset{(2)}{-\,\overline{u_i\theta}\frac{\partial\Theta}{\partial x_i}} \underset{(3)}{-\,\frac{1}{2}\frac{\partial}{\partial x_i}\overline{u_i\theta^2}} \underset{(6)}{-\,\kappa\,\overline{(\frac{\partial\theta}{\partial x_j})^2}} \qquad (1.55)$$

where again the condensation or evaporation term has been omitted. Molecular transport terms are neglected, as will be discussed in Chapter 3, Equation (3.19).

A similar variance equation may be derived from Equation (1.31), which is left as an exercise.

In the set of variance and covariance equations there are three types of terms that need special attention before the set of equations can be closed. (a) The triple correlations (terms (3) in (1.52), (1.53), (1.54), and (1.55)), which represent the divergence of the turbulent transport of variance or covariance; (b) The pressure terms (4) in (1.52), (1.53) and (1.54); (c) the molecular terms (6) in (1.52), (1.53), (1.54) and (1.55). The closure problems related to these terms are further discussed in Chapter 3.

1.8. THE TURBULENT KINETIC ENERGY BALANCE; TEMPERATURE VARIANCE BALANCE

The equation for the turbulent kinetic energy balance is obtained by letting $i = k$ in Equation (1.52), as indicated in the previous section. The result is after dividing by 2:

$$\frac{1}{2}\frac{\partial q^2}{\partial t} = -\frac{1}{2} U_j \frac{\partial q^2}{\partial x_j} - \overline{u_i u_j}\frac{\partial U_i}{\partial x_j} - \frac{1}{2}\frac{\partial}{\partial x_j}\overline{u_j u_i^2} +$$

$$-\frac{1}{\rho_o}\frac{\partial}{\partial x_i}\overline{u_i p} - \frac{g_i}{\Theta}\overline{u_i\theta} - \varepsilon \qquad (1.56)$$

where $q^2 = \overline{u_i u_i}$ and $\varepsilon = \nu\,\overline{(\partial u_i/\partial x_j)^2}$ is the dissipation. Molecular transport has been neglected (see Equation (3.19)). There is no Coriolis term, because no work is being done by the Coriolis force on the fluid parcel. In the special case of horizontal homogeneity this equation may be written in the form

$$\frac{1}{2}\frac{\partial q^2}{\partial t} = -\overline{uw}\frac{\partial U}{\partial z} + \frac{g}{\Theta}\overline{w\theta} - \frac{\partial}{\partial z}\overline{w(\frac{1}{2}u_i^2 + \frac{p}{\rho_o})} - \varepsilon \qquad (1.57)$$

where $x = x_1$ is chosen to be in the direction of the mean wind. This equation is of special importance because of the central role it plays in atmospheric turbulence. The first term on the right-hand side, the shear production term, represents the rate at which the mean flow contributes to the turbulent kinetic energy; the second term is the buoyancy production term; the third term is a combined transport and pressure term (this term does not produce or dissipate energy); and the last term represents the viscous dissipation.

The two production terms contribute energy in different ways. The shear production term contributes essentially to the u component. This term is most important near the surface, where it is always positive, and decreases rapidly with height because $\partial U/\partial z$ usually decreases rapidly. The buoyancy production term, which can be positive or negative depending on the direction of the heat flux, contributes on the other hand to the w component, because the buoyancy force is in the vertical. This becomes immediately clear when we consider the component equations of (1.57)

$$\frac{1}{2}\frac{\partial \overline{u^2}}{\partial t} = -\overline{uw}\frac{\partial U}{\partial z} - \frac{1}{2}\frac{\partial}{\partial z}\overline{wu^2} - \frac{1}{\rho_o}\overline{u\frac{\partial p}{\partial x}} - \frac{1}{3}\varepsilon ,$$

$$\frac{1}{2}\frac{\partial \overline{v^2}}{\partial t} = -\frac{1}{2}\frac{\partial}{\partial z}\overline{wv^2} - \frac{1}{\rho_o}\overline{v\frac{\partial p}{\partial y}} - \frac{1}{3}\varepsilon , \qquad (1.58)$$

$$\frac{1}{2}\frac{\partial \overline{w^2}}{\partial t} = \frac{g}{\Theta}\overline{w\theta} - \frac{1}{2}\frac{\partial}{\partial z}\overline{w^3} - \frac{1}{\rho_o}\overline{w\frac{\partial p}{\partial z}} - \frac{1}{3}\varepsilon .$$

We have used here that viscous dissipation occurs at the high frequency end of the spectrum, where turbulence is usually assumed to be isotropic. Consequently the three dissipation terms in (1.58) may each be supposed to be equal to 1/3 ε (see (3.20).

The v component obtains its energy from the other components through redistribution by the pressure term. The pressure terms tend to make the turbulence isotropic, especially at the higher frequencies. (See the discussion in connection with (3.26).)

Since the production terms are driving the turbulence, one would expect that the ratio of these two terms will define the local structure of turbulence in dimensionless form. This ratio is the flux Richardson number, R_f,

$$R_f = \frac{g}{\Theta}\frac{\overline{w\theta}}{\overline{uw}\,\partial U/\partial z} . \qquad (1.59)$$

Although this number properly characterizes the stratification of the flow, it is awkward to use in practice, because it contains a mixture of eddy correlations and mean gradients. By using (1.44) and (1.51), (1.59) may be written

$$R_f = \frac{K_h}{K_m}\frac{g}{\Theta}\frac{\partial\Theta/\partial z}{(\partial U/\partial z)^2} = \frac{K_h}{K_m}\,Ri \qquad (1.60)$$

where Ri is the Richarson number. This non-dimensional parameter, which was originally introduced by Richardson (1920), has been widely used as a stability parameter for the atmosphere. The advantage of this number is that it only contains mean quantities. Its disadvantage is that it is an unknown or complex function of height within the boundary layer. If we assume that the ratio K_h/K_m is also uniquely determined by Ri, then Ri and R_f are both equivalent parameters in defining the structure of turbulence, because Ri is a unique function of R_f and vice versa. When Ri = 0, the heat flux is 0 and the atmosphere is neutral; when Ri > 0 the atmosphere is statically stable, and when $Ri < 0$, the atmosphere is statically unstable.

Very near the surface the shear production term is the dominant production term, but since this term decreases with height much more rapidly than the buoyancy term, there is usually a height above which the buoyancy term becomes the dominant term. Obukhov (1946) sensed that this height would be a useful scaling length and tried to estimate it. He assumed that near the surface the logarithmic wind profile would hold, which means that $\partial U/\partial z = u_*/kz$, where u_* is the friction velocity $u_* = \sqrt{(-\overline{uw})_0}$ and k the Von Karman constant, which is further discussed in Section 2.2. After substitution and setting the ratio of the first two terms on the right hand side of (1.57) equal to one, he found

$$\frac{g}{\Theta}\frac{\overline{w\theta}_0\, kz}{u_*^3} = 1$$

where $\overline{w\theta}_0$ is the heat flux at the surface. This yields

$$L = -\frac{\Theta\, u_*^3}{g\, \overline{w\theta}_0\, k} \,. \tag{1.61}$$

This length scale is called the Obukhov length. The minus sign was introduced so that L has the same sign as Ri.

From the preceding argument it is clear that the non-dimensional height $\zeta = z/L$ can also be used as a stability parameter. It has the advantage over Ri that it is a non-dimensional height and provides for the easy comprehension of similarity relations, which were originally proposed by Monin and Obukhov (1954). (See Chapter 2, Section 2.4 for further discussion.)

Equation (1.57) may be written in non-dimensional form by dividing each term by u_*^3/kz, the result for the stationary surface layer is

$$\phi_m - \zeta - \phi_D - \phi_\varepsilon = 0$$

where (1.62)

$$\phi_m = \frac{kz}{u_*}\frac{\partial U}{\partial z} \;;\; \phi_D = \frac{kz}{u_*^3}\frac{\partial}{\partial z}\overline{w\left(\tfrac{1}{2}u_i^2 + p/\rho_0\right)} \quad \text{and} \quad \phi_\varepsilon = \frac{kz}{u_*^3}\varepsilon \,.$$

There is little observational documentation of the terms of (1.62). Probably the most complete study in this area is by Wyngaard and Coté (1971), who presented the balance for unstable stratification ($\zeta < 0$). The behavior of the terms beyond the surface layer will be discussed in the Chapters 2 and 4.

We return to Equation (1.57) and some special cases.

(a) When the buoyancy term is 0 or negligible the shear production term is the only input to the reservoir of turbulent energy. As we have seen this condition corresponds to a neutrally stratified atmosphere ($Ri = \zeta = 0$) and the turbulent convection that takes place is referred to as forced convection.

(b) When the shear production term is zero or negligibly small and the buoyancy term is positive (Ri and $\zeta \to -\infty$) the turbulent convection is referred to as free convection.

(c) It is not possible to have turbulence when the shear production is zero and the buoyant production is negative unless turbulence is transported into the region where this situation occurs. This may occasionally happen at the top of the boundary layer, where turbulent eddies penetrate a stable inversion layer.

(d) A steady state transition from turbulent to laminar flow may take place when the buoyancy term is negative and becomes important with respect to the shear production term (R_f increases). This transition takes place at a critical value of R_f, (R_{fc}). Richardson (1920) was the first to consider this problem. He assumed R_f = Ri and concluded that $R_{fc} = 1$. However, he did not include the dissipation term in his considerations. There is not much experimental information about R_{fc}. The best experimental and theoretical estimates at present suggest that $0.2 < R_{fc} < 0.25$.

(e) When the turbulence has subsided and the flow is laminar all terms in (1.57) are zero and this equation becomes meaningless. The stratified laminar flow may be analyzed for stability. Using infinitesimal perturbations, Taylor (1931) arrived at a critical Richardson number, $Ri_c = 0.25$ for the transition from laminar to turbulent flow. For $Ri < Ri_c$ the flow is unstable and will become turbulent; for $Ri > Ri_c$ the flow is stable for small perturbations and will remain laminar. The value $Ri_c = 0.25$ is generally accepted as the correct value.

The equation for the variance of potential temperature may be written similarly for the horizontally homogeneous case. Using (1.55) we have

$$\frac{1}{2}\frac{\partial\overline{\theta^2}}{\partial t} = -\overline{w\theta}\frac{\partial\Theta}{\partial z} - \frac{1}{2}\frac{\partial}{\partial z}\overline{w\theta^2} - N \qquad (1.63)$$

where N is the destruction rate of temperature fluctuations given by $\kappa\overline{(\partial\theta/\partial x_j)^2}$.

There is only one production term in this equation which is similar in character to the shear production term in Equations (1.57) and the u-component of (1.58). The similarity between σ_u and σ_θ equations suggest that the behavior of these two quantities as a function of height may be similar also. However, the role of the pressure term, which is missing in Equation (1.63), is quite important and makes the similarity less than perfect.

A comparison between the third equation of (1.58) and (1.63) is of interest because these two equations are linked through the heat flux, which is approximately constant in the surface layer and decreases almost linearly with height throughout the boundary layer.

When the heat flux is positive, the ratio of the temperature variance production term over the buoyancy production term is proportional to $-\partial\Theta/\partial z$, which decreases rapidly with height. Consequently we expect the variance of temperature to decrease with height also. Since the heat flux changes slowly with height and the correlation between temperature and the vertical velocity fluctuations also changes slowly with height if at all, the variance of the vertical velocity must increase with height in the lower part of the boundary layer.

When the heat flux is negative, the buoyancy production is negative and, consequently, we expect σ_w to decrease with height and eventually to be suppressed altogether. In this case the height of the boundary layer is much less than in the unstable case and the layer of constant heat flux (the surface layer) may become quite shallow. Nevertheless, we expect in the stable surface layer that there is a tendency for σ_θ to increase with height. Observations seem to support this notion.

In the case of free convection, similarity scaling allows for rather specific relations for σ_w and σ_θ with height. These will be further discussed in Section 2.5.

APPENDIX A. REQUIREMENTS FOR A DIVERGENCE-FREE VELOCITY FIELD

Make $\partial \underset{\sim}{u}_i/\partial x_i$ non-dimensional by dividing by a characteristic velocity V, and multiply by a characteristic length, L, i.e.,

$$\frac{\partial \tilde{u}_i}{\partial \tilde{x}_i} = \frac{L}{V} \frac{\partial \underset{\sim}{u}_i}{\partial x_i} \tag{A1}$$

where $\tilde{u}_i = \underset{\sim}{u}_i/V$ and $\tilde{x}_i = x_i/L$. The velocity distribution is then approximately divergence free if (1.2)

$$\frac{L}{V} \frac{1}{\underset{\sim}{\rho}} d\underset{\sim}{\rho}/dt << \partial \tilde{u}_i / \partial \tilde{x}_i \tag{A2}$$

where $\partial \tilde{u}_i / \partial \tilde{x}_i$ is O(1) by the nondimensionalizing.

Next we need to examine $d\underset{\sim}{\rho}/dt$. Consider the density, the pressure $\underset{\sim}{p}$ and entropy $\underset{\sim}{s}$, the state parameters. We may write

$$\frac{d\underset{\sim}{p}}{dt} = \left(\frac{\partial \underset{\sim}{p}}{\partial \underset{\sim}{\rho}}\right)_s \frac{d\underset{\sim}{\rho}}{dt} + \left(\frac{\partial \underset{\sim}{p}}{\partial \underset{\sim}{s}}\right)_\rho \frac{d\underset{\sim}{s}}{dt}$$

and because $(\partial \underset{\sim}{p}/\partial \underset{\sim}{\rho})_s = a^2$, where a is the speed of sound

$$\frac{d\underset{\sim}{\rho}}{dt} = \frac{1}{a^2} \left[\frac{d\underset{\sim}{p}}{dt} - \left(\frac{\partial \underset{\sim}{p}}{\partial \underset{\sim}{s}}\right)_\rho \frac{d\underset{\sim}{s}}{dt} \right] . \tag{A3}$$

Substitution in (A2) yields

$$\frac{L}{V} \left\{ \frac{1}{\underset{\sim}{\rho} a^2} \frac{d\underset{\sim}{p}}{dt} - \frac{1}{\underset{\sim}{\rho} a^2} \left(\frac{\partial \underset{\sim}{p}}{\partial \underset{\sim}{s}}\right)_\rho \frac{d\underset{\sim}{s}}{dt} \right\} << 1 . \tag{A4}$$

The second term of (A4), which is zero when we have an isentropic process, may be written (see Batchelor, 1967, p. 170) as

$$\frac{1}{\underset{\sim}{\rho} a^2} \left(\frac{\partial \underset{\sim}{p}}{\partial \underset{\sim}{s}}\right)_\rho \frac{d\underset{\sim}{s}}{dt} = \frac{\beta}{c_p} \left\{ \Phi + \frac{1}{\underset{\sim}{\rho}} \frac{\partial}{\partial x_i} \left(k \frac{\partial \underset{\sim}{T}}{\partial x_i} \right) \right\}$$

where $\beta = - \underset{\sim}{\rho}^{-1}(\partial \underset{\sim}{\rho}/\partial \underset{\sim}{T})_p$, $\Phi \simeq \varepsilon$ is heating due to the dissipation of kinetic energy and k is thermal conductivity. The viscous dissipation is always negligible as a heat source in the boundary layer, so this term is only significant where the heating occurs primarily through thermal conduction. This may happen very close to the earth's surface. Otherwise we consider this term negligible.

The first term of (A4) needs closer scrutiny. We have

$$d\underset{\sim}{p}/dt = \partial\underset{\sim}{p}/\partial t + \underset{\sim}{u}_i \partial\underset{\sim}{p}/\partial x_i \text{ but}$$

$$\underset{\sim}{\rho}\left(\frac{\partial \underset{\sim}{u}_i}{\partial t} + \underset{\sim}{u}_j \frac{\partial \underset{\sim}{u}_i}{\partial x_j}\right) = \underset{\sim}{\rho} F_i - \frac{\partial \underset{\sim}{p}}{\partial x_i}$$

where F_i is the body force acting on the fluid and

$$\underset{\sim}{u}_i \frac{\partial \underset{\sim}{p}}{\partial x_i} = \underset{\sim}{\rho}\, \underset{\sim}{u}_i F_i - \frac{1}{2} \underset{\sim}{\rho} \frac{d\underset{\sim}{u}_i^2}{dt} .$$

Therefore,

$$\frac{1}{\underset{\sim}{\rho} a^2} \frac{d\underset{\sim}{p}}{dt} = \frac{1}{\underset{\sim}{\rho} a^2} \frac{\partial \underset{\sim}{p}}{\partial t} - \frac{1}{2a^2} \frac{d\underset{\sim}{u}_i^2}{dt} + \frac{\underset{\sim}{u}_i F_i}{a^2}$$

and (A4) may be written

$$\frac{L}{V}\left\{\frac{1}{\underset{\sim}{\rho} a^2} \frac{\partial \underset{\sim}{p}}{\partial t} - \frac{1}{2a^2} \frac{d\underset{\sim}{u}_i^2}{dt} + \frac{\underset{\sim}{u}_i F_i}{a^2}\right\} \ll 1 . \tag{A5}$$

Each of these terms must be $\ll 1$. The first term expresses the rate of change of pressure with time, which may be normalized as follows

$$\frac{L}{V} \frac{1}{\underset{\sim}{\rho} a^2} \frac{\partial \underset{\sim}{p}}{\partial t} = \frac{L}{V} \frac{1}{\underset{\sim}{\rho} a^2} \underset{\sim}{\rho} V^2 \frac{V}{L} \frac{\partial \tilde{p}}{\partial \tilde{t}} = \frac{V^2}{a^2} \frac{\partial \tilde{p}}{\partial \tilde{t}} \ll 1$$

where $\tilde{p} = \underset{\sim}{p}/\rho V^2$ and $\tilde{t} = tV/L$. If $\partial\tilde{p}/\partial\tilde{t}$ is $O(1)$ then a sufficient condition is

$$V^2/a^2 = M^2 \ll 1 \tag{A6}$$

where M is the Mach number. Pressure oscillations may be considered which have a magnitude ρLVn where n is the frequency. Then $\partial\underset{\sim}{p}/\partial t = \underset{\sim}{\rho} LVn^2 \partial p''/\partial t''$, where $\partial\tilde{p}''/\partial t''$ is the normalized rate of change of pressure which is $O(1)$. So in this case we have

$$\frac{(nL)^2}{a^2} \frac{\partial p''}{\partial t''} \ll 1 \quad \text{or} \quad \frac{(nL)^2}{a^2} \ll 1 . \tag{A7}$$

If $n \simeq V/L$ we get (A6). For sound waves $nL = a$ which shows that these waves do not obey (A5) which we expected, because they are compression waves.

The second term of (A5) may be written as

$$\frac{V^2}{a^2}\frac{d\tilde{u}_i^2}{d\tilde{t}} \ll 1$$

but, by our normalization, $d\tilde{u}_i/d\tilde{t} \simeq O(1)$, therefore,

$$V^2/a^2 = M^2 \ll 1 \tag{A6}$$

which is the same requirement as found for the first term.

The last term of (A5) contains the specific body force F_i which usually is the specific force of gravity, g. For this term we find then

$$\frac{L}{V}\frac{\tilde{w}g}{a^2} = \frac{Lg}{a^2}\tilde{w} \ll 1$$

thus

$$\frac{Lg}{a^2} \ll 1 \quad \text{or} \quad \frac{V^2}{a^2}\frac{Lg}{V^2} = \frac{M^2}{Fr} \ll 1 \tag{A8a}$$

where $Fr = V^2/Lg$, the Froude number which is the ratio of the inertial force over the force of gravity. Also

$$\frac{Lg}{a^2} = \frac{L}{H} \ll 1 \tag{A8b}$$

where $H = a^2/g \simeq 12$ km is the scale height of the atmosphere. Dutton and Fichtl (1969) refer to this as limiting the study to shallow convection, i.e., to the atmospheric boundary layer (ABL).

In summary, for (1.3) to be valid we must have:

(1) $M^2 < 1$, i.e., $V < 100$ m s^{-1}

(2) Low frequency pressure waves $nL < a$

(3) $M^2 < F_r$

(4) The characteristics length scale $L < 12$ km.

All these conditions are usually fulfilled when considering turbulence in the ABL.

APPENDIX B. THE MAGNITUDE OF PRESSURE FLUCTUATIONS

In the boundary layer there are two types of pressure fluctuations, i.e., dynamic fluctuations caused by fluctuations in wind speed and static fluctuations caused by the passage of convective elements which contain lighter air than their surroundings.

The dynamic fluctuations p, may be related to the mean wind, U, and the fluctuations of the wind, u, as follows

$$p \simeq \rho U u . \tag{B1}$$

The dynamic fluctuations related to density fluctuations are an order of magnitude smaller because $\rho'/\rho \ll u/U$. For a mean wind speed of 10 m s^{-1} and $\|u/U\| \simeq 0.1$ we find, for example $\|p\| \simeq 10$ Pa.

The static pressure fluctuations may be crudely approximated by assuming a column of air which is Δz high and $\Delta\rho$ less dense than the environment over the point of observation. In this case

$$\|p\| \simeq g \, \Delta\rho \, \Delta z . \tag{B2}$$

For $\|\Delta\rho/\rho\| = 0.01$ and $\Delta z = 100$ m we find $\|p\| \simeq 10$ Pa, which is the same order of magnitude as the dynamic pressure fluctuations. Strong static fluctuations occur in light winds and strong dynamic fluctuations occur in strong winds, so we may assume that pressure fluctuations for most boundary layer conditions are not larger than the above estimates: $p = O(10$ Pa$)$ or $O(0.1$ mbar$)$ and therefore typically $\|p/P\| = O(10^{-4})$.

In very strong winds the dynamic pressure fluctuations may be an order of magnitude larger, and the temperature fluctuations an order of magnitude smaller, so that in this situation $\|p/P\| \sim \|\rho'/\rho\| \sim \|T'/T\|$ which means that (1.25) cannot be used and sensible heat flux measurements using the eddy correlation become unreliable.

The static pressure fluctuations are related to the largest eddies in the boundary layer so that they will only appear in the low frequency range of the turbulence variance spectrum of the wind or the temperature.

APPENDIX C. THE ENTHALPY EQUATION FOR MOIST AIR

Averaging (1.21) yields

$$\frac{\partial}{\partial t}\overline{\underset{\sim}{\rho}\,\underset{\sim m}{h}} + \frac{\partial}{\partial x_i}\overline{\underset{\sim}{\rho}\,\underset{\sim i}{u}\,\underset{\sim m}{h}} = \frac{\partial}{\partial x_i}\left(k\frac{\partial \overline{\underset{\sim}{T}}}{\partial x_i} - \overline{\underset{\sim ni}{R}}\right) + \overline{(c_{pv}\underset{\sim}{T} + b_2)D_q\underset{\sim}{\rho}\frac{\partial^2 \underset{\sim}{q}}{\partial x_i^2}} \tag{C1}$$

In order to keep things manageable, the expansion of this equation into mean and fluctuating parts has been carried out for the special but important case of steady-state and horizontal uniformity. In this case we have found (1.32), $\partial W/\partial z = 0$, thus $W = 0$, because $W = 0$ at $z = 0$.

This is not quite correct when evaporation occurs, because then a flux of water vapor passes through the surface layer. In this case we use the fact that the density is composed of a dry air component, $\underset{\sim}{\rho}_d$, and a water vapor component, $\underset{\sim}{\rho}_v$, thus

$$\overline{\underset{\sim}{\rho}\,\underset{\sim}{w}} = \overline{\underset{\sim}{\rho}_d\,\underset{\sim}{w}} + \overline{\underset{\sim}{\rho}_v\,\underset{\sim}{w}} = \overline{\underset{\sim}{\rho}(1-\underset{\sim}{q})\underset{\sim}{w}} + \overline{\underset{\sim}{\rho}\,\underset{\sim}{q}\,\underset{\sim}{w}}$$

and because

$$\overline{\underset{\sim}{\rho}_d\,\underset{\sim}{w}} = \overline{\underset{\sim}{\rho}(1-\underset{\sim}{q})\underset{\sim}{w}} = 0, \tag{C2}$$

$$\overline{\underset{\sim}{\rho}\,\underset{\sim}{w}} = \overline{\underset{\sim}{\rho}\,\underset{\sim}{q}\,\underset{\sim}{w}} \simeq \rho\,\overline{wq'}\ . \tag{C3}$$

Applying the conditions of steady-state and horizontal uniformity to (C1), substituting for $\underset{\sim}{h}_m$ and then integrating from the surface to a height z in the surface layer results in

$$\overline{\{(1-\underset{\sim}{q})\,\underset{\sim}{\rho}\,c_p\,\underset{\sim}{w}\,\underset{\sim}{T} + \underset{\sim}{q}\,\underset{\sim}{\rho}\,c_{pv}\,\underset{\sim}{w}\,\underset{\sim}{T} + \underset{\sim}{\rho}\,\underset{\sim}{w}\,b_4\}}_z$$
$$= -\,(k\,\partial T/\partial z)_o - (c_{pv}\ \rho\ T + b_2)\ D_q(\partial q/\partial z)_o \tag{C4}$$

where the radiation term and the molecular conduction and diffusion terms at height z have been neglected. The first term on the right-hand side of (C4) represents the sensible heat flux and the second term the enthalpy flux of water vapor, which corresponds to the latent heat flux.

We now expand the individual terms on the left hand side of (C4). The first term is

$$\overline{(1-\underset{\sim}{q})\underset{\sim}{\rho}\,c_p\,\underset{\sim}{w}\,\underset{\sim}{T}} = c_p\,\overline{\underset{\sim}{\rho}\,\underset{\sim}{w}\,\underset{\sim}{T}} - c_p\,\overline{\underset{\sim}{q}\,\underset{\sim}{\rho}\,\underset{\sim}{w}\,\underset{\sim}{T}}\ . \tag{C5}$$

First consider

$$\overline{\underset{\sim}{\rho}\,\underset{\sim}{w}\,\underset{\sim}{T}} = \underset{\text{I}}{\rho\,W\,T} + \underset{\text{II}}{\rho\,\overline{w\,T'}} + \underset{\text{III}}{W\,\overline{\rho' T'}} + \underset{\text{IV}}{T\,\overline{\rho' w}} + \text{TC}.$$

TC stands for triple correlation which we will neglect from hereon as well as higher correlations. The terms I and IV may be combined with (C2) and (C3) to give $\rho\, T\, \overline{wq'}$; term III is negligible, thus

$$\overline{\underset{\sim}{\rho}\,\underset{\sim}{w}\,\underset{\sim}{T}} = \rho\,\overline{w\,T'} + \rho\,T\,\overline{wq'}\ . \tag{C6}$$

Next consider

$$\overline{\underset{\sim}{q}\,\underset{\sim}{\rho}\,\underset{\sim}{w}\,\underset{\sim}{T}} = \underbrace{q\,\rho\,\overline{w\,T'}}_{\text{I}} + \underbrace{\rho\,T\,\overline{wq'}}_{\text{II}} + \underbrace{\rho\,W\,\overline{q'T'}}_{\text{III}} + \underbrace{T\,W\,\overline{q'\rho'}}_{\text{IV}} + \underbrace{\rho\,T\,\overline{wq'}}_{\text{V}} + \text{TC}\ .$$

Terms II, III, and IV are at least an order of magnitude smaller than I and V, thus

$$\overline{\underset{\sim}{q}\,\underset{\sim}{\rho}\,\underset{\sim}{w}\,\underset{\sim}{T}} \simeq q\,\rho\,\overline{w\,T'} + \rho\,T\,\overline{wq'} \tag{C7}$$

Substituting (C6) and (C7) into (C5) yields

$$\overline{(1-\underset{\sim}{q})\underset{\sim}{\rho}\,c_p\,\underset{\sim}{w}\,\underset{\sim}{T}} \simeq \rho\,\overline{wT'}\,c_p(1-q) \tag{C8}$$

The second term on the left hand side of (C4) is analogous to (C7)

$$\overline{\underset{\sim}{\rho}\,\underset{\sim}{w}\,\underset{\sim}{q}\,c_{pv}\,\underset{\sim}{T}} = \rho\,\overline{w\,T'}\,c_{pv}q + \rho\,\overline{wq'}\,c_{pv}T\ . \tag{C9}$$

The third term is $\overline{\underset{\sim}{\rho}\,\underset{\sim}{w}}\,b_4 = \overline{\underset{\sim}{\rho}\,\underset{\sim}{w}\,(1-\underset{\sim}{q})}\,b_1 + \overline{\underset{\sim}{\rho}\,\underset{\sim}{w}\,\underset{\sim}{q}}\,b_2$. Using (C2) and (C3) this becomes

$$\overline{\underset{\sim}{\rho}\,\underset{\sim}{w}}\,b_4 = \rho\,\overline{wq'}\,b_2\ . \tag{C10}$$

Combining (C8), (C9), and (C10) results in

$$\overline{\underset{\sim}{\rho}\,\underset{\sim}{w}\,\underset{\sim}{h}_m} = \rho\,\overline{w\,T'}\,(c_p + q(c_{pv}-c_p)) + \rho\,\overline{wq'}\,(c_{pv}T + b_2)\ . \tag{C11}$$

The specific enthalpy of water vapor contains the latent heat of vaporization, L,

$$\underset{\sim}{h}_v = \underset{\sim}{h}_\ell + L \tag{C12}$$

where h_ℓ is the specific enthalpy of liquid water given by

$$\underset{\sim}{h}_\ell = c_\ell\,\underset{\sim}{T} + b_3 \tag{C13}$$

where c_ℓ is the specific heat of water. Combining (C12) and (C13) with (1.19) we recognize that

$$b_2 = b_3 + \underset{\sim}{T}\,(c_\ell - c_{pv}) + L\ . \tag{C14}$$

If we always refer to water substance in the form of liquid water we may set a convenient value for b_3 at

$$b_3 = -c_\ell T_r \tag{C15}$$

where T_r is a reference temperature, e.g., 0 ^oC. Substitution of (C14) and (C15) in (C11) yields

$$\overline{\rho\, w\, h_m} = \rho\, \overline{wT'}\, (c_p + q(c_{pv} - c_p)) + \rho\, \overline{wq'}\, (c_\ell(T - T_r) + L) \tag{C16}$$

This equation shows that the total specific enthalpy flux is equal to the sensible and latent heat flux. The development given here is suggested by Frank and Emmitt (1981) in response to recent suggestions by Brook (1978), who developed an erroneous correction to the sensible heat flux.

Because $q(c_{pv} - c_p) << c_p$ and $c_\ell(T - T_r) << L$, (C16) may be simplified with an error less than 5%, to

$$\overline{\rho\, w\, h_m} = \rho\, c_p\, \overline{wT'} + \rho\, L\, \overline{wq'}\ . \tag{C17}$$

APPENDIX D. THE EKMAN SPIRAL

In the special case that there is no turbulence Equations (1.33a) and (1.33b) have a steady state analytical solution, which was first described by Ekman (1905) and applied to ocean currents. We shall here review the Ekman solutions for the atmospheric boundary layer and for the drift current in the upper ocean (see also Batchelor, 1967).

First we observe that above the boundary layer, where the viscosity can be neglected, we have the geostrophic balance,so that the velocity is equal to the geostrophic wind, $V=V_g$ and $U=U_g$,given by (1.34).

It is convenient to choose the x-axis in the direction of the geostrophic wind. The Equations (1.33) can then be written

$$fV = -\nu \frac{d^2U}{dz^2}\ , \tag{D1a}$$

$$f(U - G) = \nu \frac{d^2V}{dz^2} \tag{D1b}$$

with boundary conditions $U = V = 0$ for $z = 0$ and $U = G$, $V = 0$ for $z = \infty$. The solution to this set is most easily obtained by introducing a complex velocity $W = (U - G) + iV$. The two Equations introducing a complex velocity $W = (U - G) + iV$. The two Equations (D1) may then be combined into

$$ifW = \nu\, \partial^2 W/\partial z^2 \qquad \text{(D2)}$$

with the general solution

$$W = A\, e^{-(1+i)az} + B\, e^{(1+i)az} \qquad \text{(D3)}$$

where $a = (f/2\nu)^{1/2}$. For $z = 0$, $A = -G$ and for $z = \infty$, $B = 0$, we find, therefore,

$$U = G\,(1 - e^{-az} \cos az) \qquad \text{(D4a)}$$

$$V = G\, e^{-az} \sin az\ . \qquad \text{(D4b)}$$

When $az = \pi$, the velocity vector is in the direction of the geostrophic wind. The height of this point is $z = \pi\sqrt{(2\nu/f)} \simeq 1.7\text{m}$ for $\nu \simeq 0.15\ \text{cm}^2\ \text{s}^{-1}$ and $f \simeq 10^{-4}\ \text{s}^{-1}$. It is unlikely that this laminar solution has ever been observed in the atmosphere because the flow becomes unstable for $R = Gz/\nu > 100$. This means that a laminar Ekman spiral in the atmosphere can only exist for $G < 10^{-3}\ \text{m s}^{-1}$. However, if we assume that ν is a constant eddy viscosity of about $10\ \text{m}^2\ \text{s}^{-1}$ then this solution gives the right order of magnitude for the height of the atmospheric boundary layer.

In case the wind stress at the surface of an ocean at rest drives a surface current, the equations are

$$-\,fV = \nu \frac{\partial^2 U}{\partial z^2} \qquad \text{(D5a)}$$

$$fU = \nu \frac{\partial^2 V}{\partial z^2} \qquad \text{(D5b)}$$

When we choose the x-axis in the direction of the stress the boundary conditions are $\nu(\partial U/\partial z)_0 = \tau_0/\rho = u_*^2$ and $\partial V/\partial z = 0$ for $z = 0$; and $U = V = 0$ for $z \to -\infty$. In this case we use the complex velocity in the form $W = U + iV$ and arrive again at Equation (D2). The solution is

$$U = \frac{u_*^2}{\sqrt{(\nu f)}}\, e^{az} \cos\,(az - \tfrac{1}{4}\pi) \text{ and} \qquad \text{(D6a)}$$

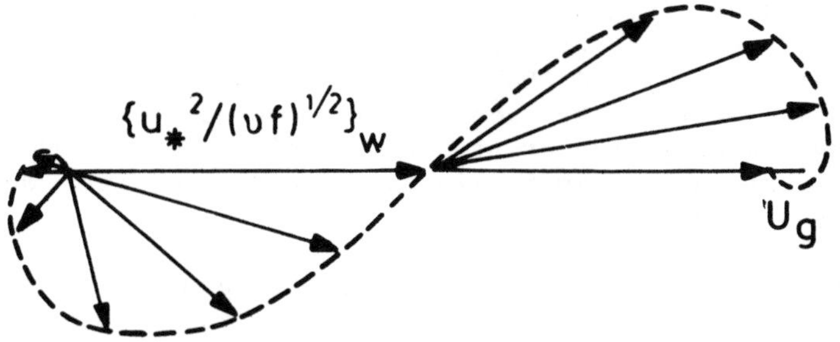

Figure 1.1. The two hodographs described by Equations (D4) and (D6). The scale of the drift velocity of the water has been normalized to appear to have the same magnitude as the wind speed.

$$V = \frac{u_*^2}{\sqrt{(\nu f)}} e^{az} \sin (az - \tfrac{1}{4}\pi) \ . \qquad \text{(D6b)}$$

As can be seen from Equations (D6) the stress is in the direction of the surface wind which makes an angle of 45° with the geostrophic wind. Similarly the stress makes an angle of 45° with the surface current. Consequently the surface current is in the direction of the geostrophic wind. Thus, if we turn the coordinate system 45°, U and V in Equations (D6) will be parallel to U and V of (D4). Furthermore by differentiating (D4) with respect to z and setting z = 0 we find that

$$G = \{\frac{u_*^2}{\sqrt{(\nu f)}}\}_a \qquad \text{(D7)}$$

where the index a refers to air.

The difference between (D4) and (D6) is that the first set of equations is in air and the second set in water. The stress, τ_0, at the surface is the same for water and air. Consequently the ratio

$$\{\frac{u_*^2}{\sqrt{(\nu f)}}\}_a \ / \ \{\frac{u_*^2}{\sqrt{(\nu f)}}\}_w$$

where the index w refers to water, determines the ratio of the velocity scales for air and water. When we assume that the ratio of the eddy viscosities $\nu_a/\nu_w \simeq u_{*a}^2/u_{*w}^2$ then we find that the drift velocity of the water surface is about G/30.

The coordinate system for the atmospheric spiral is moving with the drift velocity of water. This requires a small correction in the Equations (D2) and (D4) which has been neglected here. From Figure 1.1 we see that the ageostrophic component of the flow causes convergence in the atmosphere but divergence in the ocean when the isobars are cyclonically curved and vice versa when the isobars are anticyclonically curved.

These analytical solutions have only diagnostic value in analyzing the problem, but as such are quite useful, as will become clear when more realistic and more sophisticated solutions are discussed.

ACKNOWLEDGEMENT

The material in this chapter was borrowed in large part from the unpublished notes by the author, which were presented at the short course on boundary-layer meteorology held at Boulder, August 1978. Part of the material in Section 1.6 has been taken from John Wyngaard's notes from the same course. The material of Appendices A and B has been formulated by Carl Friehe.

2. SIMILARITY RELATIONS, SCALING LAWS AND SPECTRAL DYNAMICS

H. Tennekes

Royal Netherlands Meteorological Institute,
de Bilt, The Netherlands.

2.1. INTRODUCTION

This chapter is written from the perspective of a theoretician concerned with sensible ways of presenting experimental data. An experimental program, or a series of field observations is not complete when a stack of computer tapes full of numbers is obtained. At that point, the scientist involved faces the task of presenting his or her data in such a way that the results can be understood and employed most easily. Often (not always) this requires that the data be presented in non-dimensional form. Non-dimensional plots are independent of the system of units used, they facilitate order-of-magnitude estimates and comparisons with data obtained elsewhere, and they often contribute to the discovery of simple functional relationships.

It is easy enough to make data non-dimensional. To be a bit facetious, we might claim that, in principle, all velocities one encounters in nature can be compared with the speed of light, and all lengths with the radius of the earth or with the international astronomical unit. The problem, therefore, is not the non-dimensionalization in itself, but the art of selecting smart, sensible scaling parameters. A major theme in this chapter is: how do we scale the characteristic features of the planetary boundary layer in an intelligent way? How do we find the appropriate length, velocity, and temperature scales?

This is no mean task. As a first example, let us consider the wind field in the lowest 100 m of a neutrally stratified, horizontally-homogeneous atmosphere. For the moment, we do not worry too much about the fact that these conditions are seldom found in practice. The wind speed (U) in this 'surface layer' is observed to be of the same order of magnitude as the geostrophic wind speed, G. Therefore, U/G is of order one, and G appears to be an appropriate velocity scale for the wind field near the earth's surface. But how do we know that G is the proper choice?

The wind speed in the surface layer is a function of height. What is the appropriate scale for z? Two choices appear possible: we can either select the surface roughness length, z_0, or a length scale G/f which is related to the synoptic pressure gradient and the presence of ageostrophic velocity components in the boundary layer. We conclude that there are two length scales in this problem, a small one (z_0) and a very large one (G/f). Their ratio, the surface Rossby number G/fz_0, is a central non-dimensional parameter in planetary boundary-layer flow.

If we decide to use the roughness height (z_0) as the appropriate scale for z in the surface layer, we could write that U/G must be a function of z/z_0. This statement, however true, is also incomplete because we have no reason to assume that U/G is not a function of the surface Rossby number. But the statement that U/G is a function of both z/z_0 and G/fz_0 is so general that it serves no practical use in data presentation. What can we do instead?

Looking for an alternative approach, we note that the surface stress τ_0 must also be a function of the surface Rossby number. The characteristic velocity based on the surface stress is called the friction velocity u_*; it is defined by

$$\tau_0 = \rho u_*^2 \ . \qquad (2.1)$$

If τ_0 is a function of G/fz_0, then so is u_*, we put

$$u_*/G = F_g(G/fz_0) \ . \qquad (2.2)$$

Since we suspect that u_*/G increases as G/fz_0 decreases, we are forced to decide which of the two velocity scales (u_* or G) should be used for non-dimensional wind profiles in the neutrally stratified planetary boundary layer.

The friction velocity is not an independent, external parameter as far as the flow in the planetary boundary layer is concerned. However, it does account for the effects of the large-scale pressure field and the surface roughness. Also, since the surface stress equals the turbulent momentum flux in the air just above the surface, u_* is in some sense representative of the turbulent wind fluctuations in the lower layers. Finally, since the surface wind is observed to be aligned with the surface stress the use of u_* as a velocity scale does not involve any problems that arise from the difference in direction between τ_0 and G. Therefore, the use of u_* as a velocity scale for the wind in the lowest 100 m seems to have advantage above the use of G. This choice can be defended on theoretical grounds; those will be presented in Section 2.2.

As we have suggested above, the most appropriate scale for the height z is the roughness height z_0. Therefore, we can write the wind profile in a neutral surface layer as

$$U/u_* = f_x(z/z_0) \ . \qquad (2.3)$$

Plotting wind profiles in this form, we may expect that they will not show any explicit dependence on the surface Rossby number. This advantage does not come without cost: the friction velocity is a dependent, internal parameter of the boundary layer, so that the surface-layer similarity law (2.3) is a relationship based on internal parameterization. Through Equation (2.2), (2.3) contains an implicit dependence on G/fz_0; the problem of data representation cannot be considered solved until both F_g in (2.2) and f_x in (2.3) are known. This is the subject of the next section.

2.2. ROSSBY-NUMBER SIMILARITY

The equations of motion for steady, horizontally homogeneous turbulent flow in a neutrally stratified atmosphere without heat transfer are given by (1.33a) and (1.33b). We shall neglect here the viscous stresses $\nu\partial^2U/\partial z^2$ and $\nu\partial^2V/\partial z^2$. The amplitude of the geostrophic wind is defined as $G = \sqrt{(U_g{}^2 + V_g{}^2)}$. It is convenient to align the x-axis with the direction of the surface stress τ_0. At the surface roughness level $z = z_0$, we have

$$\tau_x = -\rho\ \overline{uw} = \tau_0 = \rho\ u_*^2, \qquad (2.4)$$

$$\tau_y = -\rho\ \overline{vw} = 0 \ . \qquad (2.5)$$

These boundary conditions suggest that it is advantageous to non-dimensionalize the Reynolds-stress components $\rho\,\overline{uw}$ and $\rho\,\overline{vw}$ occurring in (1.33a) and (1.33b) by the surface stress. That choice keeps the normalized stresses $\overline{uw}/u_*^2$ and $\overline{vw}/u_*^2$ finite, no matter how large the surface Rossby number becomes. The selection of a normalizing velocity for the velocity differences $(U - U_g)$ and $(V - V_g)$ is not quite as straightforward. Here, we need to borrow from the observed behavior of other kinds of boundary layers (Tennekes and Lumley, 1972, Chapter 5). Experience with laboratory flows suggests that u_* is the appropriate velocity scale for the ageostrophic wind components; this choice is consistent with the use of u_* as a normalizing factor for the wind profile in the bottom 100 m (See Section 2.1). The basic idea behind this is that G and its components U_g, V_g are irrelevant in the frame of reference used in (1.33a) and (1.33b). The Galilean transformation involved in taking $U - U_g$ and $V - V_g$ as dependent variables amounts to observing the planetary boundary layer from a balloon drifting along with the geostrophic wind. The wind differences observed from that balloon are caused by friction; it seems logical to scale them on the friction velocity.

Continuing this approach, we decide to absorb the Coriolis parameter f and the remaining u_* by constructing a non-dimensional height zf/u_*. This line of reasoning transforms (1.33a) and (1.33b) into the non-dimensional set

$$\frac{V - V_g}{u_*} = \frac{d(\overline{uw})/u_*^2}{dzf/u_*}, \tag{2.6}$$

$$\frac{U - U_g}{u_*} = -\frac{d(\overline{vw})/u_*^2}{dzf/u_*}. \tag{2.7}$$

In this way, all free parameters have been absorbed; equations (2.6) and (2.7) do <u>not</u> depend explicitly on G/fz_0. Furthermore, in this non-dimensional presentation all boundary conditions at the top of the boundary layer are homogeneous: both the stress components and the ageostrophic wind components are zero outside the boundary layer. The boundary conditions at the surface do pose problems, however. The normalization of the stresses makes the boundary conditions on the stress parameter-free, but the stress divergence at the surface has the components V_g/u_* and U_g/u_*, both of which presumably are functions of the surface Rossby number. This way, an explicit dependence on G/fz_0 can enter through the back door, as it were.

Since G/u_* is an increasing function of G/fz_0 (the surface stress decreases with decreasing surface roughness), the surface boundary condition on the stress divergence leads to a stress singularity in the formal limit process $G/fz_0 \rightarrow \infty$. We cannot avoid that singular behavior if we insist that (2.6) and (2.7) satisfy all of the boundary conditions at the surface. The mathematical techniques known as singular-perturbation methods (Van Dyke, 1966; Cole, 1968) suggest that the proper approach in circumstances such as these is to ignore the surface boundary conditions, and to subject the vicinity of the singular point to a separate-scale analysis.

If the system (2.6), (2.7) does not need to satisfy the surface conditions, it and its upper conditions are free of explicit parameters. Therefore, its solution is expected to be parameter-free. In particular, we expect that

$$\frac{U - U_g}{u_*} = F_x\left(\frac{zf}{u_*}\right) \quad \text{and} \quad \frac{V - V_g}{u_*} = F_y\left(\frac{zf}{u_*}\right) \tag{2.8}$$

are universal functions of the nondimensional height zf/u_*. Plotted according to (2.8), the wind profiles of a neutral, barotropic boundary layer does not depend on the surface Rossby number. This is called Rossby-number similarity of the ageostrophic wind in neutrally-stratified planetary boundary layers. It should be noted that Rossby-number similarity is not necessarily a terribly practical concept in the context of wind profile measurements in the field, since the atmospheric boundary layer is seldom stationary, neutrally stratified, and horizontally homogeneous all at the same time. Nevertheless, the conceptual framework of Rossby similarity can lead to very useful approximations (e.g., the interpolation of wind profiles, in Van Ulden and Holtslag, 1980).

One consequence of (2.8) is that the top of the boundary layer is located by this theory at a fixed value of the nondimensional height zf/u_*. Therefore, Rossby-number similarity theory suggests that the height h of neutrally stratified planetary boundary layers is given by

$$h = c\, u_*/f \,, \tag{2.9}$$

where c is an unknown constant. It should be kept in mind that this expression refers to an atmosphere without potential-temperature differences and without vertical heat flux; those introduce other parameters (to be discussed later in this chapter), which makes the issue of determining the height h much more complicated. In experimental practice, it is often preferable to use the observed boundary-layer height, if only because h tends to change appreciably during the course of a diurnal cycle, even if u_* does not.

The Surface Layer

We now turn to a scale analysis of what happens to the equations of motion near the surface. As we have seen earlier, the roughness height z_o is the appropriate length scale for the flow in the immediate vicinity of rough surfaces, and u_* is the appropriate velocity scale. On that basis, (1.33a) and (1.33b) are normalized as follows:

$$- \frac{fz_o}{u_*} \frac{V - V_g}{u_*} = - \frac{d(\overline{uw})/u_*^2}{dz/z_o} , \quad (2.10)$$

$$+ \frac{fz_o}{u_*} \frac{U - U_g}{u_*} = - \frac{d(\overline{vw})/u_*^2}{dz/z_o} . \quad (2.11)$$

The left-hand sides of these equations are quite small. Typical numbers are $U_g = 10\ m\ s^{-1}$, $V_g = 5\ m\ s^{-1}$, $u_*^2 = 0.1\ m^2s^{-2}$, $f = 10^{-4}\ s^{-1}$ and $z_o = 0.01\ m$. The left-hand side of (2.10) then is at most $5\ 10^{-5}$, and the left-hand side of (2.11) is at most 10^{-4}. It seems plausible to assume that these numbers would decrease with further increases in the value of the surface Rossby number, so that in the limit, as $G/(fz_o) \to \infty$, the left-hand side of (2.10) and (2.11) vanishes:

$$\frac{d(\overline{uw})/u_*^2}{dz/z_o} = 0 , \qquad \frac{d(\overline{vw})/u_*^2}{dz/z_o} = 0 . \quad (2.12)$$

With the boundary conditions (2.4), (2.5) on the stress, the result is that the shear stress is approximately constant for all values of z/z_o that are not too large:

$$- \overline{uw} = u_*^2 , \qquad - \overline{vw} = 0. \quad (2.13)$$

This is the basis for the theoretical concept of a 'constant stress layer' near the bottom of the planetary boundary layer. Note that this concept is a consequence of the scaling adopted here: the original equations of motion show that the shear stress changes most rapidly near the surface, but (2.12) suggests that these changes are imperceptibly slow if we use focus on very large values of G/fz_o and use a telescope that restricts our field of view to finite values of z/z_o.

With such a severely restricted field of view, we cannot see the geostrophic wind at the top of the planetary boundary layer, and the appropriate way to present wind profiles over land is a generalization to (2.3)

$$U/u_* = f_x(z/z_o), \qquad V/u_* = 0 \; . \tag{2.14}$$

The second of these is trivial because the surface stress has no y-component and because the order-of-magnitude analysis following (2.10) and (2.11) has shown that the effects of the Coriolis parameter are negligible in the lowest layers, so that there is no reason to suspect that the wind can turn with height. The region of validity of (2.14) is called the surface layer. In the surface layer, the wind is parallel to the surface stress. Also, since the effects of the Coriolis force are negligible here, (2.14) does not depend explicitly on the surface Rossby number G/fz_o. This non-dimensional wind profile, like its counterpart in the layers above, thus exhibits Rossby-number similarity. Fortunately, surface-layer similarity of the type expressed in (2.14) is not a rare occurrence in field observations. This is in contrast with the seldom-observed similarity of (2.8). Because of the relatively small effective time and length scales of the surface layer (z/u_* and z, respectively), it is not impossible to find conditions that are sufficiently stationary and homogeneous to obtain data without excessive scatter.

The Process of Matching

The surface-layer similarity law (2.14) is obviously useless at extremely large values of z/z_o. The other similarity laws, (2.8), on the other hand, cannot be extended to low values of z/z_o because of the very steep gradients there. The missing boundary conditions are now supplied by a process known in singular-perturbtation theory as 'asymptotic matching'. Matching is a technique by which one requires that the upper-level similarity laws (2.8) are identical with the surface-layer similarity law (2.14) in the double limit process $zf/u_* \to 0$ (looking down towards the surface from the middle levels in the planetary boundary layer) and $z/z_o \to \infty$ (looking upwards from the surface layer).

Because of the Galilean transformation between the first members of (2.8) and (2.14), it is convenient to do the matching of the U-component on the wind-shear component $\partial U/\partial z$. From (2.8) we get

$$\frac{\partial U}{\partial z} = f \frac{dF_x}{d\eta} \qquad (\eta = zf/u_*) \; , \tag{2.15}$$

and from (2.14), we obtain

$$\frac{\partial U}{\partial z} = \frac{u_*}{z_o} \frac{df_x}{\partial \xi} \qquad (\xi = z/z_o) \; . \tag{2.16}$$

These two have to be identical for the double-limit process $\eta \to 0$, $\xi \to \infty$. Equating (2.15) and (2.16) and multiplying both by z/u_*, we obtain

$$\frac{z}{u_*} \frac{\partial U}{\partial z} = \eta \frac{dF_x}{\partial \eta} = \xi \frac{df_x}{d\xi} . \tag{2.17}$$

The expression in the center is a function of η only, the one at the right is a function of ξ only. However, η is approaching zero and ξ is increasing beyond bounds; the only way in which the two can be the same without producing singular or trivial results is one by which <u>both</u> approach a non-zero asymptote:

$$\frac{z}{u_*} \frac{\partial U}{\partial z} = \frac{1}{k} \qquad (\xi \to \infty, \eta \to 0) . \tag{2.18}$$

Here, k is an asymptotic number; it is called the Von Kármán constant. In practical terms (2.18) states that, in a layer where z/z_0 is so large that the effects of z_0 are negligible, while zf/u_* is so small that the effects of f must be negligible, the wind shear $\partial U/\partial z$ can depend only on the friction velocity u_* and the height z. The earliest formulation of (2.18) were all in one way or another, based on the assumption that there must be a layer in which the height z has to serve as its own length scale. In fact, as our analysis shows, the existence of a layer free of explicit dependence on external scale lengths is a direct consequence of the double-limit process $z/z_0 \to \infty$, $zf/u_* \to 0$.

Integration of (2.18) yields to the well-known <u>logarithmic wind profile</u>:

$$U/u_* = \frac{1}{k} \ln(z/z_r), \tag{2.19}$$

where z_r is an arbitrary integration constant. Since the asymptotic analysis shows that this profile obeys both the surface-layer law (2.14) and the upper-level law (2.8), there are <u>two</u> alternative forms of (2.19)

$$U/u_* = \frac{1}{k} \ln(z/z_0) \qquad (z/z_0 \gg 1), \tag{2.20}$$

$$(U - U_g)/u_* = \frac{1}{k} [\ln(zf/u_*) + A] \qquad (zf/u_* \ll 1). \tag{2.21}$$

The region of validity of (2.20) and (2.21) has been called the

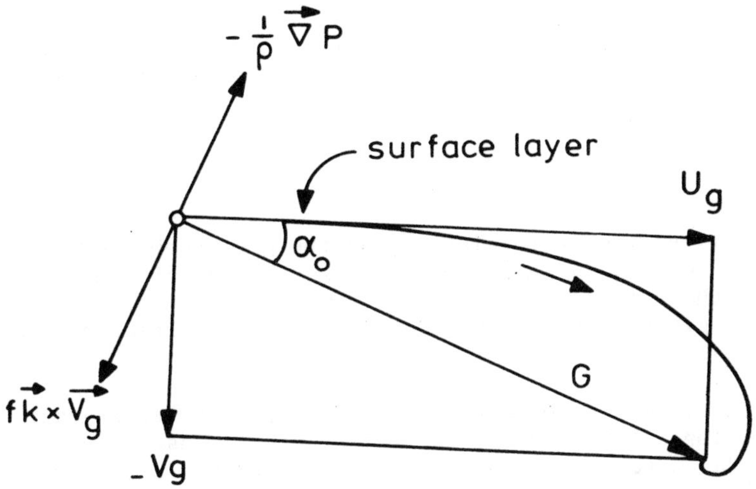

Figure 2.1: The wind spiral in a neutral barotropic boundary layer.

'inertial sublayer' (Blackadar and Tennekes, 1968), this in reference to the analogy with the inertial subrange in the turbulent kinetic energy spectrum (Tennekes and Lumley, 1972).

Taking the difference between (2.20) and (2.21), we obtain

$$\frac{U_g}{u_*} = \frac{1}{k} (\ln u_*/fz_0 - A) . \tag{2.22}$$

A corresponding relation for V_g/u_* can be found by equating the second members of (2.8) and (2.14) in the vicinity of the surface. That yields

$$\frac{V_g}{u_*} = F_y(0) = -\frac{B}{k} . \tag{2.23}$$

The angle α_o between the surface wind and the geostrophic wind is equal to $\arctan(V_g/U_g)$ (see Figure 2.1), so that

$$\alpha_0 = \arctan(-V_g/U_g) = \arctan\left(\frac{B}{k}\frac{u_*}{U_g}\right). \tag{2.24}$$

Here, u_*/U_g can be determined from (2.22); note that the angle increases with increasing surface roughness. That was to be expected.

With the definition of the amplitude G of the geostrophic wind, $G = \sqrt{(U_g^2 + V_g^2)}$, we obtain from (2.22) and (2.23)

$$\frac{G}{u_*} = \frac{1}{k}\left(\{\ln(u_*/fz_0) - A\}^2 + B^2\right)^{1/2}. \tag{2.25}$$

This is an implicit relation for the geostrophic drag coefficient u_*/G as a function of the surface Rossby number G/fz_0. The logarithmic friction law (2.25) completes the formal analysis, because it gives a formula for determining the surface stress if the external parameters G, f, and z_0 are known.

Since planetary boundary layers hardly ever satisfy all of the conditions that Rossby-number similarity imposes, the experimental data on the constants A and B in (2.22), (2.23) and (2.25) show a considerable amount of scatter. A comprehensive table of proposed values is presented by McBean (1979). Further information on these constants is given in the next section.

The Constant-Stress Layer

The logarithmic wind profile is a key feature of the inertial sublayer. That layer is found at large values of z/z_0, but small values of zf/u_*. In practice, this means that z has to be well above the top of the roughness elements, but below about one-tenth or so of the height of the planetary boundary layer. The inertial sublayer is not really a constant-stress layer. If we substitute (2.23) into (2.10), and recall that V = 0 near the surface, we obtain

$$-\frac{fB}{ku_*} = -\frac{d}{dz}(\overline{uw})/u_*^2 , \tag{2.26}$$

so that

$$-\frac{\overline{uw}}{u_*^2} = 1 - \frac{fBz}{ku_*} . \tag{2.27}$$

For $f = 10^{-4}$ s^{-1}, B = 5 (typical of the values reported in the literature), k = 0.35 (Businger et al., 1971), and $u_* = 0.3$ m s^{-1} the x-component of the Reynolds stress changes about 5% for every 10 m of increase in z. If we want the concept "surface layer" to be interchangeable with "constant-stress layer", and if we allow a 10% overall change in $\overline{uw}$, then the surface layer is only about 20 m thick in typical circumstances. In neutrally-stratified conditions, the logarithmic wind profile is often accurate up to about 100 meters; clearly, the inertial sublayer is thicker than the 'constant-stress' layer.

The Von Kármán Constant

One of the practical problems associated with the equations derived above is that the numerical value of the Von Kármán constant has to be obtained from experimental data. Rossby-numer similarity theory is an asymptotic, first-order theory; it becomes more accurate as G/fz_0 increases (meadows, prairies), but loses accuracy at relatively low values of G/fz_0 (forests, cities). The theory claims that k is a 'universal constant' in an asymptotic sense; it cannot give guidance on what deviations to expect at values of G/fz_0 encountered in practice. From this perspective, it comes as no surprise that no two researchers agree on the value of k. Numbers quoted range all the way from 0.33 (Tennekes, 1968) to 0.41 (Hinze, 1959). In micrometeorology, the value k = 0.35 obtained in the Kansas experiments of AFCRL (Businger _et al._, 1971) gained some acceptance in the seventies. Most researchers now again use k = 0.4, partly because doubts have been raised about the Kansas experiments (Wieringa, 1980).

Arguments about the correct value of k have been going on for 50 years. In 1930, Von Kármán proposed k = 0.36 (Prandtl and Tietjens, 1957, p. 73); Nikuradse's classic experiments, reported later that same year, gave a value of 0.44 (Prandtl and Tietjens, 1957, p. 74). Perhaps one should restrain the urge to expound on this issue, and state simply that

$$k = 0.4 \pm 0.04 \tag{2.28}$$

represents the state of the art.

In practice, the scatter indicated in (2.28) means that when one uses $(z/u_*)\ \partial U/\partial z = 1/k$ to determine u_* from wind-shear data (a common procedure), there is a 10% uncertainty in the result. That corresponds to a 20% error in the surface stress. For this reason alone, it is advisable to include direct stress measurements (preferably by making eddy-flux measurements with fast-response instrumentation) in all surface-layer measurements.

2.3. DIABATIC EXTENSION OF ROSSBY-NUMBER SIMILARITY

The similarity theory of neutral planetary boundary layers given above is of value, primarily because it provides a theoretical framework for boundary-layer research. However, the theory is based on steady, horizontally homogeneous flow in an adiabatic atmosphere (the potential temperature has to be independent of height not only inside the boundary layer, but also in the air aloft); those conditions are seldom, if ever, met simultaneously in the world outside laboratory windows. Rossby-number similarity provides a sound theoretical foundation, but a shaky reference for practical applications. Therefore, sound practical judgement is required of persons involved in boundary layer work.

In diabatic situations (with a non-zero surface heat flux) the similarity laws derived above have to be generalized. Since the theory is cast in terms of kinematic variables (lengths, times, velocities), it is logical to ask if the surface heat flux adds a new kinematic scale to the problem. Indeed it does: the length L, defined by (1.61) can be used to represent the effects of the surface heat flux $Q_o = \rho c_p \overline{w\theta_o}$ on the various functions involved in the theory. Because Obukhov (1946, 1971) first defined it, L is called the Obukhov length (see also Monin and Yaglom, 1971). The effects of L on wind and temperature profiles in the surface layer will be discussed in Section 2.4; here we analyze the complications that arise in the similarity laws for the atmospheric boundary layer as a whole.

If the surface heat flux differs from zero, the functions involved in Rossby-number similarity theory depend not only on the non-dimensional height zf/u_*, but also on the non-dimensional parameter Lf/u_*. Also, the height h of the planetary boundary layer, which is supposedly proportional to u_*/f in neutral cases, is likely to be a function of stability

$$hf/u_* = \Lambda(Lf/u_*) \quad . \tag{2.29}$$

This does not exhaust the potential complications. Cases with heat flux are unlikely to be steady; horizontal temperature gradients create thermal winds so that the geostrophic wind becomes a function of height, the temperature profile in the air above the boundary layer may not be adiabatic, inversions in the air aloft may limit the boundary-layer height, and so forth. All of these severely limit the practical usefulness of Rossby-number similarity theory.

It is convenient to generalize the relations occurring in the similarity theory by allowing the height h to function as an 'independent' variable. This choice forces one to make a seperate study of the way in which h is determined by initial and boundary conditions. Here, we give the principal results of the generalized Rossby-number theory. In the geostrophic drag relation (2.25) and the corresponding relation for the angle α_o between the surface stress and the geostrophic wind, the scale height u_*/f is replaced by the actual height h, the geostrophic wind components U_g and V_g are replaced by the actual components U_h and V_h at z = h (this is believed to reduce the influence of thermal wind components on the similarity laws), and the 'constants' occurring in the

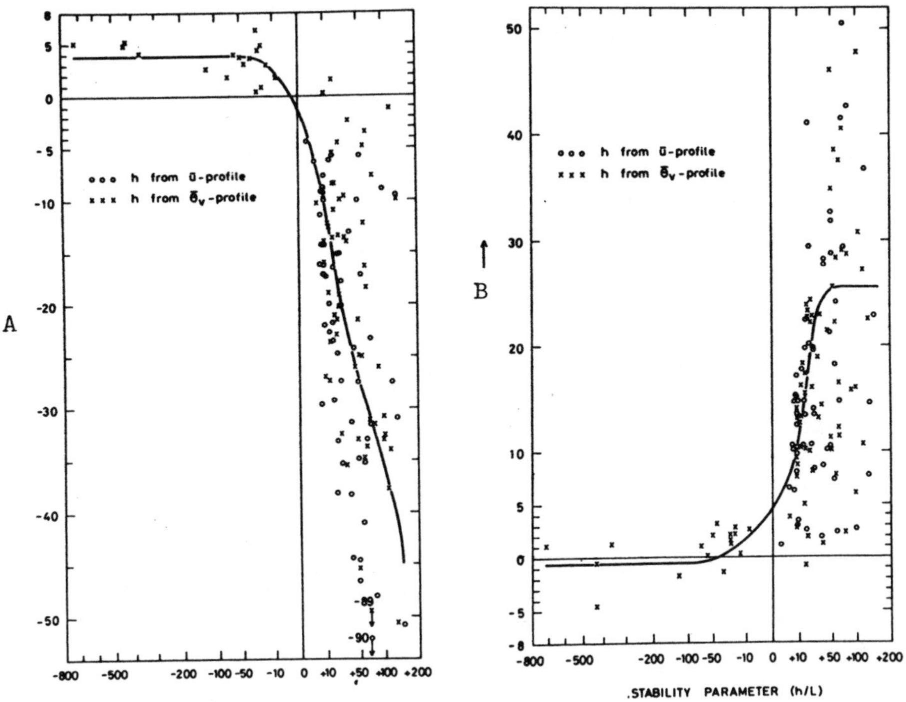

Figure. 2.2. The angle stability function B and the resistance stability function A as a function of h/L (Melgarejo and Deardorff, 1974).

relations are allowed to be functions of the stability parameter $\mu = h/L$. This yields (Zilitinkevich and Deardorff, 1974; Monin and Zilitinkevich, 1974)

$$\frac{V_h}{u_*} = -\frac{B(\mu)}{k}, \quad \alpha_o = \arctan\left(\frac{B(\mu)}{k}\frac{u_*}{U_h}\right) \tag{2.30}$$

$$\frac{(U_h^2 + V_h^2)^{1/2}}{u_*} = \frac{1}{k}\left(\{\ln(\frac{h}{z_o}) - A(\mu)\}^2 + B^2(\mu)\right)^{1/2} \tag{2.31}$$

The functions $A(\mu)$ and $B(\mu)$ are plotted in Figures 2.2 (taken from Melgarejo and Deardorff, 1974).

Considering the arbitrariness of the choices involved in (2.30) and (2.31), we should not be surprised to learn that no two investigators agree on the best way to proceed. Arya (1975) advocates a version in which $\mu = h/L$ is taken to be the independent parameter, but the surface geostrophic wind components U_{go} are used instead of the actual wind height. Clarke and Hess (1974) use the wind speed components at a fixed value of u_*/f. Zilitinkevich and Chalikov (1968) use the wind components at a fixed height, which may be different from the actual height h. In all cases, the scatter is comparable to that in the figures given here. There are indications that much of the scatter is caused by baroclinic effects (Wyngaard and Arya, 1974). For a recent summary of opinions on this issue, see McBean (1979).

2.4. MONIN-OBUKHOV SIMILARITY IN THE SURFACE LAYER

In a neutrally stratified inertial sublayer, the mean wind profile is logarithmic because the mean wind shear does not depend on an externally imposed scale length. If $z/z_o >> 1$ and $z/h << 1$, the absence of an external dynamic length forces us to use the height z, because that is the only length available. However, if the surface heat flux differs from zero, the Obukhov length has to be taken into account. The only way in which L can be used in nondimensional relations for the surface layer is through the nondimensional height $\zeta = z/L$. The adoption of ζ as the fundamental parameter in diabatic surface layers is the basic premise of what is called Monin-Obukhov similarity (Monin and Yaglom, 1971).

According to Monin-Obukhov similarity theory, the non-dimensional wind shear in the inertial sublayer must be a function of ζ

$$\frac{kz}{u_*}\frac{\partial U}{\partial z} = \phi_m(\zeta) . \tag{2.32}$$

Here, the normalization is such that $\phi_m(0) = 1$.

In diabatic conditions, the potential-temperature profile also needs to be presented in nondimensional form. We define the temperature scale θ_*, in analogy with u_*, by the relation

$$\overline{w\theta}_o = - u_*\theta_* . \tag{2.33}$$

The temperature counterpart of (2.32) then becomes

$$\frac{kz}{\theta_*}\frac{\partial \theta}{\partial z} = \phi_h(\zeta) . \tag{2.34}$$

Similarity theory has no methods to help decide on the shapes of the functions ϕ_m and ϕ_h; that information has to be provided by

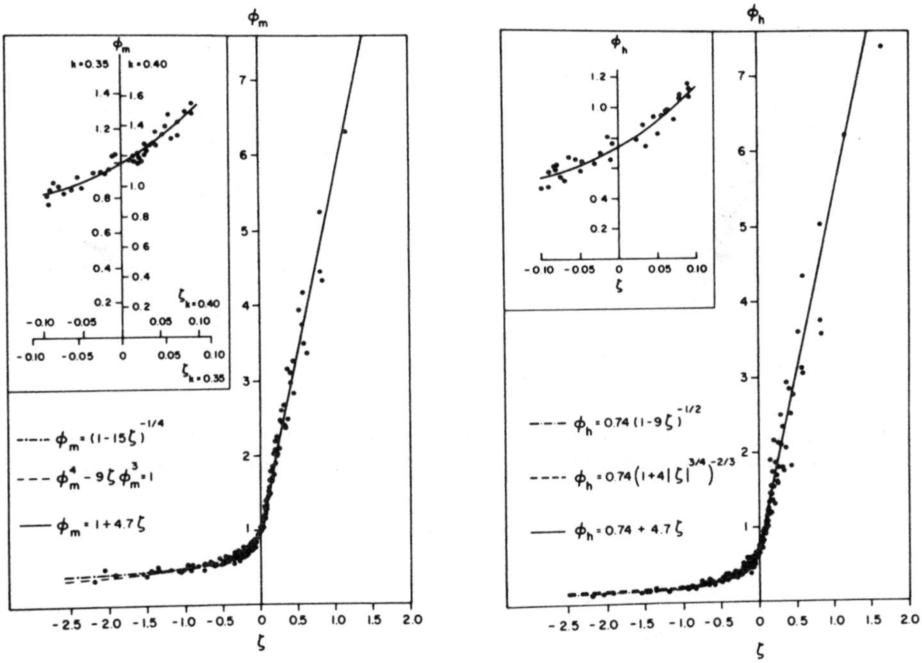

Figure 2.3. Comparison of dimensionless wind shear and of dimensionless temperature gradient observations with interpolation formulas (Businger et al., 1971).

experimentation. The data by Businger et al. (1971) are presented in Figure 2.3. Yaglom (1977) gives an authoritative review of the many different interpolation formulas for ϕ_m and ϕ_h that have been proposed in the literature.

The observed behavior of ϕ_m and ϕ_h for various values of the stability parameter ζ requires further discussion. If we define the exchange coefficients for momentum and heat, K_m and K_h, by

$$- \overline{uw}_o = u_*^2 = K_m \frac{\partial U}{\partial z} \quad \text{and} \quad -\overline{w\theta}_o = u_* \theta_* = K_h \frac{\partial \Theta}{\partial z} \qquad (2.35)$$

we obtain the following relations:

$$\alpha \equiv K_h/K_m = \phi_m / \phi_h \qquad (2.36)$$

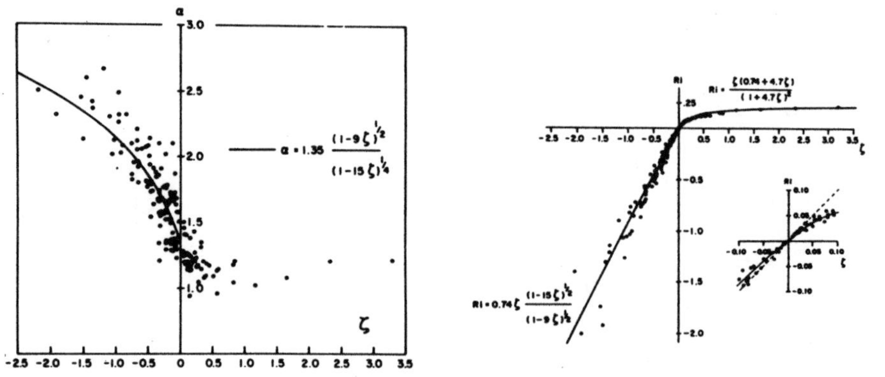

Figure 2.4. The dependence of the ratio of eddy diffusivities and the dependence of the Richardson number on stability (Businger et al. 1971).

$$\mathrm{Ri} \equiv \frac{g}{\Theta} \frac{\partial\Theta/\partial z}{(\partial U/\partial z)^2} = \zeta \frac{\phi_h}{\phi_m^2} \,. \qquad (2.37)$$

Since ϕ_h and ϕ_m are functions of $\zeta = z/L$ only, the exchange-coefficient ratio α and the Richardson number Ri (introduced in eq. (1.60)) are also functions of ζ only. The data on $\alpha(\zeta)$ and $\mathrm{Ri}(\zeta)$ obtained by Businger et al. (1971) are plotted in Figure 2.4 The neutral value of α is 1.35, corresponding to a 'turbulent Prandtl number' $K_m/K_h = 0.74$. The neutral value of ϕ_h, therefore, is 0.74 (recall that ϕ_m is defined such that its neutral value is 1).

A better appreciation for the nature of these empirical relations may be obtained by looking at their behavior in very stable and very unstable conditions. In very stable conditions ($\zeta > 0.5$, say), vertical eddy motions cannot penetrate far from the height at which they originate. This suggests that the mean wind and temperature gradients at any given height z do not depend on z if the stratification is very strong. Since the Obukhov length L is the only other length scale available, we expect that $\partial U/\partial z$ will tend to be proportional to u_*/L. The non-dimensional gradients ϕ_m and ϕ_h will then be proportional to ζ, the exchange-coefficient ratio α will approach a constant value, and the Richardson number will also aproach a constant value. All of these features are confirmed by the data presented in Figures 2.3 and 2.4

The behavior of the various curves in very unstable conditions cannot be explained with simple scaling assumptions.

The data show very clearly that K_h/K_m increases with values of $-\zeta$. According to (2.36), this implies that ϕ_m decreases more slowly than ϕ_h with increasing instability. For large values of $-\zeta$, Businger et al. (1971) suggest that

$$\phi_m \propto (-\zeta)^{-\frac{1}{4}} , \quad \phi_h \propto (-\zeta)^{-\frac{1}{2}} . \tag{2.38}$$

This implies that α is roughly proportional to $(-\zeta)^{\frac{1}{4}}$ when $-\zeta$ is large enough, and that the Richardson number Ri is roughly proportional to ζ. These features are borne out by the curves in Figures 2.4. It is worth noting that the simple expression

$$Ri = \zeta \tag{2.39}$$

is a good overall approximation for unstable conditions.

2.5. SCALING OF TURBULENCE QUANTITIES IN THE SURFACE LAYER

In the Boussinesq approximation, the equations for the mean flow in the atmospheric boundary layer do not depend explicitly on the vertical heat flux, so that it is difficult to determine how the mean wind and temperature profiles are affected by buoyant forcing. The functional dependence on the stability parameters $\zeta = z/L$ and $\mu = h/L$ thus had to be determined from experimental evidence. The study of the behavior of turbulent kinetic energy and temperature variance, however, can be guided by the conservation equations for these quantities. The information contained in these equations makes it possible to give quite specific scaling rules in a number of cases.

Since a vertical heat flux in a flow field, subject to the acceleration of gravity, leads to the conversion of potential energy into kinetic energy (or vice versa), the turbulent kinetic energy equation is a suitable starting point. For horizontally-homogeneous flows it is given by (1.57). Here we consider quasi-stationary conditions, so that $\partial q^2/\partial t = 0$.

When working with (1.57), one generally assumes that the surface heat flux and the surface stress are known, and that at all heights within the surface layer, the turbulent fluxes of momentum and heat are equal to the surface fluxes. The first real issue that arises is that the dissipation rate ε has to be parameterized in terms of velocity variance and some appropriate length scale. According to one of the most fundamental premises of turbulence theory (Tennekes and Lumley, 1972), we may write

$$\varepsilon = c_\varepsilon \, \sigma_w^3/\ell , \tag{2.40}$$

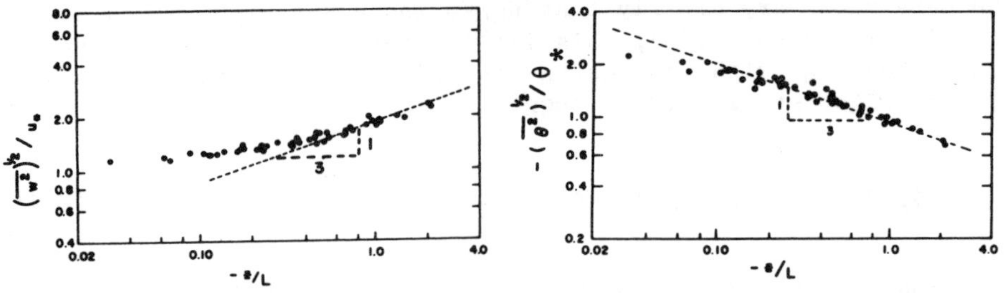

Figure 2.5. Dimensionless rms vertical velocity and temperature fluctuations under unstable conditions. The 1/3-slopes correspond to free convection in the surface layer (Wyngaard et al., 1971).

where σ_w is the standard deviation of vertical velocity and ℓ is a length which characterizes the dimensions of the most energetic eddies in the flow. The coefficient c_ε is of order one. We have chosen σ_w as respresentative of the turbulence intensity because it turns out to be rather insensitive to mesoscale disturbances associated with topography or inhomogeneous boundary conditions.

It is worth pointing out that (2.40) asserts that the rate at which turbulence loses its kinetic energy, due to internal viscous friction, is independent of viscosity. The rate at which energy is dissipated depends only on the characteristic length and velocity of the most energetic eddies. We can illustrate the usefulness of (2.40) with a simple example. In a neutrally stratified surface layer, the Reynolds stress $-\rho\,\overline{uw}$ is approximately equal to the surface stress τ_0. Since observations show that the correlation between u and w is fairly good, we may assume that σ_u and σ_w are proportional to u_*, which is independent of height. The flux divergence term of (1.57) thus should be fairly small, so that the energy budget reduces to $0 = u_*^2 \partial U/\partial z - u_*{}^3/\ell$. (Unknown constants have been absorbed into the length scale ℓ.) For a neutral surface layer, ℓ must be proportional to z (see Section 2.2). We obtain

$$0 = u_*^2 \frac{\partial U}{\partial z} - \frac{u_*^3}{kz} \; . \tag{2.41}$$

It is clear that this leads to a confirmation of the logarithmic wind profile, and the coefficient involved must the Von Karman constant k. The turbulent energy budget thus supports the scaling laws for the surface layer which were derived in Section 2.2.

As we have seen, σ_w/u_* is independent of height in a neutral surface layer. However, in diabatic conditions the Obukhov length L enters as an independent parameter. Since σ_w/u_* can depend on nondimensional parameters only, we now have to allow a possible dependence on z. We postulate that the effects of L manifest themselves as a dependence on $\zeta = z/L$. We obtain

$$\sigma_w/u_* = \phi_w(\zeta) \ . \tag{2.42}$$

This, of course, is consistent with Monin-Obukhov similarity in the surface layer.

The presence of a heat flux creates temperature fluctuations. These are made non-dimensional with the temperature scale θ_* defined in (2.33)

$$-\sigma_\theta/\theta_* = \phi_\theta(\zeta) \ . \tag{2.43}$$

Since evidence indicates that w and θ are well correlated, and since $\overline{w\theta} = -\theta_* u_*$, we may expect that $-\sigma_w\sigma_\theta \sim \theta_* u_*$, so that

$$\phi_w \, \phi_\theta = \text{constant} \ . \tag{2.44}$$

Experimental data on the behavior of ϕ_θ and ϕ_w under unstable conditions were obtained by Wyngaard et al., (1971); their results are presented in Figures 2.5. The neutral values of σ_w/u_* and σ_θ/θ_* are about 1.2 and 2, respectively; the value of $\phi_w \, \phi_\theta$ thus is nearly two.

The general form of the functions ϕ_θ and ϕ_w cannot be predicted by dimensional analysis. However, it is not difficult to make specific predictions if $-\zeta$ is large. In free convection, the wind shear and the surface stress are relatively small, but the large upward heat flux leads to rapid conversion of potential energy into kinetic energy. The flux divergence and dissipation terms in (1.57) both scale as σ_w^3/ℓ; because we are in the surface layer ℓ must scale on z. An approximate energy budget for free convection in the surface layer thus must read

$$0 = \frac{g}{\Theta}\,\overline{\theta w} - c^3 \frac{\sigma_w^3}{z} \ , \tag{2.45}$$

where c is the undetermined constant.

Since the heat flux is approximately independent of height in the surface layer, we obtain

$$\sigma_w = c^{-1} \left(z \frac{g}{\Theta} \overline{w\theta}_o\right)^{1/3}. \tag{2.46}$$

Dividing this through u_*, we find

$$\phi_w = \frac{\sigma_w}{u_*} = c_w(-\zeta)^{1/3}. \tag{2.47}$$

The corresponding prediction for σ_θ in conditions of free convection is, when we take (2.44) into account,

$$\phi_\theta = -\frac{\sigma_\theta}{\theta_*} = c_\theta(-\zeta)^{-1/3}. \tag{2.48}$$

This behavior is supported by the evidence presented in Figure 2.5. The value of c_w is about 2; the value of c_θ is approximately 0.9. This makes the product $\phi_\theta\phi_w$ roughly 1.8, which is not too different from the neutral value. This suggests that (2.44) is a fair approximation for all unstable conditions.

Dimensional analysis gives little guidance on the properties of turbulence in stable surface layers. There is some evidence that σ_θ/θ_* stays roughly constant for all positive values of ζ that have been investigated, while σ_w/u_* may increase a little with increasing stability (Lumley and Panofsky, 1964). In stable conditions, there is a tendency for the turbulent length scales to become independent of z. If that were true, the dissipation rate ε would presumably scale as $\sigma_w{}^3/L$. With $\sigma_w \simeq u_*$, the non-dimensional dissipation rate $kz\varepsilon/u_*{}^3$ would tend to become proportional to ζ. There is some evidence (Wyngaard and Coté, 1971) that indeed $kz\varepsilon/u_*^3$ increases quite rapidly with increasing (positive) values of ζ.

It should be kept in mind, however, that the vertical fluxes of heat and momentum tend to become quite small as the critical Richardson number is approached, so that it is extremely hard to get reliable data in very stable cases.

2.6. SCALING OF TURBULENCE OUTSIDE THE SURFACE LAYER

The scaling rules for the wind, temperature, and turbulence fields in the surface layer are valid only if the height z is small compared to the boundary-layer height h (the latter is often identified with the height z_1 of the first inversion in the potential-temperature profile). As soon as this condition is not satisfied, we have to generalize the relations obtained in the last two sections. In most cases, this means that the various non-dimensional functions introduced in Sections 2.4 and 2.5 become functions both of z/L and of z/h (or, alternatively, of h/L and z/h). In general, the form of such functions cannot be determined analytically.

Let us give a few examples. The non-dimensional wind shear and temperature gradient in the surface layer are represented by the functions ϕ_m and ϕ_h, respectively. Outside the surface layer, it is convenient to define

$$\frac{h}{u_*}\frac{\partial U}{\partial z} = \psi_U\left(\frac{z}{h}, \mu\right),$$

$$\frac{h}{u_*}\frac{\partial V}{\partial z} = \psi_V\left(\frac{z}{h}, \mu\right), \qquad (2.49)$$

$$\frac{h}{\theta_*}\frac{\partial \Theta}{\partial z} = \psi_\Theta\left(\frac{z}{h}, \mu\right).$$

Here, the parameter $\mu = h/L$ is the same stability parameter that was used in Section 2.3.

Very little is known about these functions other than that they, like ϕ_m and ϕ_h, tend to become very small in unstable conditions. This is why the generalized expressions of the Rossby-number similarity theory (Section 2.3) depend on experimental evidence to determine the nature of the functions involved. The scatter of the data in Figure 2.2 indicates that all kinds of complications arise in practice. The planetary boundary layer is seldom horizontally homogeneous or stationary, and vertical shear of the geostrophic wind often masks the influences of the other parameters.

Similar problems arise in the non-dimensional expressions for the standard deviations of vertical velocity and temperature. Outside the surface layer (2.42) and (2.43) may be generalized by writing

$$\frac{\sigma_w}{u_*} = \psi_w\left(\frac{z}{h}, \mu\right), \qquad (2.50)$$

$$\frac{\sigma_\theta}{\theta_*} = \psi_\theta\left(\frac{z}{h}, \mu\right); \qquad (2.51)$$

in practice, however, these relations are not very useful because other effects (advection, baroclinity, inversion erosion, and others) tend to obscure the dependence on μ.

It is fortunate that in free convection (which is much more

prevalent in the bulk of the boundary layer than in the surface layer because the height of the daytime boundary layer tends to be large) the situation improves considerably. In free convection, the mechanical production term in the kinetic energy budget (1.57) is negligible compared to the buoyant production term. If the simplified energy budget is integrated vertically from the surface to the top, the turbulent flux-divergence term vanishes (the flux is zero at the surface, and presumably also vanishes at the top of the boundary layer). The vertically-averaged energy budget for free convection thus may be approximated by

$$\int_0^h \frac{g}{\Theta} \overline{\theta w} \, dz = \int_0^h \varepsilon \, dz \, . \qquad (2.52)$$

Throughout a convective atmospheric boundary-layer, the upward heat flux has the same order of magnitude as the surface heat flux $Q_o = \rho c_p \overline{w\theta}_o$. Also, the dissipation rate ε must be estimated as $\sigma_w{}^3/h$, because the length scale involved is not the height z, but the boundary-layer height h. Therefore, (2.52) suggests that the vertically-averaged value of σ_w may be estimated as follows

$$\langle\sigma_w\rangle = c \left(h \frac{g}{\Theta} \overline{w\theta}_o\right)^{1/3} , \qquad (2.53)$$

in which c is an undetermined constant. If we compare (2.53) with the corresponding expression (2.46) for free convection in the surface layer, we see that they are identical except for the replacement of z by h.

The form of (2.53) suggests that it will be convenient to introduce a velocity scale for free convection outside the surface layer. We define the convective velocity scale w_* (Tennekes, 1970; Deardorff, 1970) by

$$w_* = \left(h \frac{g}{\Theta} \overline{w\theta}_o\right)^{1/3} . \qquad (2.54)$$

The corresponding temperature scale is

$$T_* = \overline{w\theta}_o / w_* . \qquad (2.55)$$

Recalling the definition of the Obukhov length L, we can write (2.54) and (2.55) in non-dimensional form. This yields

$$\frac{w_*}{u_*} = \left(\frac{h}{-kL}\right)^{1/3} = \left(-\frac{\mu}{k}\right)^{1/3} , \qquad (2.56)$$

$$- \frac{T_*}{\theta_*} = (\frac{h}{-kL})^{-1/3} = (- \frac{\mu}{k})^{-1/3} \quad . \tag{2.57}$$

These relations give some indication of the value of $\mu = h/L$ needed to obtain free convection. For $w_*/u_* = 2$ the kinetic energy of the convective turbulence is about $4u_*^2$. This corresponds to $h/kL = -8$, or $h/L = -3$ approximately. It is clear that it is not necessary to have very large values of $-h/L$ in order to see behavior that corresponds to free-convection rules. If the turbulent motion is predominantly of convective origin, the two-parameter expressions (2.50) and (2.51) simplify to

$$\frac{\sigma_w}{w_*} = F_w (\frac{z}{h}) \quad , \tag{2.58}$$

$$\frac{\sigma_\theta}{T_*} = F_\theta (\frac{z}{h}) \quad . \tag{2.59}$$

These simple relations have proved to be very useful in diagnostic studies of the characteristics of unstable boundary layers.

The convective velocity and temperature scales, w_* and T_*, are used also to non-dimensionalize other properties of convective boundary layers. The vertical distributions of the terms in the kinetic budget, for example, are best presented in such a way that their values are normalized with w_*^3/h. The height z is non-dimensionalized with the boundary-layer height h. The convective scales can also be used to normalize the spectra of turbulent velocity and temperature fluctuations in unstable boundary layers (see also Section 4.2).

The convective velocity scale w_* has been introduced into air pollution research by Willis and Deardorff (1976, 1978 and 1981). They have performed a series of laboratory simulations of diffusion from continuous point sources in conditions corresponding to free convection. Obviously, scaling was necessary to make their results applicable to the real atmosphere. Willis and Deardorff did this in a brilliant, creative way. They argued that the height h of the mixed layer and the convective velocity scale w_* are the ruling parameters, and that, as a consequence, $t_* = h/w_*$ determines the unit of time in these circumstances. In other words, t_* is the ruling time scale; all other times have to be measured in terms of t_*. For example, the travel distance x between the source and the point of measurement corresponds to a travel time x/U if the mean wind speed is U. The travel time has to be normalized with t_*. Therefore, the appropriate non-dimensionalization of x is, according to Willis and Deardorff,

$$X = \frac{x/U}{t_*} = \frac{x/U}{h/w_*} = \frac{xw_*}{hU} \ . \qquad (2.60)$$

This application of convective scaling has been quite useful, as we shall see in Chapter 5.

2.7. CORRELATION FUNCTIONS AND SPECTRA

In this context it seems appropriate to introduce correlation functions and their Fourier transforms, spectra, by paraphrasing Taylor's classical paper on diffusion by continuous movements (Taylor, 1921). We will extend the subject matter of that paper somewhat.

Imagine the motion of infinitesimal (mass-less) particles or pollutant molecules tossed around by turbulent velocity fluctuations. If the particle leaves the origin at t = 0, its position Y at time t is given by

$$Y(t) = \int_0^t V_y(t')\, dt' \ . \qquad (2.61)$$

We have considered only motion along the y-direction and we assume that the mean velocity in this direction is zero ($\overline{V}_y = 0$).

Averaging (2.61) over many realizations (i.e., a long series of releases of particles), we find that the mean value of $\overline{Y}$ is zero if the velocity fluctuations and the boundary conditions satisfy certain reasonable conditions (Tennekes and Lumley, 1972, Chapter 7). A more interesting equation is obtained by multiplying (2.61) with $V_y(t)$

$$Y(t)V_y(t) = Y(t)\frac{dY}{dt} = \frac{d}{dt}(\tfrac{1}{2}Y^2) = \int_0^t V_y(t)V_y(t')\, dt' \ . \qquad (2.62)$$

Taking the average of (2.62) over many particles, we obtain

$$\frac{d}{dt}(\tfrac{1}{2}\,\overline{Y^2}) = \int_0^t \overline{V_y(t)V_y(t')}\, dt' \ . \qquad (2.63)$$

Both sides of this equation have the dimensions of a diffusivity ($m^2\ s^{-1}$). In fact, the value of (2.63) at large values of t is the conventional eddy diffusivity.

If we want to solve (2.63) we need to know more about the covariance appearing under the integral. If the turbulent motion is statistically stationary and homogeneous, the correlation funtion R_L^y is an even function of the time difference $\tau = t - t'$. We put

$$\overline{V_y(t)V_y(t')} = R_L^y(\tau) = \overline{V_y^2}\,\rho_L^y(\tau)\,. \tag{2.64}$$

The subscript L refers to the fact that these are Lagrangian correlations, measured along the trajectories of moving particles. Note that the correlation coefficient satisfies $\rho(0) = 1$.

Subsitution of (2.64) into (2.63) yields

$$\frac{d}{dt}(\tfrac{1}{2}\overline{Y^2}) = \int_0^t R_L^y(\tau)\,d\tau = \overline{V_y^2}\int_0^t \rho_L^y(\tau)\,d\tau\,. \tag{2.65}$$

When (2.65) is integrated with respect to time, there results

$$\tfrac{1}{2}\overline{Y^2} = \int_0^t (t-\tau)\,R_L^y(\tau)\,d\tau = \overline{V_y^2}\int_0^t (t-\tau)\,\rho_L^y(\tau)\,d\tau\,. \tag{2.66}$$

Equations (2.65) and (2.66) define turbulent dispersion in terms of correlation functions. Of particular interest is the behavior of these equations for large values of t. If $\rho_L(\tau)$ goes to zero as $\tau \to \infty$ (which, fortunately, is nearly always the case if it is measured properly), the integral in (2.65) approaches a constant value if t is large enough. This is called the Lagrangian integral scale:

$$T_L = \int_0^\infty \rho_L(\tau)\,d\tau\,. \tag{2.67}$$

In practice, one often finds that $\rho_L(\tau) < 0.01$ for $\tau > 3\,T_L$. The integral thus converges rapidly toward its asymptotic value. For $\tau \gg T_L$, the dispersion rate in (2.65) can be approximated by

$$\frac{d}{dt}(\tfrac{1}{2}\overline{Y^2})_{t\to\infty} = \overline{V_y^2}\int_0^\infty \rho_L^y(\tau)\,d\tau = \overline{V_y^2}\,T_L^y\,. \tag{2.68}$$

The eddy diffusivity defined by (2.68) is the product of the turbulent velocity variance and the integral scale (correlation interval).

Evaluation of the behavior of (2.66) for $t \to \infty$ requires a bit more care. The factor $(t - \tau)$ inside the integral is proportional to the triangular filter $(1 - \tau/t)$. If $t >> T_L$, the filter is ineffective, because its value remains nearly one for all values of τ at which $\rho_L(\tau)$ differs appreciably from zero. Therefore, (2.66) becomes

$$(\tfrac{1}{2}\, \overline{Y^2})_{t\to\infty} = \overline{v_y^2}\, t \int_0^\infty \rho_L^y(\tau)\, d\tau = \overline{v_y^2}\, T_L^y\, t \,. \tag{2.69}$$

This result is also obtained if the approximate dispersion rate (2.68) is integrated with respect to time. We note that (2.68) and (2.69) demonstrate the well-known parabolic behavior of diffusion in homogenous turbulence.

For some purpose it is convenient to use a simple analytical approximation for the correlation coefficient. A straightforward exponential function has many attractive features and only one minor disadvantage (Neumann, 1978; Tennekes, 1979). We put

$$\rho_L(\tau) = \exp(-\tau/T_L) \,. \tag{2.70}$$

Substituting this approximation into the general Equations (2.65) and (2.66), we obtain

$$\frac{d}{dt}\,(\tfrac{1}{2}\overline{Y^2}) = \overline{v_y^2}\, T_L^y\, \{1 - \exp(-t/T_L^y)\} \,, \tag{2.71}$$

$$\tfrac{1}{2}\overline{Y^2} = \overline{v_y^2}\, [\, T_L^y\, t - (T_L^y)^2\, \{1 - \exp(-t/T_L^y)\}] \tag{2.72}$$

It is worth noting that the asymptotic form (2.68) has exponential accuracy for $t > T_L$, while the fractional difference between (2.72) and its asymptote (2.69) is hyperbolic in normalized time (t/T_L).

Now the time has come to introduce spectra. A spectrum is the Fourier transform of a correlation function. In the context of this discussion, cosine transforms are appropriate (Monin and Yaglom, 1975). We define the transform pair $R_L(\tau)$, $E_L(\omega)$ by

$$R_L(\tau) = \int_0^\infty E_L(\omega) \cos \omega\tau \, d\omega \,, \tag{2.73}$$

$$E_L(\omega) = \frac{2}{\pi} \int_0^\infty R_L(\tau) \cos \omega\tau \, d\tau \; . \qquad (2.74)$$

The transform of $R_L(\tau)$, $E_L(\omega)$ is called the energy spectrum because $E_L(\omega)\,d\omega$ is the contribution to variance made by fluctuations in the interval of width $d\omega$ centered at ω.

For $\tau = 0$, (2.73) becomes

$$R_L^y(0) = \overline{V_y^2} = \int_0^\infty E_L(\omega)\, d\omega \; . \qquad (2.75)$$

This confirms that (twice) the kinetic energy is obtained if the spectrum is integrated over all frequencies of interest.

For $\omega = 0$, (2.74) becomes

$$E_L(0) = \frac{2}{\pi} \int_0^\infty R_L^y(\tau)\, d\tau = \frac{2}{\pi} \overline{V_y^2}\, T_L^y \; . \qquad (2.76)$$

The factor $\overline{V_y^2}\, T_L^y$ was encountered before, in the asymptotic diffusion Equations (2.68) and (2.69). Apparently, the value of the spectrum at the origin defines the eddy diffusivity.

This raises the question how different frequency components contribute to the dispersion of pollution. In order to answer this question we return to the general equation for $\frac{1}{2}\overline{Y^2}$, (2.66). According to Parseval's theorem (see Tennekes and Lumley, 1972, Chapter 6) the convolution of $R_L(\tau)$ with the triangular filter $(1 - \tau/t)$ is equal to the convolution of $E_L(\omega)$ - which is the Fourier transform of $R_L(\tau)$ - with the Fourier transform of the filter function

$$\tfrac{1}{2}\overline{Y^2} = \int_0^t (t-\tau)\, R_L^y(\tau) d\tau = \tfrac{1}{2} t^2 \int_0^\infty E_L(\omega) \left(\frac{\sin \omega t/2}{\omega t/2}\right)^2 d\omega \; . \qquad (2.77)$$

The filter spectrum has its major passband around zero frequency; the side bands are of no concern here.

If t is very large, the filter is very narrow, because its first zero occurs at $\omega t/2 = \pi/2$, i.e., when $\omega = \pi/t$. In that case, the filter 'selects' $E_L(0)$, while discarding the contributions made by higher frequencies. The integral in (2.77) then can be approximated as

$$\tfrac{1}{2}\overline{Y^2} = \tfrac{1}{2}t^2\, E_L(0) \int_0^\infty \left(\frac{\sin \omega t/2}{\omega t/2}\right)^2 d\omega$$

$$= \tfrac{1}{2}\, t^2\, E_L(0)\, \frac{2}{t}\, \frac{\pi}{2} = \overline{V_y^2}\, T_L^y\, t\ . \qquad (2.78)$$

Here we have used (2.76) to substitute for $E_L(0)$, in order to verify that the result agrees with the asymptotic expression (2.69). It does, and we conclude that diffusion for large times depends on the behavior of the spectrum near the origin. We found before that the eddy diffusivity is proportional to the value of the spectrum at the origin; here we find another facet of the same gem.

Integrals of the type (2.77) occur in all kinds of averaging problems. An instrument with a finite response length modifies the true spectrum in a similar way, and averaging experimental data over finite time intervals also leads to integrals that are similar to (2.77). This is illustrated elegantly in pages 11 through 16 of Pasquill's 1974 classic book on atmospheric diffusion.

We now turn to the behavior of (2.77) at small values of time. If t is very small, the first zero of the filter occurs at very large frequencies. Since the numerical value of the filter at $\omega = 0$ equals 1, the first approximation for (2.77) at $t \ll T_L$ is

$$\tfrac{1}{2}\overline{Y^2} = \tfrac{1}{2}t^2 \int_0^\infty E_L(\omega)\, d\omega = \tfrac{1}{2}\, \overline{V_y^2}\, t^2. \qquad (2.79)$$

The point here is that all frequency components in the turbulence contribute to the dispersion without attenuation.

2.8. INERTIAL SUBRANGES

For a very short time interval after the instant of release, turbulent dispersion follows the equation we have just discussed, i.e., (2.79). In this equation the evolution of turbulence plays no role whatsoever, because the initial stage of dispersion is determined by the velocity fluctuations at the source at the instant of release.

How is (2.79) modified bv the evolution of the eddies that carry the pollutant? In principle, this is clear from (2.77): as time proceeds, the averaging filter becomes narrower and begins to remove energy from the turbulence at the high-frequency end. This means that the first correction term in (2.79) should be negative: the dispersion slows down relative to its inital stage. The spectral behavior described above also suggests that the correction term should be related to the dynamics of high-frequency eddies. It is in this way that we want to enter the world of small-scale turbulence.

A convenient approach is to perform a series expansion on the approximation (2.72) (Tennekes, 1979). This yields, for $t \ll T_L$,

$$\tfrac{1}{2}\overline{Y^2} = \tfrac{1}{2}\overline{V_y^2}\, t^2 - \frac{1}{6}\,\overline{V_y^2}\, t^3/T_L^y + \ldots . \tag{2.80}$$

How is this behavior related to the small scale dynamics of turbulence? In order to answer this question, we have to make a small detour.

We need to introduce the Lagrangian structure function, which is defined by

$$D_L^y(\tau) = \overline{\{V_y(t_o) - V_y(t_o + \tau)\}^2} \, . \tag{2.81}$$

The structure function is related to the correlation function by

$$D_L^y(\tau) = 2\,\overline{V_y^2} - 2\,R_L^y(\tau), \tag{2.82}$$

where R_L^y is defined by (2.64). It is advantageous to use the structure function because it is a reasonably good high-pass filter at small values of τ. We can see this by substituting (2.73) and (2.75) into (2.82). That yields

$$D_L^y(\tau) = 2\int_o^\infty (1 - \cos\omega\tau)\, E_L(\omega)\, d\omega. \tag{2.83}$$

Clearly, if $\tau \ll T_L$, the first maximum of $(1 - \cos\omega\tau)$ occurs somewhere in the high-frequency end of the energy spectrum $E_L(\omega)$. Equally important is the fact that the filter equals zero at zero frequency: large-scale energy is removed from the integration, and $D_L(\tau)$ depends only on small-scale parameters if $\tau \ll T_L$.

The theory of the inertial subrange is based on the hypothesis that the small-scale behavior of turbulence (exclusive of features that are directly affected by viscosity) is determined exclusively by the energy dissipation rate per unit mass (which is given the symbol ε) and the independent variable or variables concerned. Applying this hypothesis to the small-scale behavior of $D_L(\tau)$, we claim that $D_L(\tau)$ is a function of ε and τ at values of τ that are small compared to T_L and that no other variables or parameters enter the picture. Considering the dimensions of $D_L(\tau)$, ε, and τ, we find that this functional relationship must have the form (Monin and Yaglom, 1975, p. 359).

$$D_L(\tau) = \beta\varepsilon\tau \ , \tag{2.84}$$

where β is an undetermined non-dimensional coefficient. This is the inertial subrange in the Lagrangian structure function. The corresponding inertial subrange in $R_L(\tau)$ is

$$R_L^y(\tau) = \overline{v_y^2} - \tfrac{1}{2}\beta\varepsilon\tau \ . \tag{2.85}$$

The second term in (2.85) must be related to the second term in (2.80). We can find out what is going on by subsituting (2.85) into the general dispersion equation (2.66) and carrying out the integration. That yields

$$\tfrac{1}{2}\overline{Y^2} = \tfrac{1}{2}\overline{v_y^2}\, t^2 - \frac{1}{12}\beta\varepsilon\tau^3 + \ldots \ . \tag{2.86}$$

If we compare this with (2.80), we conclude that

$$\varepsilon = \frac{2}{\beta}\,\overline{v_y^2}/T_L^y \ . \tag{2.87}$$

This is a most interesting result. It shows that the dissipation rate of turbulence is determined by its large-scale dynamics. Energy is dissipated at a rate proportional to the energy available and, reciprocally, to the evolution time of the energetic eddies (which is represented by the Lagrangian integral scale, T_L). As we have seen in (2.40), this is one of the fundamental premises of turbulence theory (Tennekes and Lumley, 1972).

In the inertial subrange, the Lagrangian energy spectrum $E_L(\omega)$ can be a function of ε and ω only. Dimensional considerations require that

$$E_L(\omega) = \gamma\ \varepsilon\ \omega^{-2} \ , \tag{2.88}$$

where γ is another non-dimensional constant. This behavior, valid only if $\omega T_L > 1$, can be compared with the Fourier transform of the exponential correlation function (2.70), which reads

$$E_L(\omega) = \frac{2\,\overline{v_y^2}\,T_L}{\pi(1+\omega^2 T_L^2)} \,. \qquad (2.89)$$

At high frequencies (2.89) shows the same dependence on ω as (2.88) thus demonstrating that the exponential approximation to the Lagrangian correlation function is consistent with the postulated form of the Lagrangian spectrum in the inertial subrange (Tennekes, 1979).

The inertial subranges met in experimental practice exhibit a different power law. Usually, turbulent velocity fluctuations are measured in conditions that allow Taylor's hypothesis (the frozen-turbulence approximation) to be applied. The measured time series then are converted into spatial statistics. A streamwise Eulerian correlation function, for example, is defined by

$$R_E(r) = \overline{v^2}\,\rho_E(r) = \overline{v(x_o)v(x_o + r)}, \qquad (2.90)$$

where, in practice, r is determined as $U\tau$, U being the mean wind speed and τ the time delay in the velocity record at a fixed location, x_o.

The Eulerian correlation function and its counterpart, the (one-dimensional) energy spectrum, constitute a Fourier transform pair similar to (2.73) and (2.74)

$$R_E(r) = \int_0^\infty E(k) \cos kr\, dk \,, \qquad (2.91)$$

$$E(k) = \frac{2}{\pi} \int_0^\infty R_E(r) \cos kr\, dr \,. \qquad (2.92)$$

In the inertial subrange, $E(k)$ is a function of ε and k only, and by dimensional considerations there results for the Eulerian spectra

$$E(k) = \alpha\, \varepsilon^{2/3}\, k^{-5/3} \,. \qquad (2.93)$$

This form is sometimes used to determine ε from measured spectra. For further study of inertial-range dynamics, the reader is referred to Tennekes and Lumley (1972) or Monin and Yaglom (1975).

ACKNOWLEDGEMENT

The material in this chapter was borrowed in large part from the unpublished lecture by the author at the short course on boundary-layer meteorology held at Boulder, August 1978. The material of Sections 2.7 and 2.8, however, is taken primarily from other sources (Tennekes, 1977 and 1979).

3. BOUNDARY-LAYER MODELING

J.C. Wyngaard

N.C.A.R., Boulder, U.S.A.

3.1. THE CALCULATION OF BOUNDARY-LAYER STRUCTURE

Certainly one cannot hope for analytical solutions to the equations of motion in the turbulent boundary layer, but one might ask: Why is it necessary to model them before solving them numerically? Why can't we solve them directly on today's large, fast computers?

The answer lies in the basic nature of turbulent flow, which contains a range of 'eddy sizes', or scales of motion, all being coupled through the nonlinearity of the equation of motion (1.15). The smallest scales are dynamically essential, being responsible for the viscous dissipation. Any numerical solution of (1.15) must resolve these dissipative eddies, whose length scale η is, from Kolmogorov's (1941) arguments,

$$\eta = (\nu^3/\varepsilon)^{\frac{1}{4}} \tag{3.1}$$

where ν is the kinematic viscosity and ε is the rate of dissipation of turbulent kinetic energy per unit mass. Measurements reveal that $\varepsilon \sim q^3/\ell$, where q and ℓ are velocity and length scales characteristic of the energy-containing eddies. Thus the largest ($\sim\ell$) and smallest ($\sim\eta$) eddies have a length scale ratio

$$\frac{\ell}{\eta} = \left(\frac{q\,\ell}{\nu}\right)^{\frac{3}{4}} = R_\ell^{\frac{3}{4}} \tag{3.2}$$

where R_ℓ is a Reynolds number for the dominant turbulent eddies.

Taking typical boundary-layer values of $q \sim 1\ ms^{-1}$ and $\ell = 300$ m, we find that $\ell/\eta \sim 3\ 10^5$. Thus we would have to resolve a range of eddy scales from 300 m down to 1 mm in a numerical solution of (1.15). The grid must be three-dimensional, since the fluctuations are, and solving the equation over a 10 km x 10 km volume would require roughly 10^{20} grid points. This is clearly impossible.

Thus we must set our sights lower; we cannot hope to solve numerically the equation for the boundary-layer velocity field. However, we might not want all that detail anyway; averaged flow fields will do just as well for many purposes.

Let us investigate what happens when we average the equation of motion. We represent the velocity field $\tilde{u}_i$ by its average plus a deviation,

$$\tilde{u}_i = U_i + u_i \tag{3.3}$$

without yet defining the nature of the averaging process. We do assume averaging and differentiation commute, however. The averaged equation of motion is given by (1.26a). The non-linear term in this equation reads $\partial(\tilde{u}_i\tilde{u}_j)/\partial x_j$. Using our decomposition (3.3), one can write

$$\overline{\tilde{u}_i\tilde{u}_j} = \overline{U_iU_j} + \overline{U_iu_j} + \overline{u_iU_j} + \overline{u_iu_j}. \tag{3.4}$$

Now let us consider specific types of averaging. A mathematically convenient one is the average over an infinite ensemble of realizations of the flow, called the ensemble average. This has the properties

$$\overline{U_iU_j} = U_iU_j \quad \text{and} \quad \overline{U_iu_j} = 0 = \overline{u_iU_j} \tag{3.5}$$

so that the nonlinear term becomes

$$\tilde{u}_i\tilde{u}_j = U_iU_j + \overline{u_iu_j}. \tag{3.6}$$

Thus ensemble averaging gives an equation for the averaged field U_i, but this equation has new terms - the Reynolds stresses $\overline{u_i u_j}$, named after O. Reynolds who discovered them in 1895.

Providing we can deal with the Reynolds terms, averaging does bring advantages. It gives an equation which governs only the mean velocity, not the intricate details of the randomly fluctuating velocity. We would expect that mean velocity field variations would have length scales no smaller than ℓ, so the numerical grid spacing need only be of order ℓ, not η. Thus many fewer grid points would be needed to solve the mean field equations over a given region than would be required for the unaveraged equations.

Sometimes it is more convenient to use volume averaging. For example, one can average over a grid volume $\Delta x \, \Delta y \, \Delta z$:

$$\{\underset{\sim}{u}_i(x_i,t)\} = \frac{1}{\Delta x \, \Delta y \, \Delta z} \int_{z-\frac{1}{2}\Delta z}^{z+\frac{1}{2}\Delta z} \int_{y-\frac{1}{2}\Delta y}^{y+\frac{1}{2}\Delta y} \int_{x-\frac{1}{2}\Delta x}^{x+\frac{1}{2}\Delta x} \underset{\sim}{u}_i(\xi,\eta,\zeta,t) \, d\xi d\eta d\zeta \quad (3.7)$$

This averaging operator does not have the properties (3.5) of the ensemble average. Thus, in this case we simply write the nonlinear term as

$$\{\underset{\sim}{u}_i \underset{\sim}{u}_j\} = \{U_i\}\{U_j\} + R_{ij} \quad (3.8)$$

where R_{ij} is the generalized Reynolds stress; again, it must be specified before the averaged equations can be solved.

Volume and ensemble averages have different properties. We have pointed out one important difference, that ensemble averages follow (3.5) while volume averages as defined by (3.7) do not. Another important difference is that ensemble averaging completely removes randomness, while volume averaging does not. That is, while the velocity field $\underset{\sim}{u}_i(x_i,t)$ is random, the ensemble-averaged field $U_i(x_i,t)$ is non-random, since it is averaged over an infinite ensemble of realizations. The volume-averaged field $\{U_i(x_i,t)\}$, however, will still be random, although our intuition suggests its randomness will decrease in some sense as the averaging volume increases.

Another way of looking at the latter difference is that while the ensemble averaging operation removes all the turbulence, the volume averaging operator removes only those components of turbulence having spatial scales smaller than the averaging scale. In a statistically homogeneous, stationary, unbounded turbulent flow, the volume average converges to the ensemble average in the limit as the averaging volume increases. Thus one can think of the volume-averaged field as being the sum of the ensemble-averaged field and the large-scale turbulence.

Let us return to the discussion of the averaged equations. We saw that averaging introduces Reynolds terms. Much of the turbulence research since Reynolds' time has dealt with specifying these terms so that the averaged equations can be solved; this is sometimes called the 'closure' problem. It has turned out to be one of the more difficult problems in all of physics, and from a fundamental standpoint it must still be considered unsolved.

We can now answer our opening questions. Computers are not nearly large or fast enough to solve the equations for boundary-layer fields, nor is there any prospect that they will ever be. Averaging the equations greatly eases the computational problem, but introduces Reynolds terms which have not yet been theoretically specified in closed form. In practice it is necessary to use approximate models for the Reynolds terms, and hence to deal with boundary-layer models, not with the exact boundary-layer equations.

In this short chapter we cannot give a comprehensive summary of the history and state of the art of boundary-layer modeling. Even if it were possible, it would be of limited value because models are changing quite rapidly, and any such survey will quite likely be out of date in a few years. For that reason we will content ourselves with a review of the fundamental aspects of two broad classes of boundary layer models: ensemble-average and volume-average models.

3.2. ENSEMBLE-AVERAGE MODELS

If we ensemble average the equations of motion we obtain Equation (1.26a). We cannot solve this for the ensemble-averaged velocity field U_i without first closing (1.26a) by specifying the Reynolds stress $\overline{u_i u_j}$. It is now well established that the Reynolds stress is important in mean-flow dynamics and cannot be ignored. In this section we will discuss two approaches to specifying the Reynolds stress: first-order and second-order closure.

First-Order or Eddy-Diffusivity (K) Closure

This type of closure was introduced in Chapter 1. To illustrate it further we consider the equations for mean velocity in a horizontally-homogeneous case, which we considered in Section 1.3.

$$\frac{\partial U}{\partial t} + \frac{\partial}{\partial z}\overline{uw} = f(V-V_g) ,$$

$$\frac{\partial V}{\partial t} + \frac{\partial}{\partial z}\overline{vw} = f(U_g-U) \tag{3.9}$$

where we use the geostrophic wind components and the Coriolis parameter introduced in Equation (1.34). We have neglected the viscous terms since, as we saw in Chapter 1, they are always small compared with the turbulence (Reynolds stress) terms. In the simplest case, the time-change terms in (3.9) are much smaller than the others and can be neglected, giving a balance between friction, Coriolis, and pressure-gradient forces. In the earliest attempts to solve (3.9), the turbulent stresses $\overline{uw}$ and $\overline{vw}$ were modeled after the viscous stresses in a Newtonian fluid, but with the proportionality factor a much larger 'eddy' viscosity K_m;

$$\overline{uw} = -K_m \frac{\partial U}{\partial z}$$
$$\overline{vw} = -K_m \frac{\partial V}{\partial z} . \tag{3.10}$$

The subscript m, for momentum, differentiates this K from that for a scalar such as temperature, which can be different.

The solution of (3.9) and (3.10) with K_m = constant was given in (D4a) and (D4b) of Chapter 1. A close examination shows the solution to be unphysical, even when modified to allow the boundary-layer depth to be externally determined; its principal defect is that its wind shear is proportional to the geostrophic wind shear even when K_m is arbitrarily large. Observations suggest instead that the wind profile in this baroclinic case approaches the flat, 'well-mixed' idealization if the convection is sufficiently strong.

Recall from Chapter 1 that K_m is a property of the flow, not of the fluid, and that we expect K_m to scale with $q\ell$, the product of turbulent velocity and length scales. Since both q and ℓ vary with height z above the surface, we might try to solve (3.9) and (3.10) for a z-dependent K_m. Near the surface under neutral conditions, K_m behaves as

$$K_m = \frac{-\overline{uw}}{\partial U/\partial z} = k u_* z . \tag{3.11}$$

Ellison has solved (3.9) and (3.10) by using the K_m profiles of (3.11), as discussed by Krishna (1980). However, this solution also has the unphysical behavior of the K_m = constant solution in baroclinic conditions. It seems that reasonable solutions require a K_m which also decreases to small values at the boundary-layer top and, hence, has a mid-layer maximum.

Buoyancy has a strong influence on K. We can examine these

effects in a rough way by considering the equation for the fluctuating velocity u_i, given by (1.26b). Turbulence acts to keep the leading terms in this equation of order q^2/ℓ in root-mean-square (Tennekes and Lumley, 1972). The buoyancy term will be of this order if

$$\sigma_\theta = (\overline{\theta^2})^{\frac{1}{2}} \sim \frac{q^2\Theta}{\ell\, g} \quad . \tag{3.12}$$

For typical values of q (1 ms^{-1}) and ℓ (300 m), we find from (3.12) that buoyancy and inertia forces are of the same order if $\sigma_\theta \sim 0.1°C$. Seldom are temperature fluctuation levels much smaller than this, and so we expect the boundary layer will seldom be in a dynamically neutral state.

When the surface is warmer than the air, so the heat flux is upward, buoyancy acts as a source of turbulent kinetic energy; in the opposite case it is a sink. We can crudely examine the sensitivity of different size eddies to buoyancy by considering, in isolation, a single eddy with velocity scale q and size ℓ in a background vertical potential temperature gradient $\partial\Theta/\partial z$. The eddy motion will generate a temperature fluctuation of order $\ell\partial\Theta/\partial z$ and hence induce buoyancy forces of order $(\ell g/\Theta)\partial\Theta/\partial z$. Since inertia forces are of order q^2/ℓ, the ratio of buoyancy and inertia forces is

$$\frac{g\ell^2\partial\Theta/\partial z}{\Theta\, q^2} = Ri_\ell \tag{3.13}$$

which is an eddy Richardson number. Thus, the largest eddies feel the strongest buoyancy effects.

We conclude that in convective situations the energy input by buoyancy is felt at the largest scales of turbulent motion, with both q and ℓ being increased over neutral values. In stably-stratified conditions, the energy extraction due to buoyancy occurs principally also at the largest scales, with q and ℓ decreased. In both cases K is directly influenced, since $K \sim q\ell$. Thus, since our estimate (3.12) implies a low threshold for buoyancy effects, we should be cautious in using neutral K values.

We can now quickly sketch the expected behavior of K, without differentiating between momentum and scalar transfer. Near the surface $q \sim u_*$, $\ell \sim z$, so $K \sim u_* z$. Stability effects become important near $z \sim \|L\|$, enhancing K under unstable conditions and suppressing it under stable conditions. At the boundary-layer top, the outer edge of the continuously turbulent region, K approaches zero; thus K has a maximum in mid-layer. This characteristic K-profile creates a tendency for strong mean gradients to occur only near the boundary-layer top and near the surface. Mean gradients within the boundary layer tend to be smaller, particularly under unstable conditions when it is often called the 'mixed layer'.

A bewildering array of K-profiles exists in the literature. Wipperman's monograph (1973), for example, mentions several dozen. This proliferation reflects both the difficulty of determining K analytically and the lack of data useful for determining it empirically. Let us try to put the history of K specifications in perspective before we discuss current practice.

Very simple K-profiles were used before computers were available, since the mean equations had to be solved analytically. Ekman's (1905) classical paper on the mean flow spiral used a constant K, for example. Most of the early work also dealt with the neutrally-stratified case. From the beginning, some have preferred to express K in terms of a mixing length ℓ_m, such that

$$K = \ell_m^2 \left[\left(\frac{\partial U}{\partial z}\right)^2 + \left(\frac{\partial V}{\partial z}\right)^2 \right]^{\frac{1}{2}} \tag{3.14}$$

and to specify the behavior of ℓ_m rather than K.

Experiments gradually revealed the behavior of K in the surface layer, including its dependence on stability. There are now extensive surface-layer measurements of the dimensionless wind shear ϕ_m and temperature gradient ϕ_h, which are related to K_m and K_h by

$$\phi_m = \frac{kz}{u_*}\frac{\partial U}{\partial z} = \frac{kzu_*}{K_m} \quad \text{and} \quad \phi_h = \frac{kz}{\theta_*}\frac{\partial \Theta}{\partial z} = \frac{kzu_*}{K_h} . \tag{3.15}$$

Less is known about K values aloft, partly because of the lack of good data; measurements there are much more difficult and expensive to make. The direct measurement of K_m in a convective case where $K_m \sim 10^2\ m^2s^{-1}$ and $\overline{uw} \sim 10^{-1}\ m^2s^{-2}$ would require the measurement of $\partial U/\partial z$ values of the order of 0.1 m s^{-1} per 100 m, which is extraordinarily difficult. In addition, turbulence structure aloft is influenced by many more processes than in the surface layer. Horizontal inhomogeneity, baroclinicity, entrainment through the capping inversion, gravity waves, and non-stationarity, all probably affect K-distributions in the upper portions of the boundary layer.

There is some evidence on the behavior of K_h in the convective case, however. With strong convection $\partial\Theta/\partial z$ can change sign above the surface layer and remain slightly positive over most of the mixed layer. This implies negative K_h values. Deardorff (1966) suggested avoiding these negative values by using the modified flux-gradient relation

$$\overline{w\theta} = -K_h \left(\frac{\partial \Theta}{\partial z} - \gamma_c\right) \tag{3.16}$$

with K_h positive and $\gamma_c \sim 0.7 \times 10^{-3}\ ^{o}C\ m^{-1}$.

It should now be clear that eddy diffusivity, being a flow property, cannot be accurately prescribed a priori without knowledge of the flow structure. Unfortunately, in most problems one does not know this structure in sufficient detail before doing the calculations. There are basically three approaches to the prescription of K:

(1) to prescribe K values;
(2) to prescribe K-profile shapes;
(3) to prescribe K-dynamics.

The first (and oldest) approach is still useful since constant-K models can often be solved analytically. This can give useful insight into problems involving boundary-layer response to various types of forcing.

The second approach allows K to depend on height, stability, and other parameters, some of which can be found from the solution as it proceeds. For example, Brost and Wyngaard (1978) found through a second-order model that K-profiles in the steady, horizontally-homogeneous, stable boundary layer could be parameterized as

$$K = \frac{k\, u_* \, z(1 - z/h)^{1.5}}{1 + \beta\, z/L} \,. \tag{3.17a}$$

Businger and Arya (1974) developed a first order model for the same conditions using a K-profile that asymptotically adjusts to the accepted expression for the stable surface layer as well as its first derivative. The expression is

$$K = \frac{k\, u_* z \exp(-fz\|V_g\|/u_*^2)}{1 + \beta\, z/L} \,. \tag{3.17b}$$

The β in (3.17) follows from surface layer profiles. For the Kansas experiment $\beta = 4.7$. (see Figure 2.3).

The two expressions (3.17) give similar shapes and can be

used in a K-model, in this second approach. The parameters u_*, L, $\|V_g\|$ and h in (3.17) could be estimated from the calculated mean profiles. As another example, Pielke and Mahrer (1975) prescribed a K profile for the convective case but calculated the depth h from a rate equation, so the magnitude of K was, in effect, determined by the solution as it proceeded.

The third approach, prescribing K-dynamics, is most conveniently done by carrying, along with the mean field equations, additional equations describing the evolution of the turbulent fluxes. This is the second-order-closure approach.

Second-Order Closure

This can be viewed as a logical outgrowth of first-order closure. The principal problem in first-order closure is finding a rational basis for parameterizing K, and second-order closure confronts this by dealing directly with the conservation equations for the Reynolds terms.

A procedure for generating differential equations for stress, heat flux, and other turbulence moments has existed since Reynolds (1895) introduced his mean-turbulent decomposition. Reynolds derived the turbulent energy equation; the shear stress equation came later (Chou, 1945). As we will see, these equations also cannot be solved exactly because they contain unknown moments of yet higher order. Many 'closure approximations' for these equations have been proposed over the past 20 years, and the large-scale digital computer has made it feasible to solve these approximate moment equations numerically. This activity has intensified to the point that simplified second-moment approaches are now fairly widely used as an alternative to eddy-diffusivity models in both research and engineering applications.

We derived the Reynolds-flux conservation equations in Chapter 1. The stress equation is given by (1.52) and the temperature-flux equation by (1.53). The meaning of the terms in these equations was also discussed there.

In boundary-layer applications, we can eliminate any radiative coupling in the equation for the temperature-flux on the basis of Townsend's (1958) calculations. The leading terms in (1.53) are of order $\theta q^2/\ell$, whereas the radiative term (see (1.27)) is of order $\beta\theta q$, where β^{-1} is a radiative time scale. Thus, the relative order of this term is $(\beta q\theta)/(\theta q^2/\ell) = \beta\ell/q$, which we might call T_f, the 'flux Townsend number'. Townsend shows that the maximum value of β is $16\,\sigma\, T^3\kappa'/c_p$, where σ is the Stefan-Boltzmann constant, κ' is the mean absorption coefficient, and c_p is the specific heat at constant pressure. A calculation using values typical in the boundary layer shows that T_f is quite small, so that the radiative term can be neglected as was done in (1.53).

The production, advection, buoyancy, and Coriolis terms in (1.52) and (1.53) involve covariances of the turbulence field as well as the mean values U_i and Θ. These terms can be considered known if the model set contains equations for each of these covariances and for U_i and Θ. However, the remaining terms - molecular, pressure covariances, and transport - are not directly related to the covariances nor to the mean fields and must at this point be regarded as unknowns. This is the "closure" problem in the second-moment approach - specifying these unknown terms so that the set can be solved. This closure problem, which exists in moment equations of all orders (we met it first in (1.26a) for the mean, or first moment) is the price we pay for having moment equations. If we had none, we would have to generate moments by averaging the results of three-dimensional time dependent solutions, much as we determine them from field measurements.

Let us interpret the molecular, pressure covariance, and transport terms physically. The molecular terms in (1.52) and (1.53) can be written

$$\nu\left(\overline{u_k\frac{\partial^2 u_i}{\partial x_j^2}}+\overline{u_i\frac{\partial^2 u_k}{\partial x_j^2}}\right)=\nu\frac{\partial^2\overline{u_iu_k}}{\partial x_j\partial x_j}-2\nu\overline{\frac{\partial u_i}{\partial x_j}\frac{\partial u_k}{\partial x_j}}$$

$$\nu\,\overline{\theta\frac{\partial^2 u_i}{\partial x_j^2}}=\nu\frac{\partial^2\overline{\theta u_i}}{\partial x_j\partial x_j}-\nu\frac{\partial}{\partial x_j}\left(\overline{u_i\frac{\partial\theta}{\partial x_j}}\right)-\nu\overline{\frac{\partial u_i}{\partial x_j}\frac{\partial\theta}{\partial x_j}} \tag{3.18}$$

$$\kappa\,\overline{u_i\frac{\partial^2\theta}{\partial x_j^2}}=\kappa\frac{\partial^2\overline{\theta u_i}}{\partial x_j\partial x_j}-\kappa\frac{\partial}{\partial x_j}\left(\overline{u_i\frac{\partial\theta}{\partial x_j}}\right)-\kappa\overline{\frac{\partial\theta}{\partial x_j}\frac{\partial u_i}{\partial x_j}}\;.$$

In each of (3.18) the first term on the right represents molecular diffusion. To proceed further, we use scaling arguments of the type discussed in the Tennekes and Lumley (1972) text. Scaling the first of (3.18) as an example, we have

$$\nu\frac{\partial^2\overline{u_iu_k}}{\partial x_j\partial x_j}\sim\nu\frac{q^2}{\ell^2}\sim\frac{\nu}{q\,\ell}\frac{q^3}{\ell}\sim R_\ell^{-1}\frac{q^3}{\ell}\;. \tag{3.19}$$

Since R_ℓ is very large in the boundary layer (recall that in the example (1.47), $R_\ell \sim 10^7$), (3.19) shows that molecular diffusion is negligible. Analogous arguments hold for the remaining members of the set (3.18), and the results is that only the extreme right terms in (3.18) need be retained. These surviving terms, which involve derivative moments (not derivatives of moments) and hence are dominated by the smallest-scale turbulence, can be simplified by assuming the small-scale turbulence structure is isotropic. This implies that

$$2\nu \overline{\frac{\partial u_i}{\partial x_k} \frac{\partial u_j}{\partial x_k}} = \frac{2}{3} \delta_{ij} \varepsilon \quad \text{and} \quad \overline{\frac{\partial u_i}{\partial x_k} \frac{\partial \theta}{\partial x_k}} = 0 \tag{3.20}$$

where ε is a scalar. Note that, from (3.18) and (3.20), the molecular term in the turbulent kinetic energy ($\frac{1}{2}$ q^2) equation is

$$\nu \overline{\left(\frac{\partial u_i}{\partial x_k}\right)^2} = \varepsilon. \tag{3.21}$$

Thus, we can interpret ε as the (positive definite) molecular destruction rate of turbulent kinetic energy per unit mass.

The tensor isotropy arguments that lead to (3.20) are very useful in dealing with fine structure statistics. Although we cannot discuss them in detail here, we can make them physically plausible. Isotropy implies that tensor quantities are independent of rotation, translation, or reflection of the coordinate axes. Consider the tensor $\overline{(\partial u_i/\partial x_k)(\partial \theta/\partial x_m)}$ under coordinate reflection. Since it has an odd number of indices, it changes sign; thus it must be zero, which implies each of its components is zero, so that $\overline{(\partial u_i/\partial x_k)(\partial \theta/\partial x_k)} = 0$ as written in (3.20). Similar arguments hold for the first of (3.20).

Equation (3.20) has two important consequences: only the diagonal elements of $\overline{u_i u_k}$ (the energy components, and not the shear stresses) are destroyed by molecular effects; and temperature flux $\overline{\theta u_i}$ is not affected at all. These are implications of local isotropy; they have not been extensively tested in detail, but the local isotropy assumption is generally consistent with what fine-structure data do exist.

Consider next the pressure covariance terms in (1.52) and (1.53). As an aid to their interpretation, we will examine the equations for shear stress and vertical temperature flux in a horizontally homogeneous, near-neutral surface layer. By (3.20) and the preceding arguments we can drop the molecular terms; Coriolis terms are negligible there, as can be quickly verified by scaling them; and under the assumed near-neutral conditions we can drop the buoyancy terms. Measurements (Wyngaard and Coté, 1971;

Wyngaard et al., 1971) show that transport terms are also small, so the equations reduce to, approximately

$$\frac{\partial}{\partial t}\overline{uw} = -\overline{w^2}\frac{\partial U}{\partial z} - \frac{1}{\rho_o}\left(\overline{w\frac{\partial p}{\partial x}} + \overline{u\frac{\partial p}{\partial z}}\right),$$

$$\frac{\partial}{\partial t}\overline{w\theta} = -\overline{w^2}\frac{\partial \Theta}{\partial z} - \frac{1}{\rho_o}\overline{\theta\frac{\partial p}{\partial z}}. \qquad (3.22)$$

The right-hand sides of (3.22) are of the order of their production terms, or q^3/ℓ and $q^2\theta/\ell$, respectively; thus the left-hand sides can be neglected if the scale τ of time changes is large compared with ℓ/q. For typical mid-surface-layer values of $\ell \sim 20$ m, $q \sim 1$ ms^{-1} we require for 'quasi-steady state'

$$\tau >> \ell/q \sim 20 \text{ s} \qquad (3.23)$$

which is usually satisfied. In this case (3.22) reduces to

$$\overline{w^2}\frac{\partial U}{\partial z} = -\frac{1}{\rho_o}\left(\overline{w\frac{\partial p}{\partial x}} + \overline{u\frac{\partial p}{\partial z}}\right) \text{ and } \overline{w^2}\frac{\partial \Theta}{\partial z} = -\frac{1}{\rho_o}\overline{\theta\frac{\partial p}{\partial z}} \qquad (3.24)$$

which means that (in quasi-steady state) the pressure covariances destroy stress and temperature flux at the same rate as they are produced by the mean shear and the mean temperature gradient. Although we found this result for a special case, it is thought that pressure covariances play this role in any turbulent flow with stress and/or heat flux.

Pressure covariances also transfer energy among the turbulent velocity components. To illustrate this, we consider the turbulent kinetic energy budget, Equation (1.56). From incompressibility ($\partial u_i/\partial x_i = 0$) and horizontal homogeneity we can write, in the neutral surface layer where transport effectively vanishes

$$\overline{u_i\frac{\partial p}{\partial x_i}} = \overline{u\frac{\partial p}{\partial x}} + \overline{v\frac{\partial p}{\partial y}} + \overline{w\frac{\partial p}{\partial z}} = \frac{\partial \overline{pu_i}}{\partial x_i} = \frac{\partial \overline{pw}}{\partial z} = 0, \qquad (3.25)$$

since there are mean gradients only in the vertical direction. This means that the pressure covariances in the component kinetic energy equations must sum to zero. These equations become, in the idealized neutral surface layer

$$\frac{1}{2}\frac{\partial}{\partial t}\overline{u^2} = -\overline{uw}\frac{\partial U}{\partial z} - 2P - \frac{1}{3}\varepsilon$$

$$\frac{1}{2}\frac{\partial}{\partial t}\overline{v^2} = P - \frac{1}{3}\varepsilon \qquad (3.26)$$

$$\frac{1}{2}\frac{\partial}{\partial t}\overline{w^2} = P - \frac{1}{3}\varepsilon .$$

Note from (3.26) that the pressure terms (P) are the only source of σ_v and σ_w; σ_u is produced by shear, and loses energy by pressure transfer (-2P) to the other components.

Evidence over the past decade indicates that in the unstable surface layer the pressure term $\partial\overline{pw}/\partial z$ (we will call it pressure transport) is a significant source term in the turbulent kinetic energy budget. The first evidence (Wyngaard and Coté, 1971) was indirect; the measured shear and buoyant production, transport, and dissipation had a substantial imbalance and it seemed that nothing but pressure transport could account for it. Other experiments now support this inference. McBean and Elliott's (1975) direct measurements of the pressure covariance confirm that pressure transport is a source of turbulent kinetic energy in the unstable surface layer. The $\overline{pw}$ profile starts from zero near the (rigid) surface, then becomes increasingly negative with increasing height under unstable conditions. Perhaps this corresponds to low pressures in the heated, rising elements.

Note from (3.25) that intercomponent transfer does not sum to zero when the pressure transport is non-zero. Data indicate that this energy input goes principally into the horizontal velocity components; perhaps this is due to entrainment into rising convective elements and/or the surface constraint on the large convective eddy structure, which forces their motions to be horizontal in the surface layer.

The other transport terms remain to be discussed. Each second moment equation has one; it represents the local gain, or loss, due to the divergence of the turbulent flux of the second moment in question, but integrates to zero over the entire flow. We will call these terms simply turbulent transport. There are flows with negligible turbulent transport; measurements show that it is small in the near-neutral surface layer, for example.

Turbulent transport is often very important in convectively-driven turbulence, however. As an example, consider the region just below the capping inversion in a convective boundary layer. If there is negligible wind shear (quite common with light geostrophic winds) and if the flow is quasi-steady, the turbulent kinetic energy budget reads

$$\frac{1}{2}\frac{\partial}{\partial t} q^2 \simeq 0 = -\frac{\partial}{\partial z}(\tfrac{1}{2}\overline{wu_i^2}) - \frac{1}{\rho_o}\frac{\partial}{\partial z}\overline{pw} + \frac{g}{\Theta}\overline{\theta w} - \varepsilon . \qquad (3.27)$$

In this region $\overline{\theta w}$ is negative, since entrainment brings down warmer air from above the inversion base. Thus the first two terms in (3.27) represent a net energy source, and the final two a balancing sink

$$\frac{\partial}{\partial z}\left(\tfrac{1}{2}\overline{wu_i^2} + \frac{1}{\rho_o}\overline{pw}\right) = \frac{g}{\Theta}\overline{\theta w} - \varepsilon . \qquad (3.28)$$

In this case both turbulent and pressure transport are probably important. The divergence of the total energy flux (the flux of kinetic energy, $\tfrac{1}{2}\overline{wq^2}$, plus the pressure work $\overline{pw}$) is the sole source of turbulent kinetic energy in this example.

Although turbulent transport is rather difficult to measure because of its long averaging time requirements (Wyngaard, 1973) it is now clear that it is a significant loss term in the turbulent kinetic energy budget in the unstable surface layer. There $\overline{wu_i^2}$ is positive and increases with z; it reaches a maximum in mid-layer, and decreases above that. Thus turbulent transport extracts energy from the unstable surface layer and sends it aloft against the turbulent kinetic energy gradient.

In summary, we have found that the stress equation (1.52) and the temperature flux equation (1.53), which are the foundation of second-order closure models, have three types of terms - molecular destruction, pressure covariance, and transport - which are important and must be known before a closed set of equations is obtained. In a coming section we will discuss current methods of dealing with these terms. First, however, we will show through several examples that under simplifying assumptions, the second moment equations can give some intriguing insights into the behavior of eddy diffusivity.

Consider again the near-neutral, quasi-steady, horizontally homogeneous surface layer. We showed in the previous section that the $\overline{uw}$ and $\overline{w\theta}$ equations in this limit reduce to (3.24), which expresses a balance between production by mean gradients and destruction by pressure covariances. A simple model of these pressure covariances is

$$\frac{1}{\rho_o}\left(\overline{w\frac{\partial p}{\partial x}} + \overline{u\frac{\partial p}{\partial z}}\right) = \frac{\overline{uw}}{\tau_1} \quad \text{and} \quad \frac{1}{\rho_o}\overline{\theta\frac{\partial p}{\partial z}} = \frac{\overline{\theta w}}{\tau_2} . \qquad (3.29)$$

Our previous scaling arguments indicate the time scales τ_1 and τ_2 must be of the order ℓ/q which is an 'eddy turnover' time characteristic of the energy-containing range of the turbulence. Thus from (3.24) and (3.29) we have

$$\overline{uw} = -\tau_1 \overline{w^2} \frac{\partial U}{\partial z} = -K_m \frac{\partial U}{\partial z}$$
$$\overline{\theta w} = -\tau_2 \overline{w^2} \frac{\partial \Theta}{\partial z} = -K_h \frac{\partial \Theta}{\partial z} \tag{3.30}$$

which are the usual eddy-diffusivity expressions.

As another example, consider the horizontal heat flux equation in this idealized surface layer

$$\frac{\partial}{\partial t} \overline{u\theta} = -\overline{uw} \frac{\partial \Theta}{\partial z} - \overline{\theta w} \frac{\partial U}{\partial z} - \frac{\partial}{\partial z} \overline{wu\theta} - \frac{1}{\rho_o} \overline{\theta \frac{\partial p}{\partial x}} . \tag{3.31}$$

The turbulent transport term is observed to be small; if we neglect it and use the approximation (3.29) for the pressure covariance, we have

$$\frac{\partial}{\partial t} \overline{u\theta} = -\overline{uw} \frac{\partial \Theta}{\partial z} - \overline{w\theta} \frac{\partial U}{\partial z} - \frac{\overline{\theta u}}{\tau_3} \tag{3.32}$$

where $\tau_3 \sim \ell/q$ is again a large-eddy time scale. Rewriting $\overline{uw}$ and $\overline{\theta w}$ in terms of K's gives, in quasi-steady state

$$\overline{u\theta} = \tau_3 \frac{\partial U}{\partial z} \frac{\partial \Theta}{\partial z} (K_m + K_h) . \tag{3.33}$$

The result (3.33) is plausible physically and can be inferred from a mixing-length argument. As we discussed in Chapter 1, in the mixing-length approach, one models the u-fluctuation as $-\ell_m \partial U/\partial z$. Extending this to temperature gives $\theta \sim -\ell_m \partial\Theta/\partial z$, and hence

$$\overline{u\theta} \sim \ell_m^2 \frac{\partial U}{\partial z} \frac{\partial \Theta}{\partial z} . \tag{3.34}$$

We expect ℓ_m^2 to be of the order of ℓ^2. From (3.33), we note that $\tau_3 \sim \ell/q$, $K_m \sim K_h \sim q\ell$, so that both (3.33) and (3.34) are equivalent to

$$\overline{u\theta} \sim \ell^2 \frac{\partial U}{\partial z} \frac{\partial \Theta}{\partial z} . \tag{3.35}$$

The result (3.33) allows us to evaluate one off-diagonal diffusivity in the expression

$$\overline{u_i \theta} = - K_{ij} \frac{\partial \Theta}{\partial x_j} . \tag{3.36}$$

From (3.33), we have

$$K_{13} = -\tau_3 \frac{\partial U}{\partial z} (K_m + K_{33}) \tag{3.37}$$

where $K_{33} = K_h$.

As a third example consider the problem, mentioned earlier and also discussed by Bradshaw (1969), of the near-neutral surface-layer response to a change in surface roughness. The appropriate form of our model $\overline{uw}$ equation is now

$$\frac{d}{dt} \overline{uw} = U \frac{\partial}{\partial x} \overline{uw} + W \frac{\partial}{\partial z} \overline{uw} = -\overline{w^2} \frac{\partial U}{\partial z} - \frac{\overline{uw}}{\tau_1} . \tag{3.38}$$

We make two further assumptions. First, we assume that $\overline{uw}$ remains proportional to $\overline{w^2}$ in the response region: $\overline{w^2} = A\, \overline{uw}$. Second, we assume that the time constant τ_1 in the response region retains the same relationship to z and $u_* = (-\overline{uw})^{\frac{1}{2}}$ that it had upstream. We can easily find this relationship by solving (3.30) for τ_1 under horizontally homogeneous conditions to find $\tau_1 = kz/(A\, u_*)$. With these assumptions (3.38) becomes

$$\frac{d\overline{uw}}{dt} = \frac{Au_*^3}{kz} (1 - \phi_m) \tag{3.39}$$

where ϕ_m is the dimensionless wind shear. Equation (3.39) says that $\phi_m \neq 1$ in the response region. In a rough-to-smooth transition, where $d\overline{uw}/dt > 0$, $\phi_m < 1$. Conversely $\phi_m > 1$ in the smooth-to-rough case.

The result (3.39) can be also interpreted in terms of eddy viscosity. If we use the relation $\phi_m = (kzu_*)/K_m$, (3.39) becomes

$$\frac{d\overline{uw}}{dt} = \frac{Au_*^3}{kz} \left(1 - \frac{kzu_*}{K_m}\right) . \tag{3.40}$$

This indicates that $K_m > ku_*z$ in a rough-to-smooth transition, and $K_m < ku_*z$ in the smooth-to-rough case; this is observed (Peterson, 1969).

As another example, consider the $\overline{w\theta}$ equation in the convective boundary layer. Observations (Kaimal et al, 1976) indicate that in mid-layer, but not too close to the inversion base, turbulent transport is small. We assume quasi-steady, horizontally homogeneous conditions and use the pressure covariance approximation of (3.29), writing (Deardorff, 1972)

$$\overline{w^2}\frac{\partial\Theta}{\partial z} = -\frac{\overline{w\theta}}{\tau_2} + \frac{g}{\Theta}\overline{\theta^2} . \tag{3.41}$$

The solution for $\overline{\theta w}$ is

$$\overline{\theta w} = -\tau_2\,\overline{w^2}\left(\frac{\partial\Theta}{\partial z} - \frac{g}{\Theta}\frac{\overline{\theta^2}}{\overline{w^2}}\right) = -K_h\left(\frac{\partial\Theta}{\partial z} - \gamma_c\right) \tag{3.42}$$

which is Deardorff's (1966) suggested form (3.16) which eliminates the need for negative K_h values in the convective ABL. In his 1972 paper, he cites considerable data on $\overline{\theta^2}$ and $\overline{w^2}$ in the mixed layer which indicate that the value of γ_c, which from (3.42) is

$$\gamma_c = \frac{g}{\Theta}\,\overline{\theta^2}\,/\,\overline{w^2} \tag{3.43}$$

is of the order of $1.0\ 10^{-3}\ \mathrm{Km}^{-1}$, in fairly good agreement with the value of $0.7\ 10^{-3}\ \mathrm{Km}^{-1}$ suggested earlier.

These examples make it clear that the second-moment equations can give useful insight into eddy viscosity behavior. Note, however, that in none of these examples did we derive any results rigorously; in each case we had to make other closure approximations. Thus, while the second-moment approach has obvious potential, its results come with some uncertainty because of its own closure approximations.

Let us now examine the second-order closure problem, that is, the problem of approximating the molecular destruction, turbulent transport, and pressure covariance terms in the second-moment equations.

The equations for the velocity variances (the diagonal elements of $\overline{u_i u_k}$) contain the dissipation rate ε, which is formally defined by

$$\varepsilon = \tfrac{1}{2}\nu \overline{\left(\frac{\partial u_i}{\partial x_j} + \frac{\partial u_j}{\partial x_i}\right)^2} . \tag{3.44}$$

In a locally isotropic field this reduces to our earlier definition (3.21). One can use the Navier-Stokes equations and (3.44) to derive a conservation equation for ε. We will not do this here, but Tennekes and Lumley (1972) derived an equation for fluctuating vorticity variance, $\overline{\omega_i\omega_i}$, a closely-related quantity. The ε equation has a large number of terms, but they can be scaled using arguments also discussed by Tennekes and Lumley (1972). The resulting ε-equation is

$$\frac{\partial \varepsilon}{\partial t} = - U_j \frac{\partial \varepsilon}{\partial x_j} - \frac{\partial \overline{\varepsilon' u_j}}{\partial x_j} - 2\nu \overline{\frac{\partial u_i}{\partial x_k}\frac{\partial u_i}{\partial x_j}\frac{\partial u_j}{\partial x_k}} - 2\overline{\left(\nu \frac{\partial^2 u_i}{\partial x_j \partial x_k}\right)^2} \tag{3.45}$$

We denote the fluctuating dissipation rate ((3.44) without the overbar) by ε'. The terms on the right of (3.45) represent advection, transport, production by vortex stretching, and molecular destruction, respectively. The largest terms are found to be production and molecular destruction. Keeping only them gives

$$\overline{\frac{\partial u_i}{\partial x_k}\frac{\partial u_i}{\partial x_j}\frac{\partial u_j}{\partial x_k}} = \nu \overline{\left(\frac{\partial^2 u_i}{\partial x_j \partial x_k}\right)^2} \tag{3.46}$$

which, although useful in interpreting turbulent fine structure (Batchelor, 1960), is not helpful here; it simply says production and dissipation balance, and that the other terms are smaller. To obtain a useful ε-equation one must retain these smaller terms, which are of the order of the production-dissipation imbalance I. This gives

$$\frac{\partial \varepsilon}{\partial t} = - U_j \frac{\partial \varepsilon}{\partial x_j} - \frac{\partial}{\partial x_j} \overline{\varepsilon' u_j} + I. \tag{3.47}$$

This equation says that ε can change with time because of advection, transport, or an imbalance between production and destruction. The terms in (3.47) are of the order of q^4/ℓ^2; since ε is of order q^3/ℓ, it follows that changes in ε occur with a time scale ℓ/q, the energy-containing range time scale.

Of the terms on the right-hand side of (3.47) only I survives when integrated over the entire flow volume; thus I alone is responsible for any time changes in volume-averaged ε. The advection and transport terms are generally important locally, however. For example, in a horizontally homogeneous, stationary boundary layer, transport balances I:

$$\partial\, \overline{\varepsilon' w}/\partial z = I \,. \tag{3.48}$$

While the ε-equation (3.47) is important in turbulence dynamics, it contains two further unknowns, the transport and imbalance terms. To make this equation useful in a second-order model we must relate these terms to known quantities such as turbulence covariances and the mean fields. One can obtain little fundamental guidance for such an effort. A few attempts can be found in the recent literature: most involve parameterizing I as proportional to $P - \varepsilon$ where P is the production rate of turbulent kinetic energy, and using a gradient diffusion approximation for transport. Rodi (1980) discusses current practice in more detail.

While it is no surprise that we met a closure problem in the ε-equations - we said earlier that every turbulence moment equation has a closure problem - this closure problem is more severe than that for $\overline{u_i u_k}$. In the form of (3.47) appropriate to a horizontally homogeneous boundary layer, each of the dominant terms on the right-hand side must be parameterized. By contrast, many of the important terms in the $\overline{u_i u_k}$ equation can be calculated exactly. Thus one expects ε to be more sensitive to closure parameterization than is $\overline{u_i u_k}$, and this has caused many workers to choose simpler approaches than (3.47) for calculating ε.

Most simpler approaches are based on the empirical observation that ε in any turbulent flow is of the order of q^3/ℓ. (See also (2.40).) One knows q^2 from the $\overline{u_i u_i}$ equation, and thus one can parameterize ε by specifying the distribution of ℓ. As an example, consider the idealized neutral surface layer. Data suggest $q^2 \sim 7.5\, u_*^2$, and $\varepsilon = u_*^3/kz$. Thus choosing $\ell \sim 7z$ gives $q^3/\ell = \varepsilon = u_*^3/kz$.

The parameterization $\varepsilon \sim q^3/\ell$ has the attractive property of providing a stabilizing, negative feedback in the turbulent kinetic energy equation. The source terms in that equation are of second order in the turbulence level; in shear-driven turbulence, for example, production is of order q^2 times the mean shear. The dissipation, in this parameterization, is proportional to q^3; thus if q^2 is larger than the mean shear can support, dissipation exceeds production and q^2 decreases with time. Conversely, if q^2 is too small, production will exceed dissipation and q^2 will grow.

The principal difficulty with this approach is that while $\varepsilon \sim q^3/\ell$ is correct in order-of-magnitude, it is difficult to tailor ℓ distributions so that it becomes an accurate representation, because ℓ is multiple-valued in the boundary layer (Wyngaard, 1980). Attempts have been made to derive a conservation equation for ℓ, but here again one meets severe closure problems. Most workers prefer to specify ℓ distributions a priori, as was done with mixing length when it was widely used.

Consider next the pressure covariances. By taking the divergence of the u_i-equation (1.26b) and using incompressibility ($\partial u_i/\partial x_i = \partial U_i/\partial x_i = 0$) we find an equation for the fluctuating pressure

$$\frac{1}{\rho_o}\nabla^2 p = -\left(\frac{\partial u_i}{\partial x_j}\frac{\partial u_j}{\partial x_i} - \overline{\frac{\partial u_i}{\partial x_j}\frac{\partial u_j}{\partial x_i}}\right) - 2\frac{\partial U_i}{\partial x_j}\frac{\partial u_j}{\partial x_i} + \\ - \frac{g_i}{\Theta}\frac{\partial \theta}{\partial x_i} - 2\,\varepsilon_{ijk}\,\Omega_j\frac{\partial u_k}{\partial x_i} \equiv f(x_i). \tag{3.49}$$

There are no time-change terms because incompressibility implies that pressure responds instantaneously to changes in the right-hand side of (3.49). This is a good approximation for our purposes.

The first term on the right-hand (3.49) is non-linear in the turbulent velocity; it represents turbulent-turbulent interactions. The remaining terms are linear in the turbulence field and represent interactions with mean shear, gravity, and rotation; they can be negligible in some situations (in shear-free, non-buoyant laboratory flows, for example). The right-hand side of (3.49) is sometimes separated into a return-to-isotropy term (the nonlinear turbulence term) and rapid terms. The rapid terms are so named because they involve external influences (mean shear, gravity, rotation) which can (in principle, in laboratory situations) be rapidly changed. In any event, external time scales can control the pressure changes due to the 'rapid' terms, while the changes due to the 'return-to-isotropy' terms occur at the internal, turbulent time scales.

Equation (3.49) is a Poisson equation. Its solution in a boundary-layer flow is (Launder et al., 1975)

$$\frac{1}{\rho_o} p(x_i) = -\frac{1}{4\pi}\int f(y_i)(1/r + 1/r^*)\, dy_i \tag{3.50}$$

where r and r^* are the lengths of the vectors $x_i - y_i$ and $x_i - y_i{}^*$. We remind here our notation, where $f(x_i)$ stands for $f(x_1,x_2,x_3)$ and dx_i denotes the integration volume $dx_1dx_2dx_3$. The image of y_i (through the surface) is denoted by y_i^*. Thus the pressure fluctuations at a point x_i are determined by events over the entire flow, with those nearest x_i having the most influence. This solution allows us to write the pressure covariance as integrals of two-point correlations. For example, the covariance in the $\overline{\theta u_i}$ equation (1.53) becomes

$$\frac{1}{\rho_o}\overline{\theta \frac{\partial p}{\partial x_i}} = \frac{1}{4\pi}\int \{\overline{\theta(x_i)\frac{\partial}{\partial x_i}(\frac{\partial u_k(y_i)}{\partial x_j}\frac{\partial u_j(y_i)}{\partial x_k})} + \tag{3.51}$$

$$2\frac{\partial^2 U_k(y_i)}{\partial x_j \partial x_i}\overline{\theta(x_i)\frac{\partial u_j(y_i)}{\partial x_k}} + 2\frac{\partial U_k(y_i)}{\partial x_j}\overline{\theta(x_i)\frac{\partial^2 u_j(y_i)}{\partial x_i \partial x_k}} +$$

$$+\frac{g_k}{\Theta}\frac{\partial^2\overline{\theta(y_i)\theta(x_i)}}{\partial x_i \partial x_k} + 2\,\varepsilon_{kjm}\,\Omega_j\frac{\overline{\partial^2 u_m(y_i)\theta(x_i)}}{\partial x_i \partial x_k}\}\{\frac{1}{r}+\frac{1}{r^*}\}\,dy_i$$

Any attempt to derive equations for these two-point correlations will generate yet more unknowns, and thus we have another closure problem.

On the basis of our earlier arguments for turbulent buoyancy and inertia forces, we would expect that the buoyancy and turbulent-turbulent terms in (3.51) are of the same order in the boundary layer, and that the mean shear term can also be important. We would also expect the Coriolis term to be appreciably smaller than the mean-shear term, since mean shear is typically an order of magnitude larger than f.

One approach to the closure problem posed by (3.51) is to make an assumption about turbulence structure which allows the integrals to be evaluated. One such assumption is that the turbulence structure is isotropic. To illustrate, we take homogeneous turbulent flow (i.e., we leave the boundary layer for this example) and consider the mean shear contribution to (3.51). Denoting that portion of the fluctuating pressure field by p^S, we have from (3.49)

$$\frac{1}{\rho_o}\nabla^2 p^S = -2\frac{\partial U_k}{\partial x_m}\frac{\partial u_m}{\partial x_k} \tag{3.52}$$

so that

$$\frac{\partial}{\partial x_i}(\frac{1}{\rho_o}\nabla^2 p^S) = -2\frac{\partial^2 U_k}{\partial x_i \partial x_m}\frac{\partial u_m}{\partial x_k} - 2\frac{\partial U_k}{\partial x_m}\frac{\partial^2 u_m}{\partial x_i \partial x_k}\ . \tag{3.53}$$

It is simplest to solve for the pressure covariances by using

Fourier-Stieltjes transform techniques (Lumley and Panofsky, 1964). We write the random variables $\partial p^S/\partial x_i$, u_m, and θ as

$$\begin{aligned}
\frac{1}{\rho_o}\frac{\partial p^S}{\partial x_i}(x_i,t) &= \int e^{i\kappa_j x_j}\, dP_i^S(\kappa_i,t),\\
u_m(x_i,t) &= \int e^{i\kappa_j x_j}\, dU_m(\kappa_i,t),\\
\theta\,(x_i,t) &= \int e^{i\kappa_j x_j}\, d\Theta(\kappa_i,t)\;.
\end{aligned}\tag{3.54}$$

It follows from (3.53) and (3.54) that

$$dP_i^S = \left(\frac{2i\,\kappa_k}{\kappa^2}\,\frac{\partial^2 U_k}{\partial x_m \partial x_i} - \frac{2\,\kappa_k\kappa_i}{\kappa^2}\,\frac{\partial U_k}{\partial x_m} \right) dU_m\;, \tag{3.55}$$

where $\kappa = (\kappa_i\kappa_i)^{\frac{1}{2}}$. Using the property

$$\frac{1}{\rho_o}\overline{\theta\frac{\partial p}{\partial x_i}^S} = \int_{\kappa_i} \overline{dP_i^S\, d\Theta^*}\;, \tag{3.56}$$

where * denotes complex conjugate, we have from (3.55)

$$\frac{1}{\rho_o}\overline{\theta\frac{\partial p}{\partial x_i}^S} = 2\,\frac{\partial^2 U_k}{\partial x_m \partial x_i}\int \frac{i\,\kappa_k}{\kappa^2}\,C_m(\kappa_i)\,d\kappa_i - 2\,\frac{\partial U_k}{\partial x_m}\int \frac{\kappa_k\kappa_i}{\kappa^2}\,C_m(\kappa_i)\,d\kappa_i, \tag{3.57}$$

where $C_m(\kappa_i)d\kappa_i(\equiv C_m(\kappa_1,\kappa_2,\kappa_3)\,d\kappa_1 d\kappa_2 d\kappa_3) = \overline{dU_m d\Theta^*}$. Now we invoke isotropy, which implies (Batchelor, 1960) $C_m(\kappa_i) = C(\kappa)\,\kappa_m$. Thus under isotropy (3.57) becomes

$$\frac{1}{\rho_o}\overline{\theta\frac{\partial p}{\partial x_i}^S} = 2\,\frac{\partial^2 U_k}{\partial x_m \partial x_i}\int \frac{i\,\kappa_k\kappa_m}{\kappa^2}\,C(\kappa)\,d\kappa_i - 2\,\frac{\partial U_k}{\partial x_m}\int \frac{\kappa_k\kappa_i\kappa_m}{\kappa^2}\,C(\kappa)\,d\kappa_i. \tag{3.58}$$

If we perform the integration over κ_i first over spherical shells (κ = constant), we see that the first integral is proportional to δ_{km} (see Lumley, 1979). Thus when multiplied by $\partial^2 U_k/\partial x_m \partial x_i$ it becomes proportional to $\partial^2 U_k/\partial x_k \partial x_i$ which vanishes by incompressibility. The second integral is odd in κ_i and vanishes directly. Thus $\overline{\theta \partial p/\partial x_i^S}$ vanishes in an isotropic field.

In the same way one finds that the turbulent-turbulent and

Coriolis contributions to $\overline{\theta \partial p/\partial x_i}$ vanish under isotropy. The buoyancy term becomes

$$\frac{1}{\rho_o} \overline{\theta \frac{\partial p}{\partial x_i}}^B = - \frac{g_k}{\Theta} \int \frac{\kappa_i \kappa_k}{\kappa^2} \phi \, d\kappa_i \tag{3.59}$$

where ϕ is the temperature spectrum, i.e.,

$$\int \phi \, d\kappa_i = \overline{\theta^2} . \tag{3.60}$$

With the assumption of isotropy ϕ depends only on κ and the integral yields

$$\frac{1}{\rho_o} \overline{\theta \frac{\partial p}{\partial x_i}}^B = - \frac{1}{3} \frac{g_i}{\Theta} \overline{\theta^2} . \tag{3.61}$$

We saw earlier that $\overline{\theta \partial p/\partial x_i}$ was non-zero (and in fact quite important) in the surface layer. Careful consideration shows that buoyancy effects alone cannot sustain $\overline{\theta \partial p/\partial x_i}$ and thus we must conclude that an isotropic model is not adequate for evaluating pressure covariances.

The next possibility that comes to mind is an expression for $\overline{\theta \partial p/\partial x_i}$ in the form of a series expansion about isotropy. The leading term would represent the isotropic contribution, which is simply the buoyancy term (3.61). Higher-order terms would represent the effects of departures from isotropy; these would presumably have important contributions from the turbulent-turbulent and mean-shear terms, and possibly from anisotropic buoyancy effects.

Lumley (1979) has developed an approach of this general nature; we will refer to it as isotropic tensor modeling. We can illustrate it by considering the mean shear contribution to $\overline{\theta \partial p/\partial x_i}$ as expressed by (3.57). We write this as

$$\frac{1}{\rho_o} \overline{\theta \frac{\partial p}{\partial x_i}}^S = 2 F_{km} \frac{\partial^2 U_k}{\partial x_m \partial x_i} + 2 G_{kim} \frac{\partial U_k}{\partial x_m} \tag{3.62}$$

where the tensors F_{km} and G_{kim} represent the integrals in (3.57). Now we assume that F_{km} and G_{kim} are isotropic tensor functions of the vector $\overline{u_j \theta}$. Thus we can write (Batchelor, 1960; Lumley, 1979)

$$F_{km} = A \overline{\theta u_k} \overline{\theta u_m} + B \delta_{km}, \tag{3.63}$$

$$G_{kim} = C\,\delta_{ki}\,\overline{\theta u_m} + D\,\delta_{im}\,\overline{\theta u_k} + E\,\delta_{km}\,\overline{\theta u_i}\,. \tag{3.64}$$

If we make the further requirement that F_{km} and G_{kim} be linear in $\overline{\theta u_j}$, then A = 0. Incompressibility implies that $\overline{\theta \partial u_k/\partial x_k} = 0$, which in turn implies that $F_{kk} = 0$. Thus B = 0. Turning to G_{kim}, we note from (3.57) that $G_{kim} = G_{ikm}$; from (3.57) and the definition of C_m that $G_{kkm} = \overline{u_m\theta}$; and from incompressibility that $G_{kik} = 0$. These conditions require that C = 2/5, D = E = -1/10, so that from (3.64) and (3.62)

$$\frac{1}{\rho_o}\overline{\theta\frac{\partial p}{\partial x_i}}^S = +\frac{4}{5}\overline{\theta u_j}\frac{\partial U_i}{\partial x_j} - \frac{1}{5}\overline{\theta u_k}\frac{\partial U_k}{\partial x_i}\,. \tag{3.65}$$

From (1.53) we see that the first term in the parameterization (3.65) would remove 80% of the direct shear production of $\overline{\theta u_i}$. The key assumption made in deriving (3.65) is that the exact expression, as represented by the integrals in (3.57), is an isotropic tensor function of $\overline{\theta u_i}$. Since the integrals were moments of the spectrum of $\overline{\theta u_i}$, this seems at least plausible. Consider next, however, the turbulent-turbulent contribution. Here the integral expression is

$$\frac{1}{\rho_o}\overline{\theta\frac{\partial p}{\partial x_i}}^{TT} = -i\int\frac{\kappa_i\kappa_k\kappa_m}{\kappa^2}\,\phi_{km}\,d\kappa_i \tag{3.66}$$

where ϕ_{km} is the cross-spectrum of $u_k u_m - \overline{u_k u_m}$ and θ. We can now denote this integral by I_{ikmkm}, note that it is a contraction of I_{ikmpq}, and write it as an isotropic tensor function of second moments - but of which one(s)? Perhaps the simplest assumption is that I_{ikmpq} depends on $\overline{\theta u_i}$, which leads to

$$\frac{1}{\rho_o}\overline{\theta\frac{\partial p}{\partial x_i}}^{TT} = \frac{C_\theta}{\tau}\overline{\theta u_i} \tag{3.67}$$

where C_θ is a constant and τ is an energy containing range time scale. Recall that this is the form we used in our example (3.29).

Let us turn now to the modeling of the pressure term in the $\overline{u_i u_k}$ equation. Part of this pressure term, $2/3\delta_{ik}(\partial\overline{pu_j}/\partial x_j)$, respresents transport. Lumley (1975) first suggested that one should subtract this part from the pressure covariance and add it to the transport term. Thus we define

$$\pi_{ik} = \frac{1}{\rho_o}\left(\overline{u_k\frac{\partial p}{\partial x_i}} + \overline{u_i\frac{\partial p}{\partial x_k}} - \frac{2}{3}\delta_{ik}\frac{\partial}{\partial x_j}\overline{pu_j}\right). \tag{3.68}$$

The isotropic tensor modeling approach gives the following parameterizations for π_{ik} (Zeman, 1981)

$$\pi_{ik}^{TT} = \frac{C_1}{\tau}(\overline{u_i u_k} - \frac{1}{3}\overline{u_m u_m}\,\delta_{ik}) = C_1\, b_{ik}\,\varepsilon\ ,$$

$$\pi_{ik}^{S} = -2\,\overline{u_m u_m}\,[\,\frac{1}{5}S_{ik} + \alpha_1(S_{ij}b_{jk} + S_{kj}b_{ji} - \frac{2}{3}S_{lj}b_{jl}\delta_{ik}) + $$

$$+\alpha_2(R_{ij}b_{jk} + R_{kj}b_{ji})]\ ,$$

$$\pi_{ik}^{B} = \frac{3}{10}\frac{g_m}{\Theta}\left(\overline{\theta u_i}\,\delta_{mk} + \overline{\theta u_k}\,\delta_{mi} - \frac{2}{3}\overline{\theta u_m}\,\delta_{ik}\right). \qquad (3.69)$$

Here, $S_{ik} = \frac{1}{2}(\partial U_i/\partial x_k + \partial U_k/\partial x_i)$ and $R_{ik} = \frac{1}{2}(\partial U_i/\partial x_k - \partial U_k/\partial x_i)$, $\tau = \overline{u_i u_i}/\varepsilon$, and $b_{ik} = \overline{u_i u_k}/\overline{u_m u_m} - \delta_{ik}/3$.

According to Zeman (1981), Lumley has used rapid-distortion results to determine $\alpha_1 = 3/7$ and $\alpha_2 = 0$. C_1 is normally taken as a free constant. The first of (3.69) is usually called the Rotta (1951) parameterization.

The set (3.69), together with our results

$$\frac{1}{\rho_o}\overline{\theta\frac{\partial p}{\partial x_i}}^{TT} = \frac{C_\theta}{\tau}\overline{\theta u_i}$$

$$\frac{1}{\rho_o}\overline{\theta\frac{\partial p}{\partial x_i}}^{S} = +\frac{4}{5}\overline{\theta u_j}\frac{\partial U_i}{\partial x_j} - \frac{1}{5}\overline{\theta u_k}\frac{\partial U_k}{\partial x_i} \qquad (3.70)$$

$$\frac{1}{\rho_o}\overline{\theta\frac{\partial p}{\partial x_i}}^{B} = -\frac{1}{3}\frac{g_i}{\Theta}\overline{\theta^2}$$

are considered to be a 'standard set' of parameterizations (Zeman, 1981). As Zeman points out, however, these parameterizations are not completely effective in the boundary layer. As examples, Zeman indicates that authors usually take the liberty of adjusting the 'invariant' constants in (3.69) and (3.70) in order to 'tune' their models to the boundary layer. Zeman and Lumley (1979) found it necessary to change the sign and magnitude of the modeling in (3.70) of $\overline{\theta\partial p/\partial x_i}^S$. Wyngaard (1975) found it necessary to increase the coefficient 1/3 in the expression (3.70) for $\overline{\theta\partial p/\partial x_i}^B$ toward 1.0 under stable conditions. Wyngaard (1980) also considers the parameterizations (3.69) and (3.70) in the context of boundary-layer data and discusses some of their unphysical properties.

There are two more points to be made with regard to pressure covariance parameterizations for boundary-layer models. The first has to do with the 'wall effects' discussed by Gibson and Launder (1978). They argue that since the formal solution (3.50) for pressure in a boundary layer has a 'wall effect' term (that involving y_i^*), the parameterizations of pressure covariances should explicitly include wall effects. They have proposed such terms but as discussed by Wyngaard (1980) it is not clear that they remedy any of the known deficiencies of existing pressure covariance parameterizations in the boundary layer. Nonetheless, there clearly are 'wall effects', particularly in the convective case where they cause departures from Monin-Obukhov surface-layer similarity (Wyngaard, 1980).

The second point concerns what is called 'realizability'. Schumann (1977) was perhaps the first to point out that parameterization for pressure covariances and other quantities in second-order models should be formulated so that the resulting second-moment field is 'realizable', i.e., has non-negative variances and correlation coefficients not greater than 1.0 magnitude. Schumann showed that the usual parameterizations (including some of (3.69) and (3.70)) do not guarantee realizability. The importance of this is clear to anyone who has computed with second-order models. While it might seem surprising that it should be raised at this relatively late date, that too is understandable; it simply has not been clear how to build realizability criteria into parameterizations. Schumann (1977) derived some realizability criteria and suggested an ad-hoc enforcement scheme. Lumley (1979) has done perhaps the most fundamental work along these lines, but the realizable forms corresponding to (3.69) and (3.70) can become very complicated and difficult to interpret. This is an area that needs much more study.

We turn now to transport. It has two contributors: pressure transport, just discussed, and divergence of the flux due to turbulent velocity, which we consider now. The flux in the $\overline{u_i u_k}$-equation equation is $\overline{u_i u_k u_j}$. Again we can get some insight by examing its conservation equation

$$\frac{\partial}{\partial t}\overline{u_i u_k u_j} = -\left[\frac{\partial U_i}{\partial x_m}\overline{u_m u_j u_k} + ..\right] - \frac{\partial}{\partial x_m}\overline{u_i u_m u_j u_k} +$$

$$+\left[\overline{u_j u_k}\frac{\partial}{\partial x_m}\overline{u_i u_m} + ..\right] - \left[\overline{u_j u_k \frac{\partial p}{\partial x_i}} + ..\right] - \left[\frac{g_i}{\Theta}\overline{\theta u_j u_k} + ..\right]$$

$$-\left[2\,\varepsilon_{imn}\,\Omega_m\,\overline{u_n u_k u_j} + ..\right] - 2\nu\left[\overline{u_k \frac{\partial u_i}{\partial x_p}\frac{\partial u_j}{\partial x_p}} + ..\right]. \qquad (3.71)$$

The + .. indicates two more terms obtained from the first by cyclic permutation of the indices i,j,k. The terms in (3.71) are, in order, shear production, transport, turbulent production, pressure interaction, buoyancy, rotation, and molecular. The transport term here is a fourth moment; even order moments of velocity and temperature (but not derivative moments) are usually observed to be Gaussian, to a good approximation. If we use a Gaussian approximation for this fourth moment, we have

$$\overline{u_i u_m u_j u_k} \simeq \overline{u_i u_m}\;\overline{u_j u_k} + \overline{u_i u_j}\;\overline{u_m u_k} + \overline{u_i u_k}\;\overline{u_m u_j} \, . \tag{3.72}$$

This allows us to combine transport and turbulent production in (3.71) to get

$$\begin{aligned} &- \frac{\partial}{\partial x_m} \overline{u_i u_m u_j u_k} + [\, \overline{u_j u_k} \frac{\partial}{\partial x_m} \overline{u_i u_m} + \,..] \\ &\simeq [\, - \overline{u_i u_m} \frac{\partial}{\partial x_m} \overline{u_j u_k} + \,...] \, . \end{aligned} \tag{3.73}$$

We cannot interpret some of the terms in (3.71) as easily as those in (1.52) and (1.53). Nonetheless let us consider (3.71) in our idealized, near-neutral surface layer, using the result (3.73) of our Gaussian approximation. Assume for the moment that the pressure (P_{ijk}) and molecular (M_{ijk}) terms in (3.71) behave roughly as their counterparts in the second-moment equations, so that

$$P_{ijk} + M_{ijk} \simeq \frac{\overline{u_i u_j u_k}}{\tau} \tag{3.74}$$

With (3.74), the $\overline{u_i u_j u_k}$-equation now reduces to

$$-\tau [\, \overline{u_i u_m} \frac{\partial}{\partial x_m} \overline{u_j u_k} + \overline{u_j u_m} \frac{\partial}{\partial x_m} \overline{u_k u_i} + \overline{u_k u_m} \frac{\partial}{\partial x_m} \overline{u_i u_j}] \simeq \overline{u_i u_j u_k} \, . \tag{3.75}$$

This is exactly the gradient-diffusion approximation used for $\overline{u_i u_j u_k}$ in many models.

Let's pursue this a bit further, still using our approximations (3.73) and (3.74). In a very unstable surface layer, or in the mixed layer during even mildly unstable conditions, the buoyant term in (3.71) will be important. Then Equation (3.71) becomes

$$\overline{u_i u_j u_k} = -\tau[\, \overline{u_i u_m} \frac{\partial}{\partial x_m} \overline{u_j u_k} + \overline{u_j u_m} \frac{\partial}{\partial x_m} \overline{u_k u_i} + \overline{u_k u_m} \frac{\partial}{\partial x_m} \overline{u_i u_j} \,]+$$

$$- \tau \left[\frac{g_i}{\Theta} \overline{\theta u_j u_k} + \frac{g_j}{\Theta} \overline{\theta u_i u_k} + \frac{g_k}{\Theta} \overline{\theta u_i u_j} \right] . \tag{3.76}$$

Now the gradient-diffusion model (3.75) for $\overline{u_i u_j u_k}$ need not hold; if the buoyant term in (3.76) is also important, $\overline{u_i u_j u_k}$ can depend on more than second-moment gradients.

Consider the situation for $\overline{w^3}$ in the unstable surface layer. Both $\overline{w^3}$ and $\partial \overline{w^2}/\partial z$ are observed to be positive there, while (3.75) would imply

$$\overline{w^3} = -3\tau \overline{w^2} \frac{\partial \overline{w^2}}{\partial z} , \tag{3.77}$$

or that they have an opposite sign. Thus the usual gradient-diffusion model fails. One reason for this failure is that the buoyant term in the $\overline{w^3}$ budget (3.73) is significant in the unstable surface layer. Wyngaard (1980) discusses the $\overline{w^3}$ budget in more detail.

As we showed in the early part of this chapter, strong buoyancy effects are the rule in the boundary layer. Thus we should expect in general that gradient-diffusion approximation (such as (3.75)) for third moments will not be accurate. In shear-driven flows, such as found in the laboratory and in many engineering applications, we would expect them to be more useful.

With this background we can discuss approaches to transport modeling, of which there are basically three. The first, and simplest, is to ignore transport. Although data for testing transport models are limited, it appears that this no-transport approximation can be useful in the stably-stratified boundary layer. In neutral shear flows, it is probably not as good an approximation, and it can fail conspicuously in the convective case, where transport is often very important.

The second approach is the gradient-diffusion approximation, which for $\overline{u_i u_j u_k}$ is (3.75). This is broadly consistent with transport behavior in shear-driven turbulence, although we still have much to learn here. Transport in convective turbulence can behave quite differently; it can carry turbulent kinetic energy up, rather than down its gradient and, in fact, can become uncoupled from the gradient.

The third approach, which has been attempted only recently, is to carry simplified rate equations for the transport terms.

This can introduce many new equations, and a new set of closure problems. It has the advantage, however, of resolving exactly the buoyancy effects on transport, since the buoyant terms are also third moments. It has been used by Lumley et al. (1978) and Zeman and Lumley (1976) to give much more realistic simulations of the convective boundary layer than models with down-gradient transport parameterizations. André et al. (1978) have also done simulations with a different version of this approach. Zeman (1981) summarizes some recent developments here, and Wyngaard (1980) discusses $\overline{w^3}$ dynamics.

In these pages we have tried simply to cover the fundamentals of second-order modeling. The interested reader will want to pursue further details in recent survey papers on the subject. Reynolds (1976), Reynolds and Cebeci (1976), and Lewellen (1977) cover contemporary shear-flow models in some detail. Launder (1976) has emphasized the modeling of heat and mass transfer, and Rodi (1980) has recently discussed models for environmental flows. Zeman (1981) has surveyed boundary-layer applications. The leading work in second-order modeling under convective conditions has been done by Lumley and his colleagues; Lumley's ideas are summarized in Lumley (1979). There is also a wealth of literature on second-order modeling in the engineering, fluid mechanics, and atmospheric science journals.

Second-order modeling of the boundary layer is still in its early stages and is evolving rapidly. A wide variety of models can be found in the literature, some based entirely on shear flow closures and some tailored specifically for buoyancy-dominated turbulence. As Zeman (1981) points out, their early success in predicting engineering flows has led to their sometimes indiscriminate use in geophysical flows. While this situation is confusing, it is well to remember its cause - that as yet there is no known way to construct a second-order model entirely from first principles. Lumley has called second-order modeling 'turbulence engineering'. Like engineering, carrying it out succesfully requires knowledge of the underlying physics, considerable working experience, access to suitable test data, a good perception of user needs, and creativity.

3.3. VOLUME-AVERAGE MODELS

We saw that the governing equations in the boundary layer must be averaged before they can be solved numerically and we just discussed the models which result from ensemble averaging. We will now discuss volume-averaging and the models it produces.

Large-Eddy Models

The averaging volume is here sufficiently small that the eddies contained entirely within it (i.e., the unresolved eddies) are of inertial-range scales and smaller. One resolves explicitly the largest (energy-containing range) eddies, those which are principally responsible for turbulent heat, momentum, and mass transfer.

We saw in the last section that ensemble-average closure is very difficult because it forces one to parameterize the effects of the entire spectral range of eddy motions - including the largest, energy-containing scales. These largest eddies in any turbulent flow are very sensitive to environmental details such as geometry and stratification. As a result, some researchers doubt whether a given second-order model may be useful over a wide range of conditions. For example, Lamb (See section 5.4) expresses reservations about second-order ensemble-average models in diffusion applications. These concerns are minimized with large-eddy models, since they calculate these environment-sensitive eddies, rather than parameterize them. Thus, while large-eddy models require much computer time, they have distinct advantages over ensemble-average models and are used today to calculate both geophysical and engineering flows.

We will examine the roots of large-eddy models by first deriving equations for the large-scale turbulent structure. We begin by decomposing each flow variable into resolvable and subgrid components. Velocity, for example, is written

$$\underset{\sim}{u}_i = \{U_i\} + u_i' , \tag{3.78}$$

where the braces represent the result of operating with a filter function. Leonard (1974) defines this filtering quite generally:

$$\{f(x_i,t)\} = \int G(x_i - x_i')\, f(x_i',t)\, dx_i' \tag{3.79}$$

where f is a flow variable, G is the filter function, and the integration is over the flow volume. We can also express the filtering operation (3.79) in Fourier terms. If we represent Fourier-transformed variables with a tilde, e.g.

$$f(x_i,t) = \int e^{i\kappa_j x_j}\, \tilde{f}(\kappa_i,t)\, d\kappa_i \tag{3.80}$$

then (3.79) becomes $\{\tilde{f}\} = \tilde{G}\,\tilde{f}$ which is an alternative form of the filter expression.

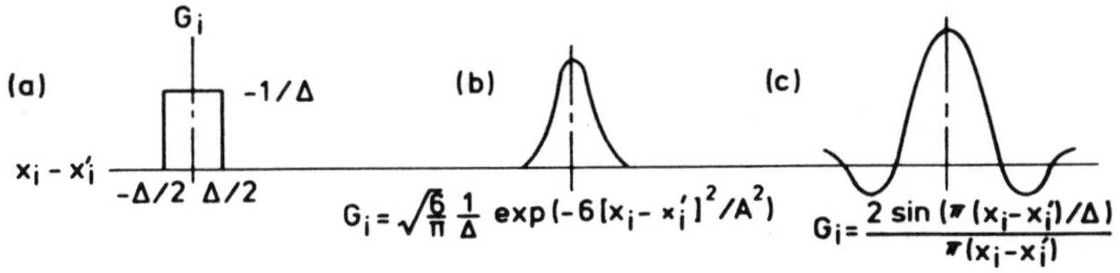

Figure 3.1. Examples of spatial filters, with $G = G_1G_2G_3$. Δ is the characteristic filter width. Filter (a) was used by Lilly (1967) and Deardorff (1973); (b) is Gaussian, and is used by the Stanford group; (c) corresponds to a truncated Fourier expansion.

Examples of spatial filters as given by Leonard are shown in Figure 3.1. The top hat filter (a) is that used in Deardorff's work, much of which is summarized in Deardorff (1973). Recent work by the Stanford group (Mansour et al., 1977) uses the Gaussian filter (b). Figure 3.1c corresponds to a truncated Fourier expansion (i.e., a step-function low pass filter). Each of these filters removes the small-scale contributions to the random variable f.

Let us consider the application of the filter (3.79) to the continuity and Navier-Stokes equations, (1.3) and (1.16) respectively. We assume that each flow variable vanishes on the boundaries, so that $\{\partial f/\partial x_i\} = \partial\{f\}/\partial x_i$ which follows from (3.79) through integration by parts. The filtered forms of the equations are

$$\frac{\partial}{\partial t}\{U_i\} + \frac{\partial}{\partial x_j}\{\underset{\sim}{u}_i\underset{\sim}{u}_j\} = -\frac{1}{\rho_0}\frac{\partial}{\partial x_i}\{P\} - 2\varepsilon_{ijk}\Omega_j\{U_k\} +$$

$$+\nu\frac{\partial^2\{U_i\}}{\partial x_j\partial x_j} + g_i = 0 \quad \text{and} \quad \frac{\partial\{U_i\}}{\partial x_i} = 0\,. \tag{3.81}$$

Equations (3.81) represent the dynamics of the large-scale motion. They cannot be solved yet, however, because they involve the unknown quantity $\{\underset{\sim}{u}_i\underset{\sim}{u}_j\}$.

We proceed as when dealing with the usual (ensemble-average) mean motion equations. We define the quantity

$$R_{ij} = \{\underset{\sim}{u}_i\,\underset{\sim}{u}_j\} - \{U_i\}\,\{U_j\} \tag{3.82}$$

and the stress tensor τ_{ij} by

$$\tau_{ij} = R_{ij} - \frac{1}{3}R_{kk}\,\delta_{ij}. \tag{3.83}$$

We introduce the modified pressure Π

$$\{\Pi\} = \{P\} + \frac{1}{3} R_{kk} \rho_0 \,. \tag{3.84}$$

With these definitions, (3.81) can be written as

$$\frac{\partial}{\partial t}\{U_i\} + \frac{\partial}{\partial x_j}\{U_i\}\{U_j\}$$

$$= -\frac{1}{\rho_0}\frac{\partial}{\partial x_i}\{\Pi\} - 2\epsilon_{ijk}\Omega_j\{U_k\} - \frac{\partial}{\partial x_j}\tau_{ij} + \nu\frac{\partial^2}{\partial x_j^{\,2}}\{U_i\} + g_i \,. \tag{3.85}$$

Eq. (3.85) is the large-eddy equation used by Lilly (1967) and Deardorff (1973). Note that it has the form of the ensemble-average momentum equation (1.26a). The 'Reynolds stress' term τ_{ij} in this case represents the effect of the unresolvable, subgrid-scale motions, and must be parameterized.

Some workers use a large-eddy equation slightly different from (3.85). Rather than using the decomposition (3.82) for R_{ij}, Leonard (1974), for example, writes

$$\{\tilde{u}_i \tilde{u}_j\} = \{\{U_i\}\{U_j\}\} + \{u'_i\{U_j\}\} + \{\{U_i\}u'_j\} + \{u'_i u'_j\} \tag{3.86}$$

Using this decomposition, we define

$$R'_{ij} = \{\tilde{u}_i \tilde{u}_j\} - \{\{U_i\}\{U_j\}\}$$

$$= \{u'_i\{U_j\}\} + \{\{U_i\} u'_j\} + \{u'_i u'_j\} \tag{3.87}$$

which differs slightly from (3.82). Define the stress tensor τ'_{ij} and the modified pressure Π' as before

$$\tau'_{ij} = R'_{ij} - \frac{1}{3} R'_{kk} \delta_{ij}, \tag{3.88}$$

$$\{\Pi'\} = \{P\} + \frac{1}{3} R'_{kk} \rho_o. \tag{3.89}$$

Then the large-eddy equation becomes

$$\frac{\partial}{\partial t}\{U_i\} + \frac{\partial}{\partial x_j}\{\{U_i\}\{U_j\}\}$$

$$= -\frac{1}{\rho_o}\frac{\partial}{\partial x_i}\{\Pi'\} - 2\epsilon_{ijk}\Omega_j\{U_k\} - \frac{\partial}{\partial x_j}\tau'_{ij} + \nu\frac{\partial^2}{\partial x_j^{\,2}}\{U_i\} + g_i \,. \tag{3.90}$$

Leonard (1974) has pointed out that the resolvable-scale inertia terms in our two forms of the large-eddy equation, (3.85) and (3.90), are different. That is, $\{U_i\}\{U_j\} \neq \{\{U_i\}\{U_j\}\}$ in general. The difference, of course, is compensated by the differences in the pressure and stress terms, since both (3.85) and (3.90) are exact.

To examine further the differences in the inertia terms, we consider the budget of large-eddy kinetic energy in flow within a finite volume. We multiply (3.85) and (3.90) by $\{U_i\}$ and integrate over the flow volume; the pressure terms become divergences and vanish on integration, and the Coriolis terms vanish indentically. The viscous terms are smaller than the remaining terms by a factor of the large-eddy Reynolds number. Thus we have, from (3.90) and (3.85) respectively,

$$\frac{1}{2}\frac{\partial}{\partial t}\overline{\{U_i\}\{U_i\}} = -\overline{\{U_i\}\frac{\partial}{\partial x_j}\{\{U_i\}\{U_j\}\}} - \overline{\{U_i\}\frac{\partial}{\partial x_j}\tau'_{ij}} \tag{3.91}$$

$$\frac{1}{2}\frac{\partial}{\partial t}\overline{\{U_i\}\{U_i\}} = -\overline{\{U_i\}\frac{\partial}{\partial x_j}\{U_i\}\{U_j\}} - \overline{\{U_i\}\frac{\partial}{\partial x_j}\tau_{ij}} \tag{3.92}$$

where the overbar represents an integral over the flow volume. We can write the triple-product term in (3.92) as a divergence also

$$\{U_i\}\frac{\partial}{\partial x_j}\{U_i\}\{U_j\} = \frac{1}{2}\frac{\partial}{\partial x_j}\{U_i\}\{U_i\}\{U_j\} \tag{3.93}$$

and hence it integrates to zero. We rewrite the remaining terms on the right sides of (3.91) and (3.92) by integrating by parts, and we have finally

$$\frac{1}{2}\frac{\partial}{\partial t}\overline{\{U_i\}\{U_i\}} = \overline{\{\{U_i\}\{U_j\}\}\frac{\partial}{\partial x_j}\{U_i\}} + \overline{\tau'_{ij}\frac{\partial}{\partial x_j}\{U_i\}} \tag{3.94}$$

$$\frac{1}{2}\frac{\partial}{\partial t}\overline{\{U_i\}\{U_i\}} = \overline{\tau_{ij}\frac{\partial}{\partial x_j}\{U_i\}}\ . \tag{3.95}$$

The right-hand sides of (3.94) and (3.95) (which are equal) represent the loss of large-eddy kinetic energy through transfer to the subgrid-scale eddies. A steady state could exist if this were balanced by energy input, for example, by buoyant forcing. This transfer loss is equal to $\{\varepsilon\}$, the volume-integrated molecular destruction rate due to the smallest eddies. Note that in (3.95) this transfer is entirely due to the subgrid-scale stresses. In (3.94), however, some of this transfer is done by the resolvable eddies. Leonard (1974) has shown that the resolvable portion is significant and has estimated it as of the order of one-third of the total transfer.

Some workers feel that because it allows this resolvable-scale energy transfer, the large-eddy equation in the form (3.90) is to be preferred over the form (3.85). This is a debatable point, however, and in any event one must parameterize the subgrid-scale stresses, which do most of the energy transfer.

Before discussing the parameterization of sub-grid scale motions, we should mention that there is an alternative to (3.79) as the underlying volume-averaging process. Recall that (3.79) represents the passage of a volume-averaging filter over the entire region occupied by the turbulence, leaving a filtered field which is defined at every point in the region. Schumann's (1975) alternative is simply to average over fixed volumes, so that the averaged field is defined only at the center of the (non-overlapping) averaging volumes. This 'control volume' averaging process has different properties from the continuous volume averaging process (3.79); in fact, control volume averaging has the same properties as ensemble averaging. Thus the first two terms on the right-hand side of (3.87), sometimes called the Leonard stresses, are zero in the case of control-volume averaging.

Although we will not discuss methods of solving the large-eddy equations here, we do have two comments on such solutions. First, although the pressure might appear to be an unknown, it is related to the velocity field through a Poisson equation, as we showed earlier. Thus (in principle) as the solution proceeds, one can calculate exactly the pressure terms in the large-eddy equations. This must be done very carefully, but suitable numerical methods are discussed in the literature (see Mansour et al. (1977), for example).

Our other point concerns the calculation of the inertia term in the case where the large-eddy equation is (3.90). There one needs $\{\{U_i\}\{U_j\}\}$. The equation solution produces directly only the filtered velocity and, hence, one has immediately available only $\{U_i\}\{U_j\}$. In early work $\{\{U_i\}\{U_j\}\}$ was generated from $\{U_i\}\{U_j\}$ by approximate means, but it can now be done exactly using fast Fourier transform techniques. In so doing, one is forced to choose explicitly an analytical form for the filter function G, rather than implicitly assuming it corresponds to the top-hat function in Figure 3.1a, for example, with Δ the three-dimensional spacing. This problem does not occur with control-volume averaging, where $\{\{U_i\}\{U_j\}\} = \{U_i\}\{U_j\}$.

We showed, through (3.94) and (3.95), that the subgrid stress term in the large-eddy equation is dynamically essential; it causes the transfer of kinetic energy to smaller scales. Thus its parameterization is a key step in developing a large-eddy model. We also showed that this stress can be written in two ways. In one way (Lilly-Deardorff) all this inertial transfer is caused by the subgrid-scale stress. In the other way (Leonard) part of this transfer is due to the resolvable-scale eddies. This complicates somewhat our discussion, since a comparison of (3.94) and (3.95) shows that the subgrid parameterizations must differ in the two approaches.

We meet another complication in parameterizing the subgrid effects. Although the large-eddy equation looks like the ensemble mean velocity equation, there is an essential difference. In the ensemble-mean velocity equation, the subgrid term - the Reynolds stress divergence - is an ensemble-average property, and therefore varies smoothly in time and space. By contrast, in the large-eddy equation the subgrid term - again a stress divergence - is a random variable; it varies randomly in time and space.

Most existing parameterizations for turbulence relate ensemble-average properties. The relation between the mean wind shear and turbulent stress in the surface layer is of this type, as are the closure parameterizations in second-order models. The usual deductions that we make from theory, and from data, apply to ensemble-average parameterizations. This is not what we need here, however; here we need parameterizations for instantaneous, randomly fluctuating variables.

Let us illustrate this by considering the simplest parameterization for τ_{ij}

$$\tau_{ij} = K \left(\frac{\partial \{U_i\}}{\partial x_j} + \frac{\partial \{U_j\}}{\partial x_i} \right) , \tag{3.96}$$

where K is an eddy viscosity. Intuitively, this seems much more restrictive than the same assumption relating ensemble averages

$$\tau_{ij} = K \left(\frac{\partial U_i}{\partial x_j} + \frac{\partial U_j}{\partial x_i} \right) . \tag{3.97}$$

Note that an ensemble averaging operation applied to (3.96) does not give (3.97) in general, since K is now a random variable; the average of the product on the right-hand side of (3.96) does not equal the product of the averages.

In spite of this formal difficulty with subgrid-scale parameterizations, there are a couple of points on the plus side.

First, we need to parameterize only the small-scale motions. There is reason to believe (Townsend, 1956; Tennekes and Lumley, 1972) that turbulent flows differ from one another principally in their large-eddy structure, and that the small-scale eddies in all turbulent flows are statistically similar. Thus, one might expect that subgrid parameterizations have some universal properties and are independent of the large-scale differences in flows. In second-order models, however, our parameterizations must cover the whole eddy range; there is evidence suggesting that such parameterizations cannot be uniformly valid in a wide range of flows. Thus while it is perhaps more difficult to get theoretical or experimental guidance for subgrid-scale parameterizations, as they are developed they might well be more broadly useful than those in second-order models.

Our second point is that there is reason to believe that the parameterizations in large-eddy models can be simpler than those in second-order models. Corrsin (1974) points out that the gradient-diffusion assumption (e.g., eddy viscosity)

> ... requires (among other things) that the characteristic scale of the transporting mechanism [mean free path in gas kinetics; Lagrangian velocity integral time scale multiplied by root-mean-square velocity, in turbulence ...] must be small compared with the distance over which the mean gradient of the transported property changes appreciably ... nearly all traditional turbulent transport problems violate this requirement ...

Note, however, that here we are parameterizing the effect of the subgrid scale motions on the resolvable scales. In the limit $\Delta/\ell \to 0$, where ℓ is the integral scale of the resolvable-scale motion, Corrsin's objections should disappear.

Recall that K, as used in mean velocity field closure, is very sensitive to buoyancy effects. In subgrid-scale closure, however, buoyancy effects generally become small in comparison with shear effects as $\Delta/\ell \to 0$. Thus the simple, K-subgrid closure can be effective even in convective turbulence.

Second-order subgrid closure has also been used. Lilly (1967) outlined the approach, and Deardorff (1973) discusses it in detail. One would expect, on the basis of the previous arguments, that it does not give the dramatic improvement over K-closure that it gives in mean field calculations. Deardorff's experience tends to confirm this.

Large-eddy simulation is the best available method of calculating the three-dimensional, time-dependent structure of large Reynolds-number turbulent flow. Its principal disadvantage is its expense. It gives, of course, a wealth of data, much of which is useful primarily to generate statistics. Deardorff (1974a, b) has used it to study the evolution and structure of a convective boundary layer. Recent applications to research flows are discussed by Ferziger et al. (1977), and the survey paper by Reynolds and Cebeci (1976) discusses the method briefly and lists a number of references to current work. Herring (1979) has given a brief but useful survey of recent developments and fundamental issues. Perhaps the most complete reference is Schumann et al. (1980), who use control-volume averaging.

Other Volume-Average Models

These include the two- and three-dimensional models often used in air-quality and regional-meteorology applications. Their distinguishing feature is an averaging volume so large that very little of the turbulence is resolved explicitly; most resides in the subgrid scales and must be parameterized.

The reader of contemporary literature might wonder whether a given three-dimensional model uses volume or ensemble averaging, since this is seldom discussed explicitly. The answer is simple when the grid scale Δ is much greater than the turbulence integral scale ℓ. In a volume-averaged model, all the turbulent transfer is then sub-grid scale and must be parameterized. In the ensemble-averaged model all the turbulent transfer must always be parameterized. Thus, if the same parameterization is used in each case, the models become identical.

What about two-dimensional models? Consider, for example, the boundary-layer flow perpendicular to a long ridge, which is sometimes calculated numerically with a two-dimensional model. If volume averaging is used, the grid volume is then infinitely long in the ridge direction. Regardless of the two-dimensional grid resolution, virtually all of the turbulent transfer will be subgrid scale, since we expect the boundary layer turbulence to be three dimensional. Thus, here again volume-average and ensemble-average models could be identical.

Consider now three-dimensional models with $\Delta < \ell$, and say we want to calculate boundary-layer structure over uniform terrain. If we use volume averaging, the calculated resolvable-scale field will be random, that is, turbulent. If we use ensemble averaging, the calculated field will be non-random and will vary smoothly with z; if the boundary conditions allow, it can reach a time independent, steady state.

For this case where $\Delta < \ell$, let us see if we can interpret the turbulent behavior of one model and the non-turbulent behavior of the other in terms of the basic equations. Suppose for simplicity that each model has a constant K-parameterization. Then the equations are of identical form, but the volume-averaged model has $K = K_v$ = constant and the ensemble-averaged model has $K = K_e$ = constant. Since K_v represents the turbulent diffusivity due to eddies of a scale smaller than Δ, we expect that $K_v \sim q_v\Delta$ where q_v is the sub-grid velocity scale (say the square root of the subgrid kinetic energy). Likewise, we expect $K_e \sim q\ell$, where q and ℓ are energy-containing range scales. Since $q \gg q_v$ and $\ell > \Delta$, we expect that $K_e > K_v$. Now in this simple example our equations of motion are simply the Navier-Stokes equations with ν replaced by K_v or K_e. Thus if q and ℓ are the velocity and length scales of the mean motion, the Reynolds numbers of the two simulations will be $R_v = q\ell/K_v$ and $R_e = q\ell/K_e$. Since $K_e \sim q\ell$, we see that $R_e \sim 1$; this will yield 'laminar' or non-turbulent behavior, since R_e is so small. However, $R_v \sim q\ell/(q_v\Delta) > 1$; in this case, R_v will most likely exceed the 'critical' value where transition and turbulence occur, and the solution will be turbulent.

ACKNOWLEDGEMENT

The material in this chapter was borrowed in large part from the unpublished lecture by the author at the short course on boundary-layer meteorology held at Boulder, August 1978.

4. OBSERVED CHARACTERISTICS OF THE ATMOSPHERIC BOUNDARY LAYER

S.J. Caughey

Meteorological Office, Belfast, BT29 4AB,
Northern Ireland

4.1. INTRODUCTION

The detailed structure of the lower layers of the atmosphere is of special interest in many respects. This is the region in which the energy input to weather systems largely occurs and it is into these layers that the gaseous waste products of heating and industrial processes are released. Much of the work on the detailed characteristics and diffusive action of atmospheric flow has been concentrated on the surface layer for the obvious reasons of accessibility and practical interest. Relatively little information exists on the rather inaccessible regions above the surface layer, more especially those beyond the reach of tall towers. However, to an increasing degree, interest is being focused on these upper regions of the planetary boundary layer as their importance to boundary layer energetics and structure is more fully appreciated.

Our knowledge of turbulence structure over land is reasonably comprehensive for heights up to a few hundred metres in convective conditions. Rather less is known of the details of turbulence in the higher regions affected by entrainment or in the vicinity of the capping inversion. In stable conditions turbulence levels are generally much lower, so instrumental difficulties are increased and interpretation of the observations is complicated by the coexistence of waves and turbulence. Much further effort is required in this area before any generalized turbulence behavior can be stated with confidence. The foregoing remarks refer to essentially cloud-free boundary layers over relatively flat and uniform terrain. The presence of various types and amounts of low cloud may radically alter the boundary-layer structure. This is a field to which surprisingly little effort has been devoted although some numerical and experimental studies of the strato-cumulus-capped mixed layer have recently been reported (Deardorff, 1980; Caughey et al., 1981) and measurements in the cloudy marine boundary layer have been discussed in the literature (Nicholls and Readings, 1979). Topographic effects may also considerably alter the structure of turbulence (Jackson and Hunt, 1975) and this area of work also requires a good deal of further experimental study (Bradley, 1980; Mason and Sykes, 1979; Sacré, 1979). In any review of such a wide ranging subject it is of course necessary to be highly selective and the material presented here is drawn almost exclusively from studies of boundary layer characteristics over land of fairly flat and uniform terrain. References to similar studies of the marine boundary layer are, however, included for completeness.

A wide range of instrumentation and supporting platforms has been employed in boundary layer studies, ranging from masts and towers to aircraft and balloons (both tethered and free flying), see, e.g., Lenschow (1970), Kaimal et al. (1966), Angell (1974), Readings and Butler (1972). For work in the higher regions, only aircraft and large tethered balloons are suitable. The latter are particularly useful since they provide a means of making simultaneous measurements at many levels and they also enable accurate positioning of instruments with respect to, for example, entrainment interfaces. Aircraft, on the other hand, can cover large distances in a short time (~100 km in 10-15 min being typical) and therefore provide more representative estimates of turbulent statistics (especially of the low frequency contributions). They are also ideally suited to investigating turbulence structure on occasions when a large degree of spatial organization is apparent (Nicholls, 1978).

In this review, the data described are heavily biased to those obtained from tethered balloon-borne instrumentation, developed over the course of the last 20 years by the British Meteorological Office. It is therefore worth briefly describing

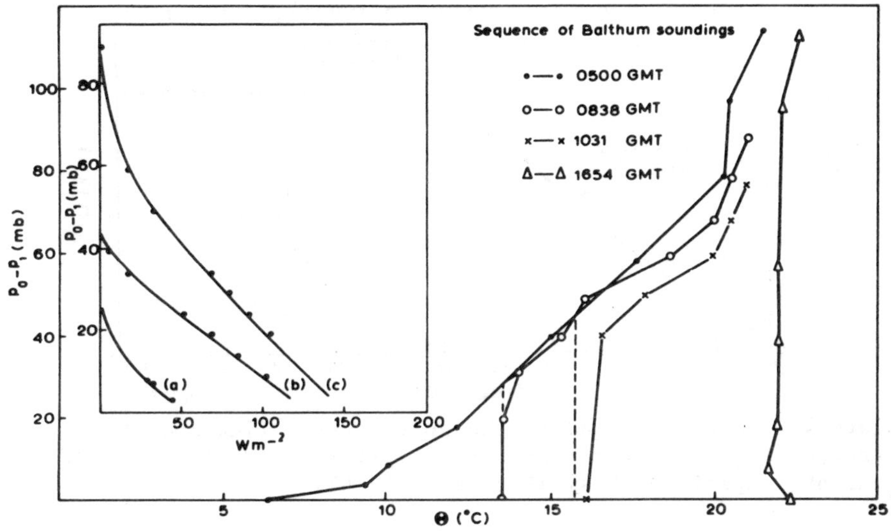

Figure 4.1. Potential temperature profiles on 24 August 1973. The inset shows the implied profiles of sensible heatflux from summation of the temperature change between (a) 0500 - 0838 GMT, (b) 0838 - 1031 GMT and (c) 1031 - 1654 GMT. The height above the surface is given by the pressure difference $P_o - P_1$ (Chorley et al., 1975).

this facility, full details of which are available elsewhere (Readings and Butler, 1972; Caughey, 1977). The turbulence probe consists of a set of sensors mounted on a vane which is attached to the balloon tethering cable in a way that leaves it free to rotate and thus keep the sensors facing into the wind. The temperature device consists of 1.8 m of 26 μm dia. platinum wire wound non-inductively on a plastic former. A double V-shaped hot wire inclinometer mounted on a damped pendulum measures the inclination of the wind vector to the horizontal plane, whilst a second inclinometer, orientated at 90° to the first, measures the instantaneous wind direction relative to the vane direction. A magnetic flux gate is used to sense the vane direction relative to the earth's magnetic field. Wind speed is measured with a multislot photoelectric anemometer fitted with an eight-cup polystyrene rotor. The balloon system can be used to position up to five turbulence probes through the lowest 1½ km of the atmosphere.

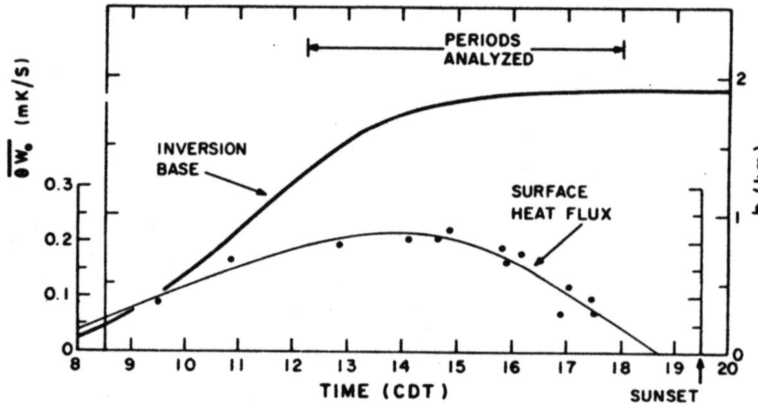

Figure 4.2. Diurnal trend in the surface heat flux and the corresponding inversion rise for a typical day during the 1973 Minnesota experiment (Kaimal et al., 1976).

4.2. CONVECTIVE BOUNDARY LAYER

Boundary Layer Development

The convective boundary layer (CBL) is generally regarded as that layer of the atmosphere most directly affected by solar heating of the earth's surface. This layer shows a strong diurnal development and in mid-latitudes over land typically reaches a height of 1-2 km by mid-afternoon (Chorley et al., 1975; Coulman, 1978). Figure 4.1 illustrates the 'warming-out' of a nocturnal inversion through surface heating. Initially, a strong surface-based inversion extends up to ~800 m with a total temperature difference of ~13°C. As warming proceeds, an adiabatic layer (depth = h, which in the CBL is usually taken as the height to the lowest inversion) develops close to the surface and slowly increases in depth with time. Ultimately a well-mixed layer extending to over 1 km is produced. Similar behavior was observed in the data obtained during the Minnesota experiment (Izumi and Caughey, 1976) and has also been observed by remote sensors such as FM-CW radars (Richter et al., 1974) and acoustic sounders (Neff, 1975; Cole et al, 1980). Referring to Figure 4.2 it can be seen that h grew rapidly in response to the steadily increasing surface heat flux (Q_o) and then slowed down as Q_o reached its maximum value. The rate of growth of h for this period agrees with the numerical model prediction of Mahrt and Lenschow (1976) which assumes a constant Q_o for the 3 hour period following local noon. As Q_o decreased through the late afternoon h began to level off to a nearly constant value, which it maintained, even after Q_o turned negative.

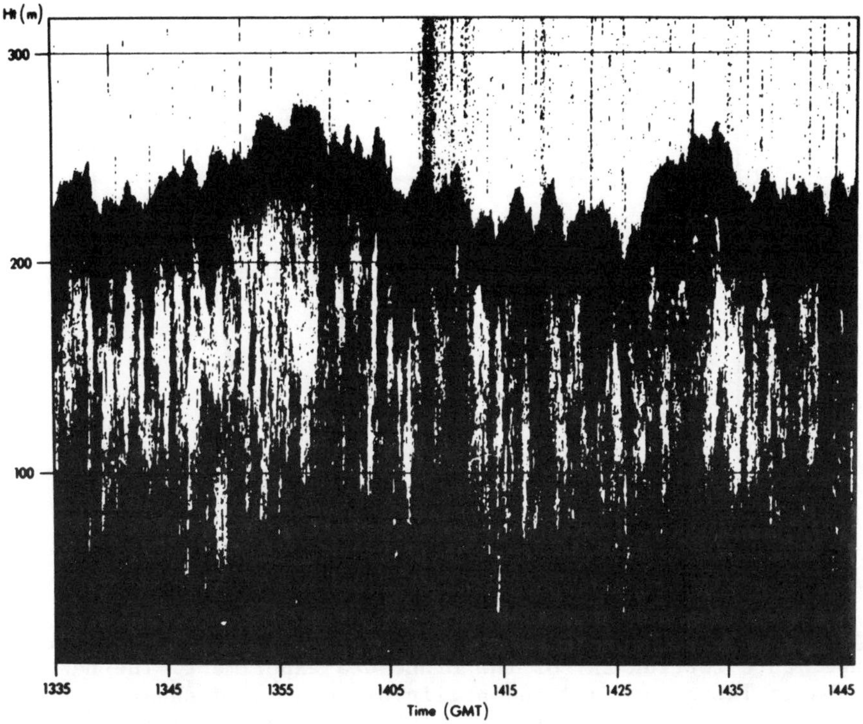

Figure 4.3. Sounder facsimile chart for the period 1335-1445 GMT on the 28 October 1975 (Crease et al., 1977).

Some very useful insights into CBL behavior can be obtained from acoustic sounding (e.g., Crease et al., 1977). An example of an acoustic echo pattern from a shallow CBL capped by a strong (~10 K) inversion with a base at ~200 m is given in Figure 4.3. The interaction of convective plumes with the region of strong temperature gradient generates an intensive echo layer. This interface represents the top of convection on this day and therefore the boundary-layer depth (h). The echo layer is distorted by a series of undulations of similar vertical scale but markedly differing along wind scale. The smaller-scale features (sometimes called 'hummocks') are clearly associated with the convection below. There is also evidence throughout this record of thermal activity, originating at the inversion, extending well down into the mixed layer. It is natural to associate this with entrainment, a process in which air is transferred across the capping inversion from the free atmosphere and mixed into the CBL. Acoustic records over longer periods provide graphic illustrations of the diurnal evolution of the CBL (e.g., Cole et al., 1980).

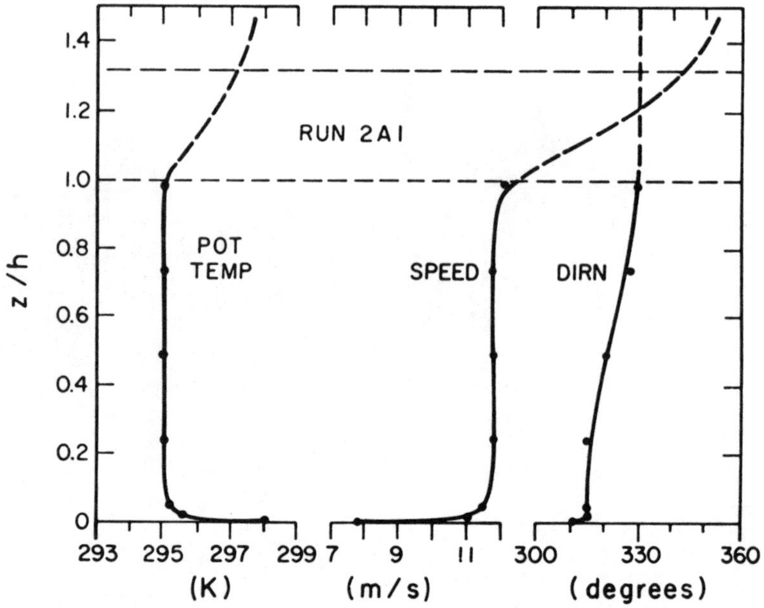

Figure 4.4. Profiles of wind speed, wind direction and potential temperature for run 2A1 of the Minnesota experiment. The near-adiabatic lapse rate and the negligible mean wind shear in the mixed layer are typical of strongly convective conditions (Kaimal et al., 1976).

The wind speed and temperature profiles observed at Minnesota (Figure 4.4) are fairly typical of daytime convective conditions. Almost all the wind shear and potential temperature gradient in the CBL are confined to a shallow region close to the ground. The sharp increase in wind speed across the capping inversion appears consistently in many daytime runs and has implications for momentum and heat transfer in the upper CBL. It is not, of course, a general feature and is often absent, especially in anticyclonic conditions.

Over land the CBL may be idealized as a multilayer structure in terms of the parameters considered relevant to the turbulence in each. Proceeding upwards from the surface we may define:

(1) The surface layer where wind shear plays a dominant role. Here Monin-Obukhov similarity applies and the controling parameters are z, τ_0, Q_0 and g/Θ. The scaling velocity and temperature for this layer are, respectively, u_* and θ_*, given by (2.1) and (2.33) respectively.

Dimensionless groups, formed with u_* and θ_*, become universal functions of ζ (= z/L). Experimental results (with the exception of u and v statistics) support this (Businger et al., 1971; Wyngaard and Coté, 1971; Kaimal et al., 1972; Busch, 1973; Garrett, 1979). (See also Section 2.4 and 2.5.) The surface layer is confined to a height range $z < \|L\|$. Since this region is the most widely studied and best understood, it will be only briefly referred to in this review.

(2) The <u>free convection layer</u> where τ_o is no longer important but height, z, continues to be the significant length scale. The governing parameters reduce to three: z, Q_o and g/Θ (see also Section 2.5). This yields a scaling velocity u_f and a scaling temperature T_f given by

$$u_f = [(Q_o/\rho c_p)\ z\ (g/\Theta)]^{1/3} \tag{4.1a}$$

$$T_f = (Q_o/\rho c_p)/u_f. \tag{4.1b}$$

According to the 'local free-convection' predictions dimensionless groups formed with u_f and T_f should be constants (Wyngaard et al., 1971a). The very unstable ($-z/L > 1$) data from the Kansas experiment support these predictions for the most part. The Minnesota data suggest an upper limit for free convection scaling of ~ 0.1 h.

(3) The <u>mixed layer</u> where the structure of turbulence is insensitive to z as well as τ_o. The boundary layer depth emerges as the controling length scale so that the scaling velocity and temperature for this layer become w_* and T_* given by (2.54) and (2.55) respectively (see also Section 2.6).

Within this region, dimensionless groups formed with w_* and T_* should be functions only of z/h. This expectation is based on model studies (Deardorff, 1972; Wyngaard et al., 1974) which show wind and temperature data scaling with w_* and T_*.

(4) <u>The entrainment interfacial layer</u> which extends roughly from 0.8 h to 1.2 h. Here the turbulence structure may be dominated by entrainment effects, the characteristics of the capping inversion and the stable atmosphere aloft. Mixed-layer scaling cannot be expected to produce universal behavior in this region and recently Wyngaard and Le Mone (1980) have developed scaling expressions for the structure parameters generated by the entrainment processes in this region.

This multi-layer idealization applies best to the typically very convective CBL's that occur over land ($h/\|L\| > 50$–100), however in the marine boundary layer h/‖L‖ may be much smaller and is not unusually ~10. This reflects the fact that surface-induced mechanical turbulence cannot be neglected at higher levels in the CBL and neither mixed-layer scaling or free convection-layer scaling are adequate. Hence, a more general approach is required which retains both τ_o and h (see Nicholls and Readings, 1979). Thus, a suitable scheme for many marine CBLs is to use θ_*, u_*, q_* (defined as in the surface layer), group the data according to h/‖L‖ and plot versus z/h.

Spectra of the Velocity Components

The main results from the Minnesota (Izumi and Caughey, 1976; Kaimal et al., 1976) and Ashchurch (Caughey and Palmer, 1976) experiments are presented within the framework of mixed layer similarity. Earlier results had shown that spectra and cospectra from the surface layer reduce to a set of universal curves that converge into a single curve in the inertial subrange, but spread out as a function of z/L at lower frequencies (Kaimal et al., 1972; Busch, 1973). Spectra from the mixed layer showed that the inertial subrange remained essentially constant with height (in contrast to the sharp decrease with height near the ground) and the spectral peaks were invariant both in their intensities and positions on the frequency scale. Extending similarity arguments to the mixed layer we can expect the mixed-layer spectra, normalized by w_*^2, to be a function of only two variables, z/h and λ/h (where λ is the wavelength approximated by U/n. The one-dimensional logarithmic u-spectrum in the inertial subrange can be expressed as

$$\frac{n\,E_u(n)}{w_*^2} = \frac{\alpha_1}{(2\pi)^{2/3}}\,\psi^{2/3}\,f_i^{-2/3} \qquad (4.2)$$

where α_1 is the spectral constant for u, ψ is the dimensionless dissipation rate ($\varepsilon T/g\,Q_o$) appropriate to the mixed layer and f_i is the dimensionless frequency nh/U for that layer; ψ and f_i are analogous to ϕ_ε and f in the surface layer formulation (Kaimal et al., 1972). By the similarity argument, ψ the ratio of kinetic energy dissipation to buoyancy production at the surface, should be a function only of z/h. This is supported by the data which show that ψ assumes a value of between 0.5 to 0.7 throughout the mixed layer. Slightly larger values are observed in the near-transition runs close to sunset. These ψ curves resemble the dissipation rate profiles of Rayment (1973), Lenschow (1974), Frisch and Clifford (1974), Volovítkaya and Ivanov (1970) and Kaimal and Haugen (1967). The Ashchurch ε data overlap well with that from Minnesota and shows that ε (and hence ψ) remains nearly constant close to the capping inversion above which it falls rapidly as turbulence levels decrease.

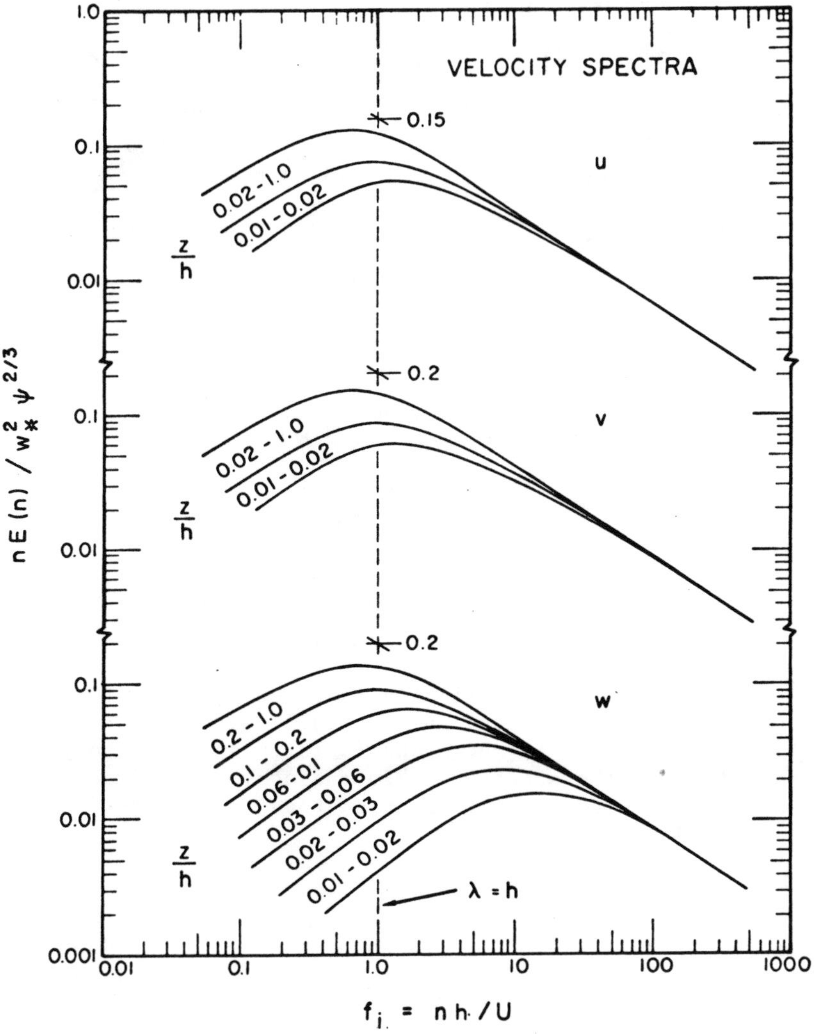

Figure 4.5. Universal curves for the velocity spectra expressed in mixed-layer similarity coordinates (Kaimal et al., 1976).

Taking α_1 to be 0.5 and rearranging terms Equation (4.2) becomes

$$\frac{n\ E_u(n)}{w_*^{\ 2}\ \psi^{2/3}} = 0.15\ f_i^{\ -2/3}. \qquad (4.3)$$

Spectral forms for the one-dimensional v and w spectra differ from

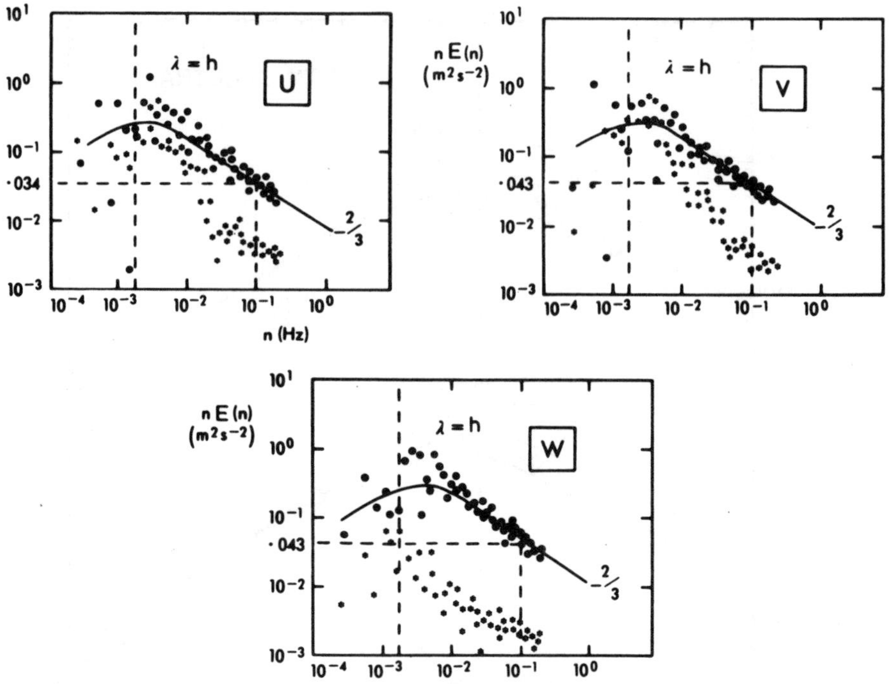

Figure 4.6. Spectra of the velocity components within the mixed layer (at 0.9 h, circles) and in the stable air aloft (at 1.4 h, stars). The line λ = h is shown and falls in approximately the same position for the spectra within and above the mixed layer because of small wind shear present on these occasions (Caughey and Palmer, 1979).

equation (4.3) by a factor of 4/3 as a consequence of isotropy and the direction of sampling (see Monin and Yaglom (1975), section 21.4) so that, for these fixed point measurements

$$\frac{n\,E_v(n)}{w_*^2\,\psi^{2/3}} = \frac{n\,E_w(n)}{w_*^2\,\psi^{2/3}} = 0.20\,f_i^{-2/3} \qquad (4.4)$$

Logarithmic-normalized spectra show systematic behavior when plotted as a function of f_i (Figure 4.5). Inclusion of the $\psi^{2/3}$ in the normalization forces all spectra to collapse into a single curve in the inertial subrange. At lower frequencies the curves seperate as a function of z/h. Of the velocity components, w shows the largest spread with height. As z/h increases from 0.01 to 1.0 the position of the spectral peak shifts to increasingly lower values of f_i, rather rapidly at first, up to z/h ~ 0.1, then more

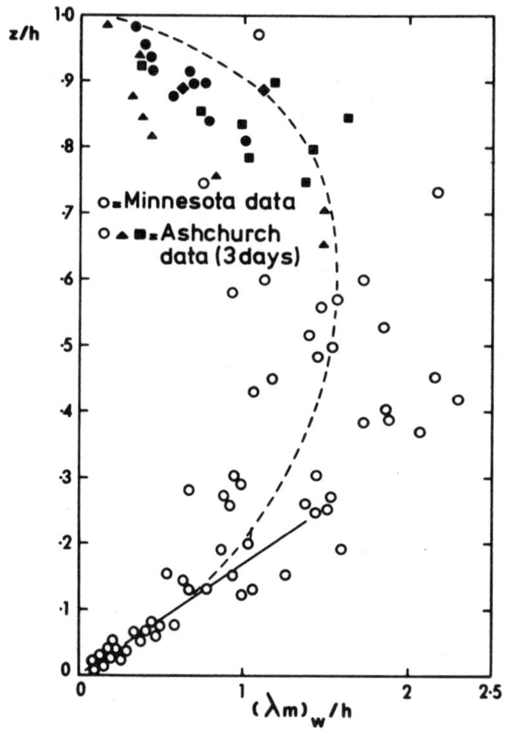

Figure 4.7. The variation of the normalized peak wavelength for the vertical velocity component with z/h. The solid line represents the free-convection prediction (4.5), whilst the dashed line represents the relationship (4.6) (Caughey and Palmer, 1979).

gradually. For u and v only two categories exist: 0.01 to 0.02 and 0.02 to 1.00.

Spectral data from the Ashchurch experiment (Caughey and Palmer, 1979) broadly confirmed the behavior observed at Minnesota and additionally emphasized the marked change in the spectral shape occurring above the CBL (Figure 4.6). The solid lines in Figure 4.6 are from Kaimal et al. (1976) and form a fairly good fit to the Ashchurch spectra, except that the spectral maximum falls at a shorter scale than $\lambda_m \sim 1.5$ h, where λ_m is the peak wave length defined by U/n_m (n_m is the frequency where the logarithmic spectrum has a maximum). Above the CBL, the small-scale spectral density collapses and only contributions from waves and undulations are apparent.

The peak wavelength, $(\lambda_m)_w$, is an important one for studies of turbulent transport in the CBL and therefore merits close examination. Figure 4.7 shows $(\lambda_m)_w$ normalized with h and plotted as a function of z/h, from the Ashchurch and Minnesota experiments. For $z < 0.1$ h, the relation is a linear one approximated by

$$(\lambda_m)_w/h = 5.9\ (z/h) \tag{4.5}$$

which is precisely the free convection limit, $z/(\lambda_m)_w = 0.17$, observed in the Kansas data (Kaimal et al., 1972). Above 0.1 h $(\lambda_m)_w$ increases more gradually and finally approaches a value of 1 - 2 h at ~0.5 h. Above this the Ashchurch data suggest that $(\lambda_m)_w$ decreases again as the capping inversion is approached. This observation relates well to laboratory studies of turbulence near boundaries (e.g., McDougall, 1979). Caughey and Palmer (1979) suggested a fairly good fit to $(\lambda_m)_w$ through the region $0.1 < z < h$ can be obtained by the relation

$$(\lambda_m)_w = 1.8\ h\ [1 - \exp(-\frac{4z}{h}) - 0.0003\ \exp(\frac{8z}{h})]. \tag{4.6}$$

Additional support for the decrease in $(\lambda_m)_w$ near a capping inversion comes from the aircraft results of Frangi (1979).

For the u and v components $(\lambda_m)/h$ shows little variation with z/h. The average value 1.3 comes close to that for w above z/h ~ 0.2. Thus, in the CBL we see a strong tendency for the peak wavelengths of all velocity components to be the same and roughly equal to ~1.5 h. These results are in general agreement with those of other investigators, e.g., Kukharets (1974) and Ivanov et al. (1973). In the surface layer, Kaimal et al. (1972) and Busch (1973) had found that λ_m for u and v did not obey Monin-Obukhov similarity. The absence of systematic behavior with z/L indicated that some length scale other than z and L controls the behavior of λ_m. It appears from the Minnesota and Ashchurch data that this length scale is h.

This is a suitable point at which to speculate on the significance of the spectral peak at $\lambda_m \sim 1.5$ h observed in the velocity spectra. This wavelength matches the horizontal length scale of large convective plumes or thermals which extend through the depth of the CBL. The existence of organized convection on this scale has been observed by, for example, Hardy and Ottersten (1969), Konrad (1970) using high power radars, Rowland and Arnold (1975) using FM-CW radars, and Frisch et al. (1975). In the Minnesota data, evidence of such structures can be found in the temperature records above heights of ~0.1 h. The traces show positive bursts spaced 4-5 min apart, the time scale being represented by λ_m in the mixed layer. The records show a high

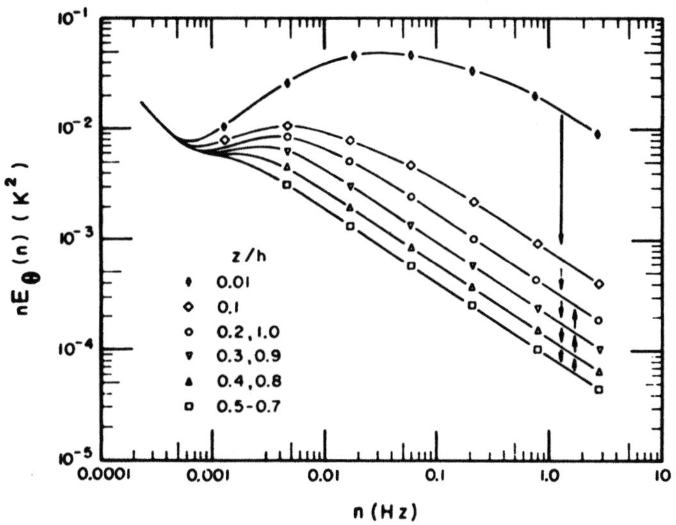

Figure 4.8. Idealized temperature spectra in the dimensional coordinates show variation in the spectral behaviour as a function of z/h (Kaimal et al., 1976).

degree of correlation between these fluctuations and the long period fluctuations in the velocity field responsible for the spectral peaks.

The prevailing convection field during the Minnesota experiment probably resembled the thermal streets in Konrad's (1970) radar observations, where individual thermals line up in the direction of the mixed-layer mean wind vector (spaced apart and with a row separation of ~3 h). Under more unstable conditions (lighter winds), there may be different patterns with thermals arranged in open rings or hexagons of diameter 5-6 h, as observed by Hardy and Ottersten (1969). More observations and analyses are required to confirm these speculations and relate the organization of convection cells to the mean structure of the lower atmosphere.

Spectrum of Temperature

Unlike velocity spectra, the spectrum of temperature cannot be conveniently generalized within the framework of mixed-layer similarity. This is due to variations in the high-frequency variance produced by entrainment effects in the upper half of the CBL. However, some generalizations can be made about spectral shapes and inertial subrange intensities (Figure 4.8). At inertial subrange frequencies, the spectral intensity drops steadily with

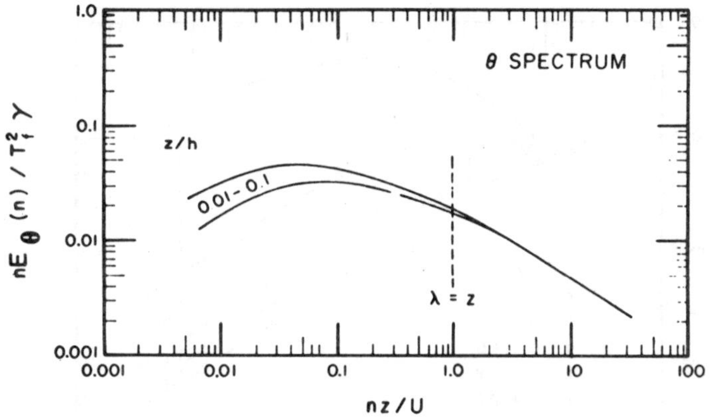

Figure 4.9. Universal curve for the θ spectrum in the range $0.01\,h < z < 0.1h$. T_f is the free-convection scaling temperature (4.1b) and γ is given by (4.8) (Kaimal et al., 1976).

height up to 0.5 h, stays at a low value between 0.5 and 0.7 h and starts to rise again above 0.7 h. This increase in the upper regions of the CBL reflects the entrainment of warm air through the capping inversion. More evidence is presented further in this section.

The only region where spectral generalization is possible is in the height range $z < 0.1$ h, where the θ spectra behave much like those observed in the Kansas experiment. Since the controling length scale in this region is z, not h, we use f, which is defined as $f = nz/U$, for the dimensionless frequency scale. The logarithmic one-dimensional temperature spectrum for the inertial subrange can be expressed as

$$n\,E_\theta(n) = \frac{\beta_1}{(2\pi)^{2/3}}\,N\,\varepsilon^{-1/3}\,z^{2/3}\,f^{-2/3} \qquad (4.7)$$

where β_1 is the spectral constant for temperature, assumed to be ~0.8 from the Kansas results, and N is the dissipation rate of the temperature fluctuations (1.63).

Substituting the value for β_1 and normalizing with the temperature scale T_f^2, gives

$$\frac{n\,E_\theta(n)}{\gamma\,T_f^{\,2}} = 0.235\,f^{-2/3} \qquad (4.8)$$

where $\gamma = N\,\varepsilon^{-1/3}\,(Q_o/\rho c_p)^{-4/3}\,(g/\Theta)^{2/3}\,z^{4/3}$. This normalization brings all θ spectra into coincidence in the inertial subrange, as seen in Figure 4.9. At mid and low frequencies they collapse into a fairly narrow band with no apparent tendency to separate according to z/h. For this composite spectrum, λ_m approximates 20 z, the free-convection limit in the Kansas spectra. As z approaches 0.1 h, λ_m approaches ~1.5 h, the characteristic wavelength found in the velocity components. Above 0.1 h, λ_m shows a tendency to increase slightly with height up to ~0.5 h and to decrease again above 0.7 h, a behavior noted in the Ashchurch velocity spectra.

The function γ is essentially a dimensionless form of the structure parameter C_T^2. We can write

$$C_T^2 = 4\,\beta_1\,N\,\varepsilon^{-1/3} \sim 3.2\,N\,\varepsilon^{-1/3}. \qquad (4.9)$$

The Kansas data provide the relationship between C_T^2 and surface layer parameters for local free convection (Wyngaard et al., 1971)

$$C_T^2 \sim 2.67\,(Q_o/\rho c_p)^{4/3}\,(g/\Theta)^{-2/3}\,z^{-4/3}. \qquad (4.10)$$

From Equations (4.8 - 4.10) we have $\gamma \sim 0.83$ for the free-convection layer and this was found to be valid for the Minnesota data as well. Indeed, γ is found to stay constant up to ~0.5 h, so that the $z^{-4/3}$ decrease in the inertial subrange intensity continues well above 0.1 h (Kaimal et al., 1976).

The above empirical relation can be used to collapse the inertial subrange temperature spectra into a single curve in a plot similar to Figure 4.9. However, no systematic trend with z/h emerges at the low frequency end.

Temperature spectra obtained at Ashchurch (Caughey and Palmer, 1979) also confirmed the main results from the Minnesota data. Within the CBL, the spectral shape exhibited a well-defined inertial subrange whereas, in the stable air aloft, the spectral level of the high-frequency fluctuation decays rapidly (see Figure 4.10). Spectra from this region resemble those from the intermittently turbulent nocturnal boundary layer (Caughey and Readings, 1975) and behave in an analogous manner to those obtained in laboratory studies by Willis and Deardorff (1974).

Cospectra of Heat Flux and Stress

The cospectra of heat flux and stress measured near the ground (z < 0.1 h) in the Minnesota experiment follow the same universal forms observed in the Kansas data, but cospectral shapes depart significantly from this as z exceeds 0.1 h. The patterns at these higher levels are not consistent enough to justify

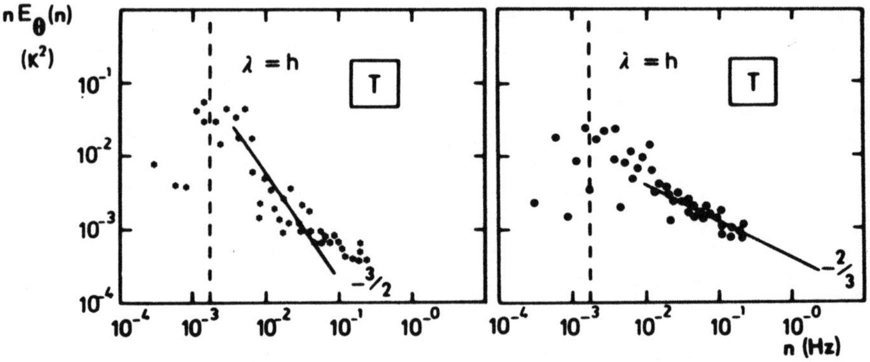

Figure 4.10. As Figure 4.6 for temperature (Caughey and Palmer, 1979).

development of universal forms, although individually they reveal some interesting perspectives into the nature of turbulent flow in the CBL.

In Figure 4.11 we see how the range of frequencies contributing to the upward transport of heat near the ground narrows down to a relatively small band centered near 0.003 Hz at a height of ~610 m. At 1219 m (close to the inversion base) there is still a small amount of upward flux but the net flux is now downwards. Transport of air downward through the capping inversion (i.e., entrainment) is responsible for this negative flux.

In the height range $z < 0.2\ h$ (i.e., below ~300 m for the Minnesota data) the flux is almost entirely upwards and the composite curves for logarithmic cospectra normalized by the local heat flux $\overline{\theta w}$, show a small but systematic variation with height in the first 150 m when plotted as a function of the dimensionless frequency $f\ (= nz/U = z/\lambda)$ (Caughey and Kaimal, 1977). In the inertial subrange they all converge to a single curve as in the Kansas results and follow the same empirical relation

$$\frac{n\ C_{w\theta}\ (n)}{\overline{w\theta}} = 0.14\ f^{-4/3}. \tag{4.11}$$

At lower frequencies, the 4 and 32 m curves fall within the narrow cospectral band defined by scatter in the unstable Kansas spectra. Above 32 m, the cospectral curves move away from the Kansas results, whilst above 150 m a more abrupt change occurs as the inertial subrange cospectrum breaks away from the surface layer form represented by Equation (4.11).

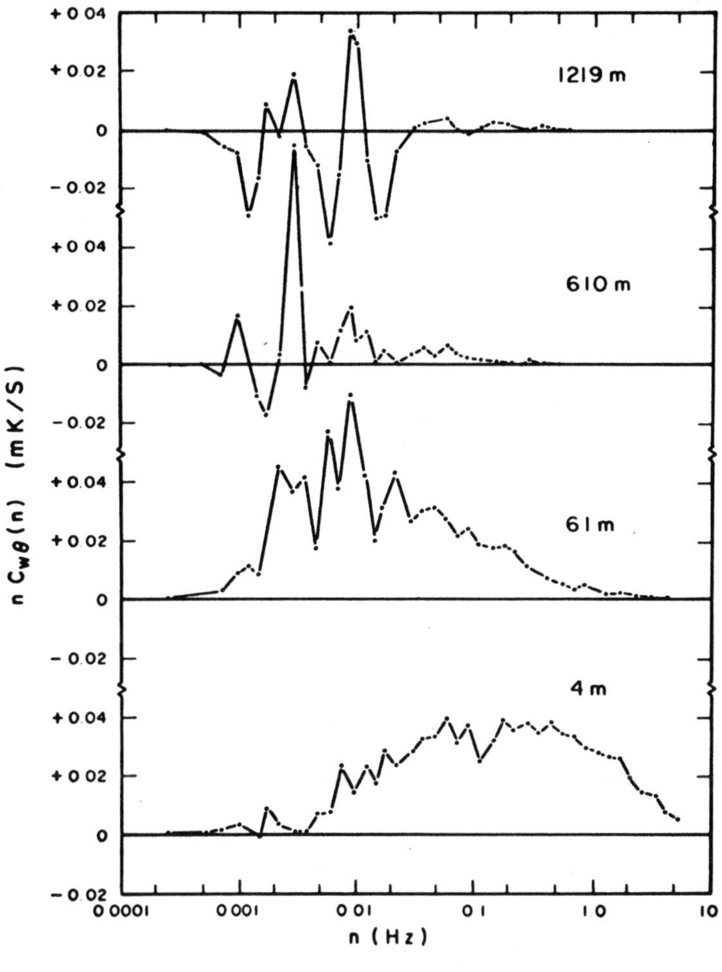

Figure 4.11. Cospectra of heat flux at four heights during run 2A1 of the Minnesota experiments (Kaimal et al. 1976).

A plot of $(\lambda_m)_{w\theta}$ (i.e., the wavelength corresponding to the cospectral peak) versus height shows that $(\lambda_m)_{w\theta}$ increases between 4 and 150 m, with a tendency to level off at a value of ~1.5 h, which is close to the limiting value in the temperature and velocity spectra (Caughey and Kaimal, 1977).

The consequences of entrainment are very apparent in the profile of sensible heat flux (Figure 4.12). These resemble those observed in other studies (Lenschow, 1974; Pennel and Le Mone, 1974), laboratory experiments (Willis and Deardorff, 1974) and

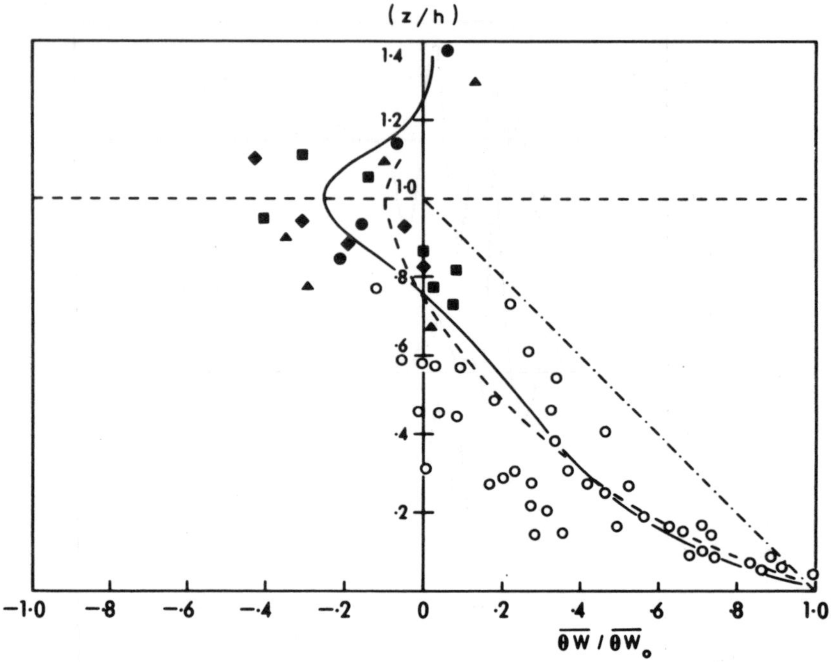

Figure 4.12. Normalized heat flux data from the Ashchurch and Minnesota experiments. The Ashchurch data are coded as solid symbols which represent data on different days. The Minnesota data are shown by open circles. The dashed line represents Lenschow's (1970) aircraft results. The solid line is a 'by-eye' fit to the Ashchurch + Minnesota data points (Caughey and Palmer, 1979).

model calculations (Wyngaard and Coté, 1974). The combined Minnesota and Ashchurch heat flux data give an indication of the behavior of the heat flux through the entire CBL. An extensive zone, between 0.6 and 1.0 h is shown to be affected by entrainment. Above the inversion, the heat flux falls rapidly to zero, but a sharp cut-off is not observed because of the large scale undulations of the CBL top during the Ashchurch experiment (typically 0.1 to 0.2 h). The ratio of the entrained to the surface-heat flux Q_e/Q_0 (where Q_e is the value of the heat flux at z = h) must be regarded as rather tentative since Q_0 has been obtained by indirect means (integration of temperature profiles) and Q_e may also be in error, due to difficulty in measuring the entrained heat flux associated with a rapidly rising inversion. Nevertheless, the range of values observed, −0.2 to −0.4, falls close to those previously reported. Carson (1973) noted values of similar magnitude in his analysis of the O'Neill data whilst Rayment and Readings (1974) found $Q_e/Q_0 \sim -0.4$ during the period of inversion rise and ~ -0.25 throughout the whole period of measurements. Similar values have been observed in laboratory models (Deardorff et al., 1969; Willis and Deardorff, 1974) and in the aircraft results of Lenschow (1970, 1974). The data indicate quite a rapid decrease in the heat flux with height through the lowest layers, so that at around 0.4 h the ratio Q/Q_0 has fallen to ~0.3, with a substantial spread of order 0.1. Above 0.7 h, significant differences are to be expected since the entrained heat flux will depend upon factors such as the atmospheric stability above h and wind shear across the inversion (Carson and Smith, 1974), in addition to the surflace flux. Simple mixed-layer scaling therefore cannot produce universal behavior in this region of the boundary layer and entrainment interfacial scales (not yet fully formulated) need to be used.

The Minnesota experiment provided an excellent opportunity to observe in detail the progression of heat flux at different heights as the CBL disintegrated shortly before sunset. Much recent work has been devoted to the study of the evolution of the CBL between sunrise and noon (e.g., Neff, 1975; Zilitinkevich, 1975; Tennekes, 1975; Chorley et al., 1975; Mahrt and Lenschow, 1976), yet little is known about the details of its dissolution near sunset. A height/time plot of the heat flux (15 min averages) showed that this breakdown is rapid. During this period the level of zero heat flux, located normally at ~0.6 h, makes a quite rapid descent to the surface. This occurs almost an hour before the local sunset. Following this event a surface-based inversion begins to develop in line with the conventional view of nocturnal layer build up and a gradual intensification of the downward heat flux near the surface occurs.

The mean components of the stress, $\overline{uw}$ and $\overline{vw}$, normally decrease with height in the CBL. However, in the Minnesota experiment they behaved anomalously, showing an increase with height and being of opposite sign. In the barotropic, non-entraining CBL one would expect $\overline{uw}$ to decrease monotonically with height approaching zero at z = h and $\overline{vw}$ to vanish both at the surface and at z = h (see Section 2.2). The observed profiles were strongly curved and behaved quite differently. Large stress values of this magnitude may arise from several causes, for example, baroclinicity (Arya and Wyngaard, 1975). The baroclinity needed to explain these observations requires positive horizontal temperature gradients of ~1°C/15 km, which were certainly not present on the synoptic scale during the Minnesota experiments. Gradients of this order may have been present on the 1-2 km scale, because of albedo variations introduced by differences between harvested and ploughed areas. Whether such gradients existed on larger scales is not known. Deardorff (1973) has shown that a wind profile such as that observed in Figure 4.4 can generate large stresses through the entrainment of momentum in the upper CBL.

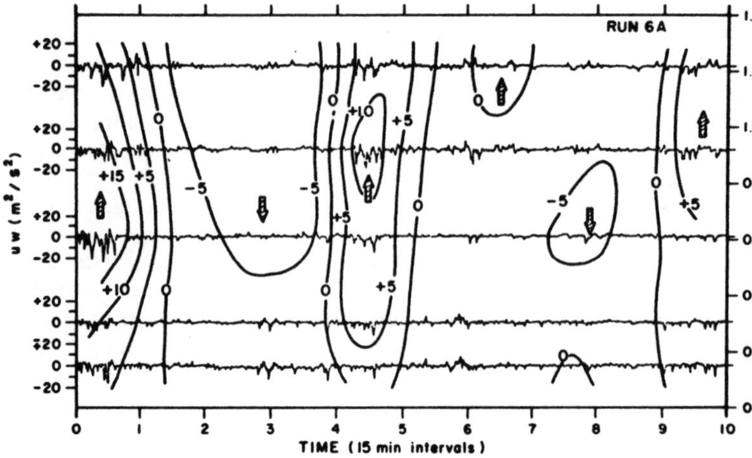

Figure 4.13. Instantaneous uw values at five levels observed during the 150 min. period covering the runs 6A1 and 6A2 of the Minnesota experiment. Isopleths of wind inclination angle (15 min averaged) at 5° intervals are shown superimposed on the uw fluctuations. Each division on the scale represents 15 min (Kaimal et al., 1976).

Wyngaard's (1975) higher-order closure model was run on some of the Minnesota data. Agreement was fairly good, except that the stress maximum occurred near the inversion and not within the CBL.

Another possible explanation for the large stresses is that the averaging period was too short, i.e., that there exist large contributions outside the recorded frequency range 0.001 - 5 Hz. For $\overline{uw}$ this would mean the presence of a large upwards transport of momentum at very low frequencies. Observed stress profiles in the marine CBL do however, show the expected decrease with height (Nicholls and Readings, 1979).

The data obtained in this experiment provide some interesting details of the momentum transfer process. The largest contributions to the stress cospectra come from frequencies in the range 0.002 - 0.1 Hz, the energy containing region of the velocity spectra. The uw traces for the 2.5 hour period covering Runs 6A1 and 6A2 with the isopleth of 15 min averaged wind inclination observed by the probes added are given in Figure 4.13. These suggest the presence of well-defined longitudinal roll vortices with the updraft regions spaced ~1 hour apart. Most of the vertical momentum transfer occurs in these updraft regions. Within these areas the amplitude of the uw bursts increases with height. This increase continues to a height of ~0.5 h. The

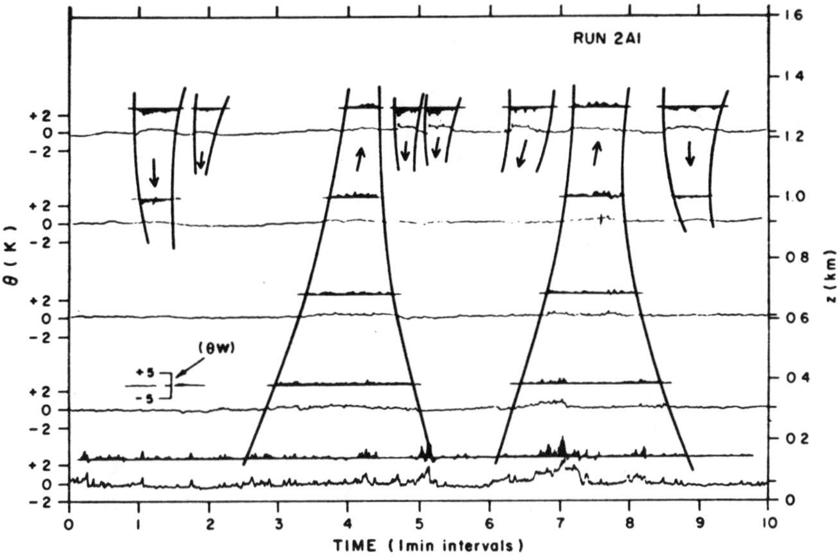

Figure 4.14. Temperature fluctuations observed during the 10 min. period from 1321 to 1331 CDT toward the end of run 2A1 of the Minnesota experiment. The top of the figure represents the inversion base for this period. The instantaneous vertical heat flux at each level is indicated by the shaded fluctuations traced above it. The scale for θw is in K ms^{-1} (Kaimal et al., 1976).

downdraft region on the other hand, is much more diffuse and occupies a much longer time interval than the updraft region. In this area the amplitude of the uw fluctuations decreases rapidly with height and remains small through most of the CBL depth.

Examining the structure of the uw bursts in the updraft region we find that it consists of smaller discrete bursts with vertical continuity through the depth of the CBL. These small bursts correspond to the thermals also evident in the Minnesota data (see Figure 4.14). It thus appears that at Minnesota, the bulk of the downward momentum transport in the CBL was carried out by thermals in the updraft regions between longitudinal roll vortices.

Entrainment

As previously noted, entrainment effects were important to the CBL structure observed at Minnesota and Ashchurch and are probably generally so. The temperature and heat flux traces (Figure 4.14) provide clear evidence of this process. Only portions of the θw traces where significant heat flux exists, are

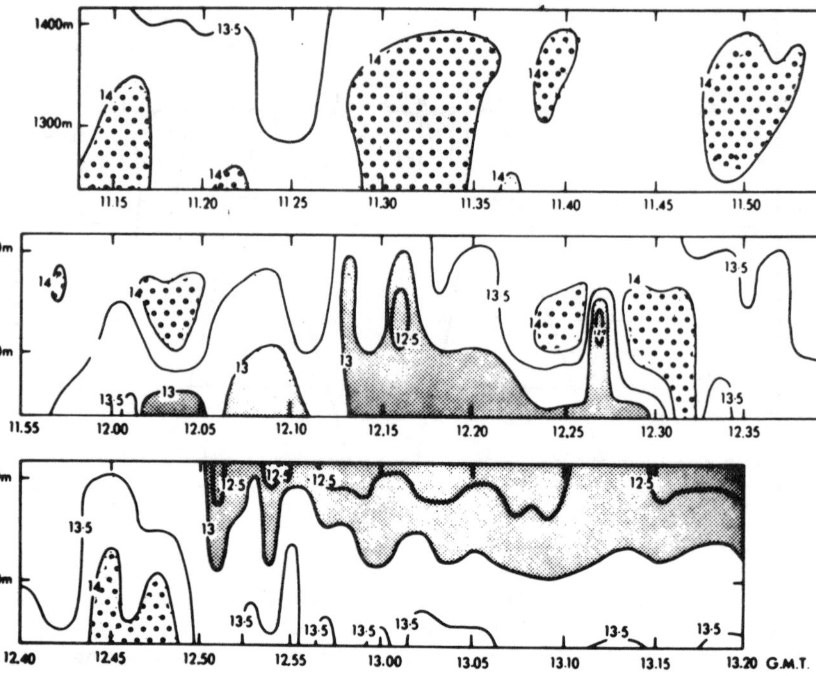

Figure 4.15. Height time profile of temperature contours (°C) drawn to 1 min mean values from the turbulence probes during the period 1113-1320 GMT. Temperatures >14°C are dotted, <13°C are shaded (Palmer et al., 1979).

shown and these are shaded to distinguish them from the temperature traces. The regions of upward flux are thermals, or groups of thermals, which originate near the surface-shear layer. Between these are regions of negative heat flux but these are also characterized by positive bursts in the temperature trace. They are suggestive of entrained warm air descending in the form of discrete plumes and have a characteristic 'saw-tooth' appearance with the largest discontinuity at the leading edge, not the trailing edge as is found near the surface (Kaimal and Businger, 1970). This is not surprising in view of the higher wind speeds above the capping inversion. Direct evidence of entrained air entering the CBL also comes from acoustic sounder studies, as noted earlier (see Figure 4.3 and Crease et al., 1977).

Some detailed information on the characteristics of turbulence near an entrainment interface has come from the Ashchurch experiment (Palmer et al., 1979; Caughey and Palmer, 1979). This experiment involved the use of high power Doppler radar in addition to the balloon-borne turbulence probes. Shown in

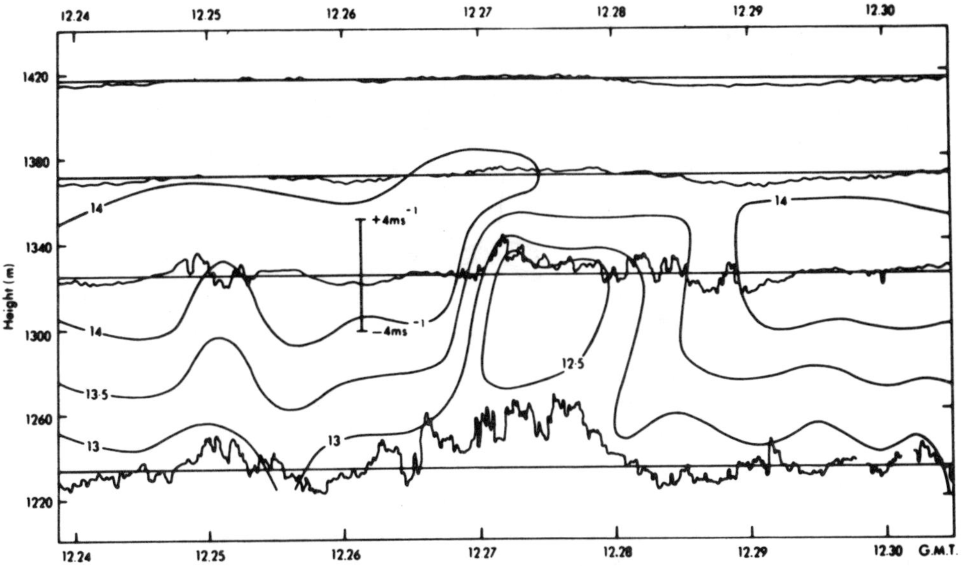

Figure 4.16. Height-time profile of temperature contours (^{o}C) drawn to 15s mean values and traces of vertical velocity from the turbulence probes over the period 1224 to 1230 GMT (Palmer et al., 1979).

Figure 4.15 is a section of data obtained as the inversion rises along the probes, so that they move into the CBL. The first section of data begins with the probes located well into the stable air, after which cooler and turbulent air begins to appear at the lowest level (heavy stipling in Figure 4.15). Further deepening of the CBL results in the probes moving into a fully turbulent region. The midle section of data illustrates the distortions present on the interface (~50-100m at 1215 GMT and again at 1225-1230 GMT). A closer examination of the latter undulation is given in Figure 4.16. At 1325 m the velocity trace clearly shows the change from non-turbulent to turbulent conditions and at the same time, the temperature dropped by ~1.8^{o}C in less than 30 s, indicating the upward movement of air from the mixed layer. A rough estimate of the displacement, d, of the interface may be made from the following relation, which is obtained by equating the loss of kinetic energy of the thermal to the work done in penetrating the stable layer, $d \sim w/\omega_B$, where,

$$\omega_B = \{(g/\Theta)\ \partial\Theta/\partial z\}^{\frac{1}{2}} \tag{4.12}$$

Taking in the present example $w \sim 4$ m s^{-1}, $\Theta \sim 296$ K, $\partial\Theta/\partial z \sim 0.025$ K m^{-1} we obtain $d \sim 130$ m, which is of the same order as that observed.

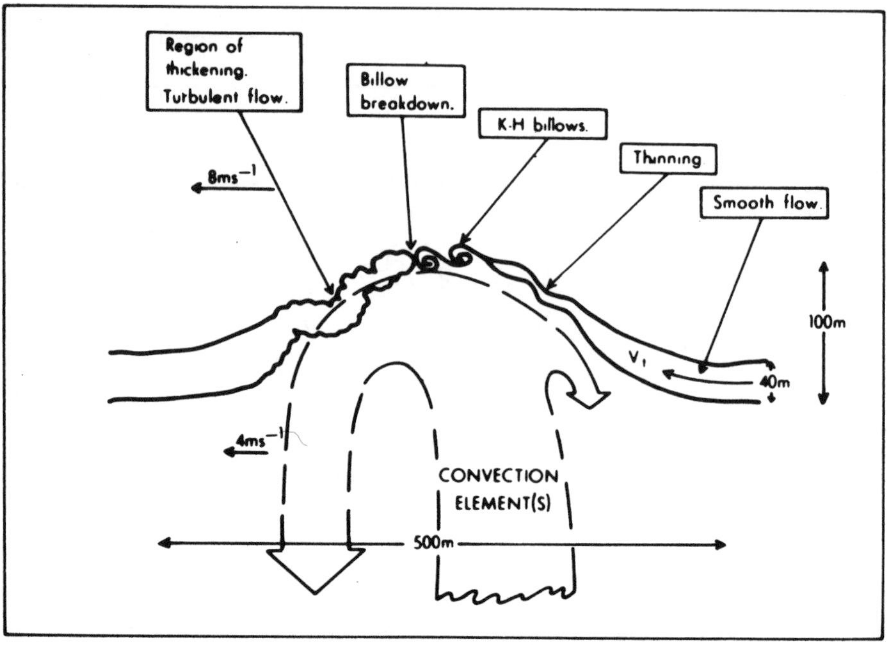

Figure 4.17. Schematic illustration of the generation of convective hummocks on a capping inversion with consequent thinning and the formation of K-H instabilities (Rayment and Readings, 1974).

Undulations on an inversion marking the top of the CBL have also been observed by Rayment and Readings (1974). These features were also ~100 m in amplitude (see Figure 4.17) and apparently were the main vehicles by which air was entrained across the interface on that occasion. These authors interpreted the observations to indicate the presence of small-scale Kelvin-Helmholtz (K-H) billows forming in the region of enhanced shear near the crests of the undulations. On occasions with somewhat lower stability in the free atmosphere, the whole undulation may become unstable and result in the engulfment of large volumes of air into the CBL as envisaged by Carson and Smith (1978). This process may have been occurring during the Ashchurch experiment, since the radar data indicated quite large volumes of entrained air near the CBL top. Much more effort is required to elucidate the varied entrainment mechanisms that doubtless occur in developing CBLs. A discussion of the dependence of the entrainment velocity on inversion strength has been given by Mahrt (1979).

The rate of rise of the capping inversion and the potential temperature step across it, $\Delta\Theta_1$, may be used to estimate the entrained heat flux, Q_e, from the relation given by Ball (1960).

$$Q_e = - \frac{c_p}{g} \Delta\Theta_i \frac{\delta(P_o - P_i)}{\delta t} \tag{4.13}$$

where g is the acceleration due to gravity, P_i the pressure at the inversion base and P_o the surface pressure. For the Ashchurch experiment the potential temperature step $\Delta\Theta_i$ was estimated directly from the turbulence probe records and the rate of rise of the inversion, $\delta(p_o - p_i)/\delta t$, from the radiosonde ascents. In their previous study, Rayment and Readings (1974) estimated entrained fluxes in the range 100-200 W m^{-2} but eddy correlation values from the turbulence probe were only ~1/4 of these. Two possible explanations were put forward for this disagreement. Firstly, the inversion base was rising steadily with time and therefore the turbulence probe was moving into a region of smaller heat flux further from the base; secondly, measurements were available for only a short 20 min period and therefore spectral losses were probably significant. In the Ashchurch study, the inversion base was also rising with time but by combining observations from different probes it was possible to improve on these shortcomings and provide flux estimates for the height range $0.95 < z/h < 1$.

The magnitude of the entrained heat fluxes obtained by this method fall in the range 30-170 W m^{-2} and are probably typical of those associated with a rising inversion in fair weather conditions. Direct eddy correlation estimates compared quite favourably with those obtained from Ball's relationship. These results serve to confirm the magnitude of the entrained heat flux and its importance to CBL development.

Variances, Dissipation Rates and Structure Parameters

When combined, the Ashchurch and Minnesota data sets provide a much more complete description of the behavior of the variances of the velocity components and temperature than was previously available. The dimensionless vertical velocity variance increases from ~0.1 near the surface to reach a rather broad maximum of ~0.4, centred near the middle of the layer (see Figure 4.18a). A free-convection prediction based on the Kansas data (Wyngaard et al., 1971) may be recast in mixed layer similarity form as

$$\sigma_w^2/w_*^2 = 1.8\ (z/h)^{2/3} \tag{4.14}$$

(see, e.g., Kaimal et al., 1976). The data indicate that this behavior is followed up to ~0.1 h. The dashed line in Figure 4.18a is from Willis and Deardorff (1974) and represents an average of their S1 and S2 cases (these refer to different temperature

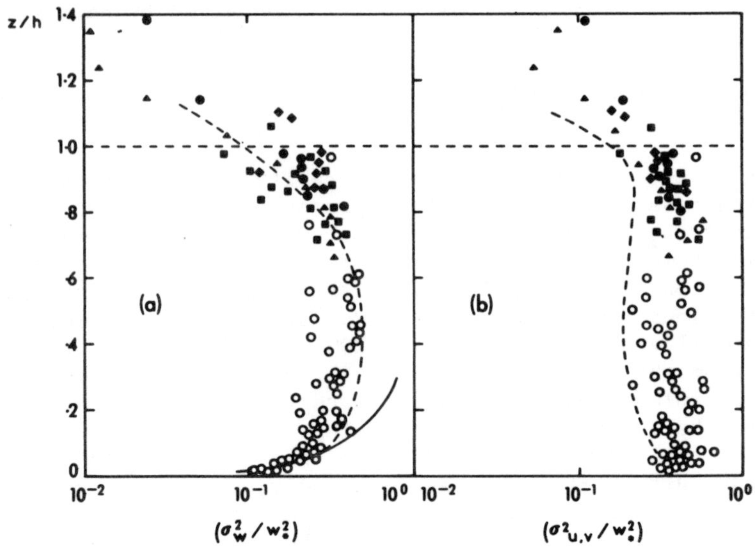

Figure 4.18. (a) Normalized vertical variance. The solid line is the free convection prediction (4.14). (b) The average of the normalized horizontal velocity components. Data coded as in Figure 4.12 (Caughey and Palmer, 1979).

gradients in the stable region above the convective layer, S2 gradient ≃ 2 x S1 gradient). The fit to the atmospheric data is quite good near the top and bottom of the CBL but the laboratory data indicate rather larger values than the atmospheric data in mid-regions. Kaimal et al. (1976) commented on the absence of any appreciable decrease in the vertical velocity variance near the inversion base in the Minnesota experiment. This was reflected in the spectral behavior, which showed no appreciable change in spectral intensity or shape above 0.1 h. However, the combined data set now suggests a slow decrease in variance, reaching ~0.1 to 0.2 w_*^2 at h. This is in excellent agreement with the aircraft results surveyed by Willis and Deardorff (1974) and the more recent aircraft results of Frangi (1979) and Guillemet et al. (1979).

The numerical results of Deardorff (1974) support a lower value for the maximum variance, i.e., they indicate $(\sigma_w^2)_{max} \sim 0.44\ w_*^2$, but this is reached at a rather lower level in the model, i.e., at 0.33 h, than the 0.5 to 0.6 h observed in the atmosphere. Near h, water tank and numerical results also indicate $\sigma_w^2 \sim 0.1\ w_*^2$, in excellent agreement with the atmospheric data. Above the CBL, the numerical results show a steadily decreasing value in much the same manner as the Ashchurch data. At

$z \sim 1.5$ h, $\sigma_w{}^2$ has decreased to $\sim 0.01\ w_*{}^2$. The finite value of $\sigma_w{}^2$ above h is associated with entrainment and, at the higher levels particularly, with internal waves. These show up clearly in the temperature contours and have periods in the range 5-10 min (Palmer et al., 1979).

The longitudinal and lateral components of air motion are on occasion significantly contaminated by balloon motion effects (Haugen et al., 1975), which can produce over-estimates of the variance by ~ 10-30%, depending on the separation between the balloon and turbulence probe. However, the data are still expected to indicate the general form and approximate magnitude of these variances in the CBL. Since the observed magnitudes of the variances did not differ significantly only their average, $\sigma^2_{u,v}$ is presented here, (Figure 4.18b). The dashed line is from Willis and Deardorff (1974) and again represents an average of the S1 and S2 cases. This falls substantially lower than the atmospheric data but has the same general shape and indicates a very slight maximum in the variance near 0.8 h, which is also just indicated in the atmospheric data. The difference in magnitude of the variances may be due to the fact that in the atmosphere significant u and v fluctuations occur at large scales, i.e., in excess of 2 h, through the complete depth of the CBL (Kaimal et al., 1976), whereas this represents the maximum horizontal length scale achieved in the laboratory model. Aircraft data (summarized by Willis and Deardorff, 1974) are in excellent agreement with the Ashchurch and Minnesota results as are those of Lenschow et al. (1980) from the marine CBL during AMTEX. An average value of $\sigma_{u,v}{}^2 \sim 0.4\ w_*{}^2$ in the mixed layer may be compared with the conclusions of Panofsky et al. (1977) based on surface-layer data. They found that the behavior of the horizontal wind components could be approximated by

$$\sigma_{u,v} \sim u_* \left(12 + 0.5\ h/\|L\|\right)^{1/3} \qquad (4.15)$$

where u_* is the friction velocity. Since the second term in parentheses usually dominates the first in very unstable conditions (for the Minnesota data this term averages ~70) Equation (4.15) may be recast with the aid of (2.56) as $\sigma_{u,v}{}^2 \sim 0.31\ w_*{}^2$. This indicates that the mixed-layer observations agree well with the unstable limit prediction from the surface-layer data.

The data from Deardorff's (1974) numerical simulation support the atmospheric results in the absence of any maximum in horizontal velocity variance near h, as was indicated in other work by Deardorff (1972). In these earlier simulations, a lid was imposed at h and clearly the change in behavior observed when the lid is removed, suggests that most of the kinetic energy of overshooting updrafts is absorbed in forcing entrainment and

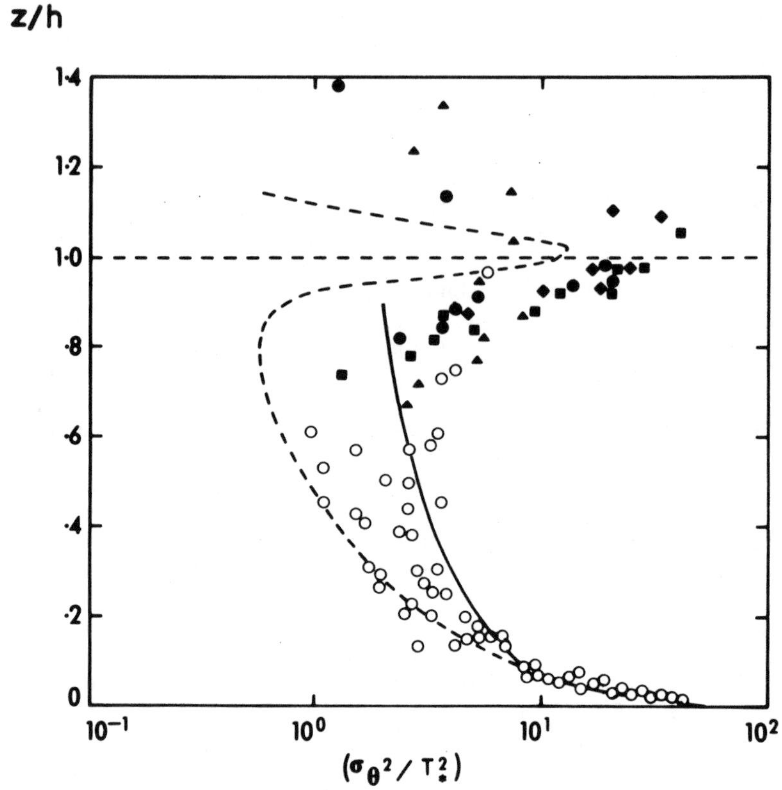

Figure 4.19. The normalized temperature variance. The solid line represents the free convection prediction (4.16) and the dashed line the S1 case from Willis and Deardorff (1974) (Caughey and Palmer, 1979).

internal waves rather than in conversion to horizontal kinetic energy. A relatively rapid decrease of variance occurs above the CBL but at 1.4 h the horizontal variance is considerably greater than that of the vertical component.

The normalized temperature variance (Figure 4.19) decreases rapidly with height from about 60 near the surface to reach a rather broad minimum of 2-3 in the center of the layer. Some indication of an increase with height in the Minnesota data is strongly supported by the Ashchurch data, which shows a well defined maximum, in the range 20-30 associated with the inversion. This behavior relates quite well in shape and magnitude to the numerical results of Deardorff (1974). Aircraft results (summarized by Willis and Deardorff, 1974) also indicate a value

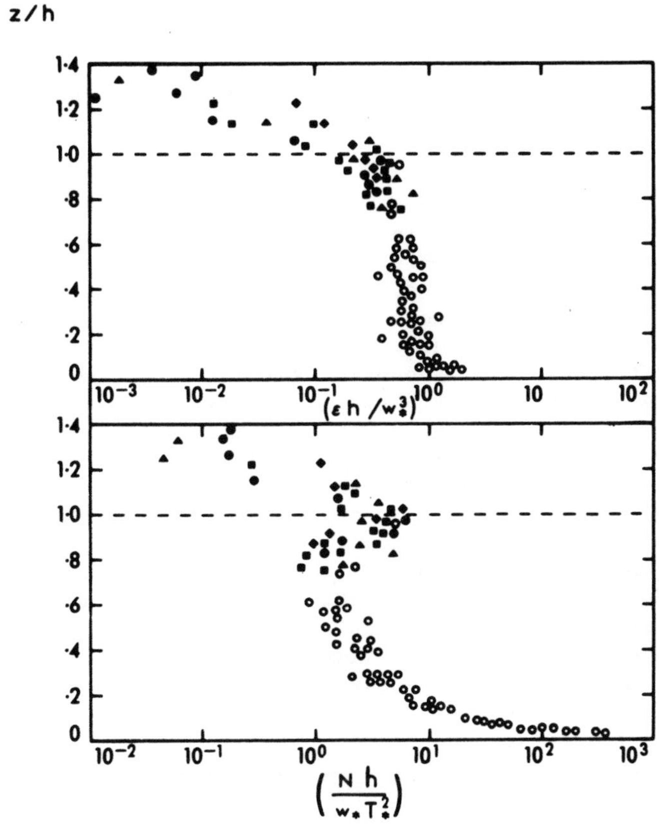

Figure 4.20. Variation of the normalized dissipation rates of turbulence kinetic energy and temperature variance with height (Caughey and Palmer, 1979).

for $\sigma_\theta{}^2 \sim 2\text{–}3\ T_*{}^2$ in the middle of the CBL and are in excellent agreement with the present results. More recent aircraft data of Frangi (1979) and Guillemet et al. (1979) are also in excellent agreement with the Ashchurch and Minnesota data.

The solid line in Figure 4.19 represents a free convection prediction (based on the Kansas results) rewritten in mixed-layer notation as

$$\sigma_\theta^2 / T_*^2 = 1.8\ (z/h)^{-2/3} \qquad (4.16)$$

and this is a reasonable fit up to ~0.1 h. The temperature variance near the inversion base is associated with entrainment and hence cannot be scaled with $T_*{}^2$. This is clearly exhibited by the numerical data of Deardorff (1974) in which the maximum variance associated with the inversion increases substantially as the day progresses. The lack of separation in the Ashchurch data in this region, most probably reflects the similarity in conditions on each of the days. In this regard, the scatter in the data presented by Guillemet et al. (1979) at heights above 0.8 h, is perhaps noteworthy.

Of interest for the turbulence budget equations is the behavior of the dissipation rates of turbulence kinetic energy, ε, and of one-half the temperature variance, N, in the CBL. The combined Ashchurch and Minnesota data sets are shown in Figure 4.20 and a marked difference in behavior of the two quantities is indicated by the present data. The normalized dissipation rate, $\varepsilon h/w_*{}^3$, maintains a fairly constant value through the bulk of the CBL with some indication of a slight decrease near the top. Above the CBL the data show a very rapid decrease in the intensity of the smaller-scale fluctuations - by roughly three orders of magnitude at 1.4 h. The data support the relationship (2.40) with ℓ equal to $(\lambda_m)_w$. The $C_\varepsilon \sim 0.26$, in fair agreement with Hanna's (1968) value of 0.29 and the more recent estimate by Wamser and Muller (1977), for a wide range of stabilities, of 0.32. The decrease of N with height is very rapid and continues up to 0.6 h. From 0.7 h to h, a substantial increase occurs and the results indicate an elevated maximum associated with the inversion. This behavior emphasizes the importance of heat entrainment into the CBL and is apparent in the highest-level observations by Caughey and Rayment (1974). Above the inversion the high frequency temperature fluctuations also decrease very rapidly.

The structure parameters $C_T{}^2$ and $C_V{}^2$ are of importance in studies of acoustic and optical propagation in the atmosphere. An asymptotic prediction for $C_V{}^2$ in the surface layer is

$$C_V{}^2\ h^{2/3}/w_*{}^2 \sim 1.5 \qquad (4.17)$$

and this forms a reasonable fit to the data above ~0.1 h (Figure 4.21). As noted by Kaimal et al. (1976) $C_T{}^2$ provides by far the best fit to the Kansas free convection predictions in the lower part of the CBL. In mixed-layer scaling the expected variation is (Figure 4.21)

$$C_T{}^2 h^{2/3}/T_*{}^2 = 2.67\ (z/h)^{-4/3}. \qquad (4.18)$$

However, as mentioned earlier, mixed-layer scaling is not expected to be applicable in the vicinity of the capping inversion where large values of $C_T{}^2$ are generated by entrainment (see also Kukharets and Tsvang, 1979). For this layer, Wyngaard and Le Mone (1980) have suggested an interfacial-layer temperature scale which seems applicable to the interfacial Ashchurch data. Other

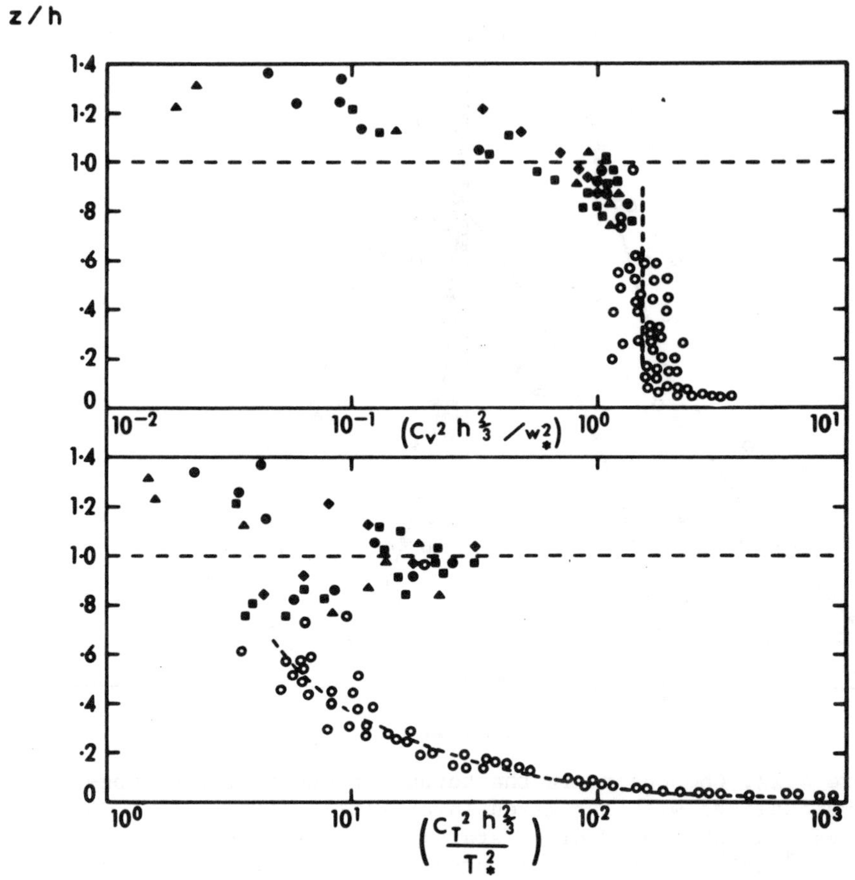

Figure 4.21. Variation of the normalized structure parameters for velocity and temperature with z/h. The dashed lines represents the free convection predictions (4.17) and (4.18) (Caughey and Palmer, 1979).

observations of C_T^2 in the marine CBL have been given by Fairall et al., (1980).

Turbulence Kinetic Energy Budget

The Minnesota data provided a good opportunity to examine the characteristics of some of the terms in the turbulence kinetic budget (Caughey and Wyngaard, 1979). Only a relatively small number of attempts have been made to measure the budget equation components above the surface layer. Instrumented aircrafts have been used to directly record some of the terms by, for example,

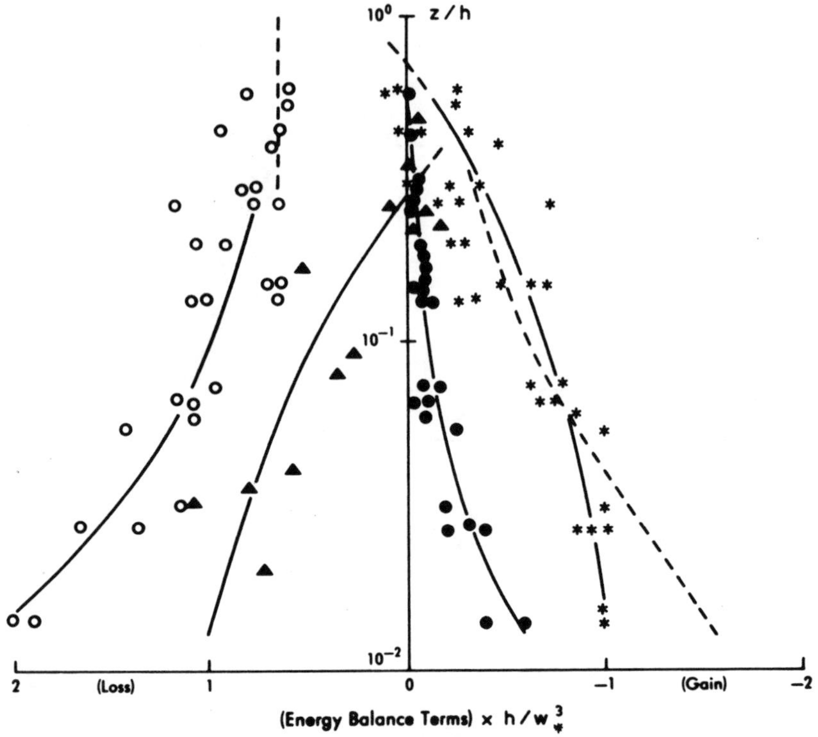

Figure 4.22. Components in the turbulent energy budget nondimensionalized with the mixed layer similarity scales w_* and h plotted against z/h. Coding; stars: buoyancy production; dots: shear production; triangles: turbulent transport; circles: dissipation; dashed line: residual (Caughey and Wyngaard, 1979).

Zubkovskiy and Koprov (1970) and Lenschow (1970). Results from these experiments indicate substantial variations in the relative importance of the turbulent kinetic energy budget components through the CBL. Rayment and Caughey (1977) demonstrated that the relative magnitudes of the budget equation terms at 91 m are significantly different from those in the surface layer.

Under the assumption of horizontal homogeneity the equation that describes the budget of turbulence kinetic energy is given in Section 1.8, Equation (1.57), where the meaning of the different terms in this equation is also discussed. Equation (1.57) is nondimensionalized with the boundary-layer height, h, and velocity scale, w_*, to examine the Minnesota data for consistent behavior on the various days. The term on the left-hand side of Equation (1.57) is the rate of change of $\frac{1}{2}q^2$ and was negligible during the Minnesota runs.

The mixed layer data were found to be much more scattered than the surface layer statistics, due no doubt to the longer averaging times required to obtain stable estimates of mixed layer quantities (Wyngaard et al., 1974). Only those runs with small wind shear were selected for analysis. Dissipation rates (see Figure 4.22) fall off steadily with height to become generally $< w_*^3/h$ above 0.1 h. The asymptotic surface layer prediction is $\varepsilon h/w_*^3 \sim 0.66$ and this falls near the lower bound of the data at heights > 0.2 h. Normalized shear production in these runs decreased rapidly, becoming negligible above 0.3 h. Only rather sketchy information exists on the behavior of the turbulence transport term, which appears to decrease from ~1 at ~0.01 h (in approximate balance with normalized buoyancy production) to become negative above ~0.3 h. Normalized buoyancy production decreases steadily with height and shows signs of becoming a loss above 0.5 to 0.6 h. The general form of these quantities agrees well with those observed in the marine CBL during AMTEX (Lenschow et al. 1980), except that the normalized dissipation rates are smaller. However, as reported by Caughey and Wyngaard (1979), the Minnesota dissipation rates appear anomalously large. The AMTEX budget is quite well balanced in mid-layer, while at small z/h there is some evidence that a gain term is needed for balance. This is consistent with the results of Wyngaard and Coté (1971), McBean and Elliott (1975), Rayment and Caughey (1977) and Caughey and Wyngaard (1979), who propose that this gain might arise from pressure transport.

4.3. STABLE BOUNDARY LAYER

The study of the stable boundary layer (SBL) is inherently more difficult and consequently less well advanced than the CBL. This is because buoyancy forces act to suppress turbulence so that fluctuation levels are generally much lower and therefore more difficult to measure. Additionally, the atmosphere can support wave motion and the co-existence of waves and turbulence complicates interpretation of the data.

General Characteristics of the SBL

A typical set of profiles from the Minnesota experiment is given in Figure 4.23. Notable features are the large gradients of the measured parameters throughout the SBL depth (see also Mahrt et al. 1979). The dashed line indicates the SBL top on this occasion and is taken as the height, h, at which the heat flux has fallen to ~5% of its surface values (see Caughey et al., 1979). The wind profile also indicates the presence of a low-level nocturnal jet (see, for example, Thorpe and Guymer, 1977). It is

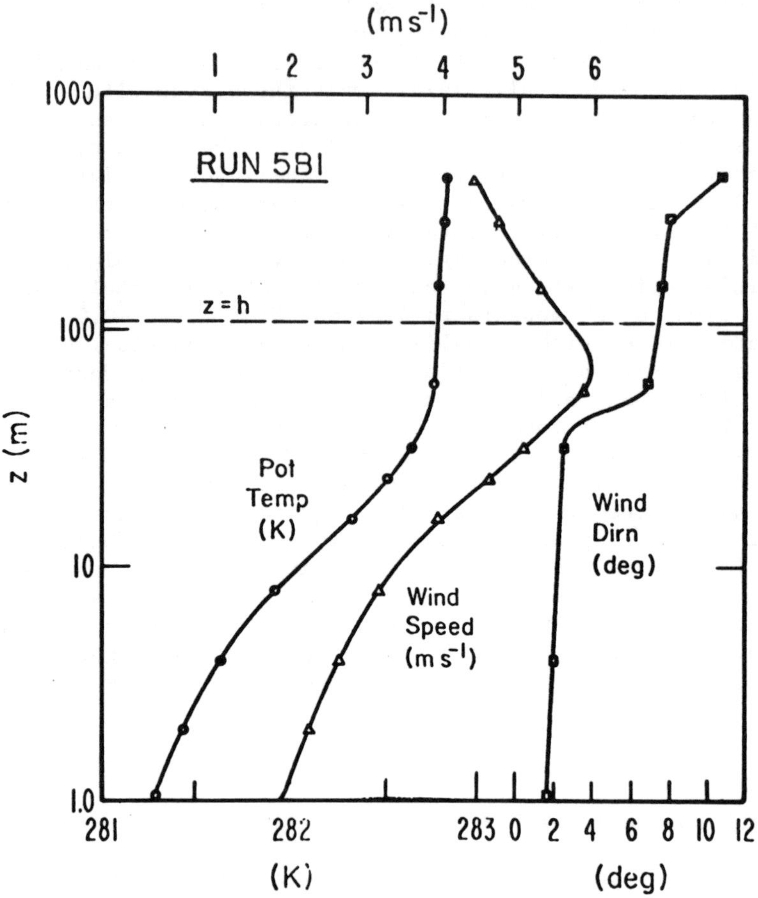

Figure 4.23. Typical mean profiles of wind speed, wind direction and potential temperature showing their relationship to the boundary depth, h (Caughey et al., 1979).

important to note that there is no simple relationship between the SBL depth, h, and the depth of the surface inversion layer. As this layer deepens and becomes more intense, h decreases, i.e., the depth of significant turbulent exchange becomes confined to a shallow layer close to the ground (see Wyngaard, 1975). Internal waves in the stable layer may however, become unstable and locally reduce the Richardson number to 0.25 with consequent production and decay of patches of turbulence, which may be the only contributors to turbulent exchange on these occasions.

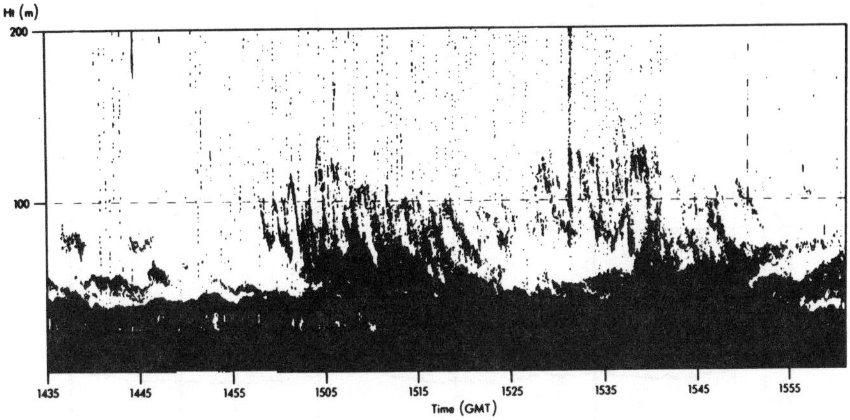

Figure 4.24. Sounder facsimile chart for the period 1435-1600 GMT on the 14 November 1975 illustrating the presence of breaking waves in the vicinity of the 100 m level (Crease et al., 1977).

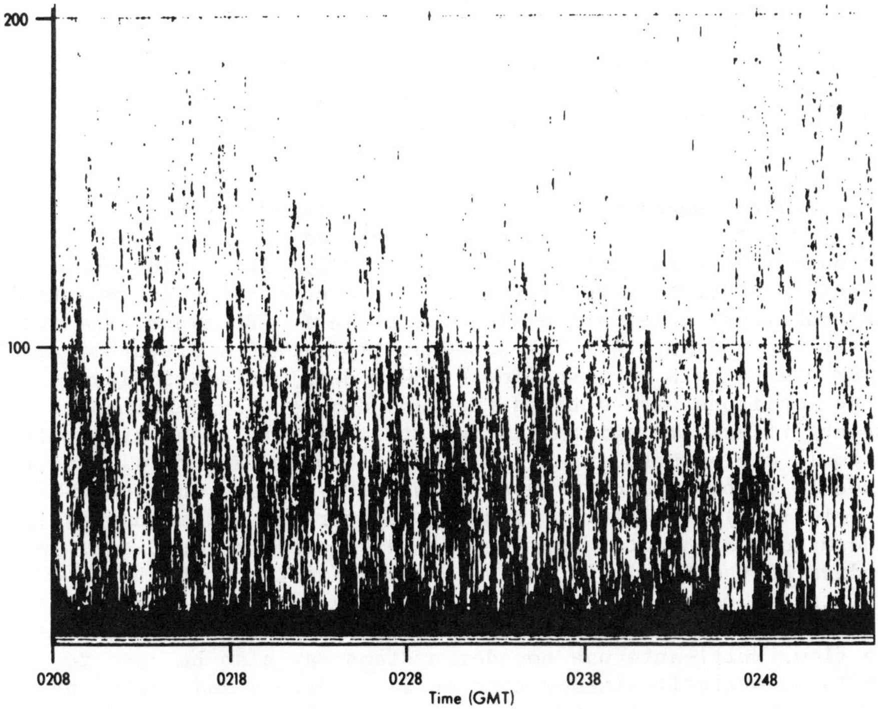

Figure 4.25. Acoustic echoes from a stable boundary layer in the presence of strong surface winds (Crease et al., 1977).

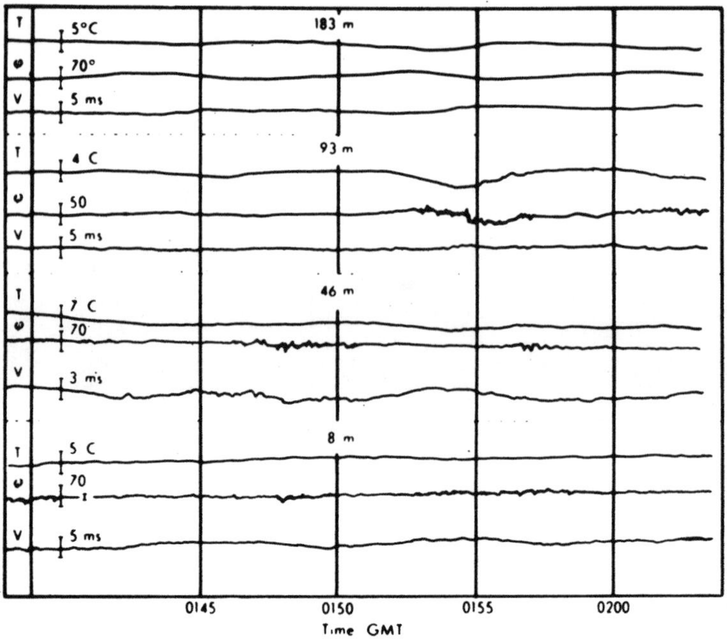

Figure 4.26. Time histories of temperature (T), inclination of the wind to the vertical plane (ϕ) and total wind speed (V) at the indicated heights (Caughey and Readings, 1975).

Acoustic sounding provides valuable information on the structure of the SBL and indeed may be used on some occasions to estimate h (see Asimakopoulos et al., 1975). The variety of structure present in records from stable conditions is remarkable and doubtless reflects the wide variation in SBL types. The record presented in Figure 4.24 is somewhat typical of a disturbed stable layer overlying a very shallow (~30 m deep) adiabatic layer. Large amplitude wave motions (perhaps due to Kelvin-Helmholtz instabilities) are evident in the stable layer, generating patches of small scale turbulence. The record shown in Figure 4.25 is from a weakly stable boundary layer with fairly strong surface wind (5 - 10 ms^{-1}). The 'striations' probably represent mechanically generated eddies acting on the stable temperature gradient. In studies of the SBL it is clearly very valuable to obtain a qualitative pictorial representation from acoustic sounding in addition to the more usual highly detailed direct measurements of the flow. Multi-antennae sounder systems may also be used to obtain quantitative information on wave motions and turbulent characteristics in the SBL (see, for example, Eymard, 1978; Caughey et al., 1980; Moulsley et al., 1981).

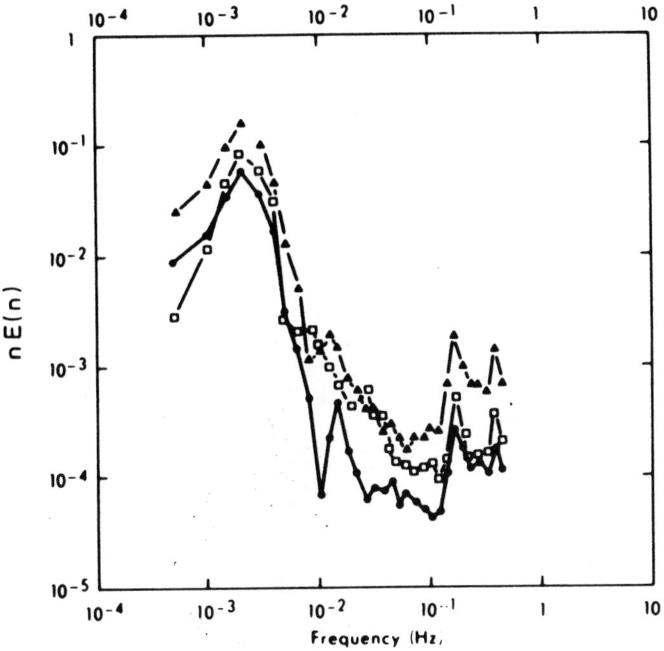

Figure 4.27. Power spectra of the vertical velocity (dots), horizontal wind speed (triangles) and temperature (squares), units are m^2s^{-2} and $^oC^2$ (Caughey and Readings, 1975).

Waves and Turbulence

As noted earlier, the interpretation of atmospheric data obtained in stable conditions may, well be complicated by the coexistence of turbulence and internal waves. These two phenomena have quite different properties and it is important to distinguish between them if erroneous deductions are to be avoided. Various ways of accomplishing this have been proposed (Stewart, 1969; Bush, 1969), one of which is based on the characteristics of the spectra of the temperature and velocity fields and has been used by Caughey and Readings (1975) in an observation of waves and turbulence in the SBL. In this study four levels of turbulence measurements were available and the time histories are given in Figure 4.26. Between 0140 and 0224 GMT all the traces exhibit wave-like characteristics, and at 46 m and 93 m associated bursts of turbulence occur whilst, at other, times these levels are essentially non-turbulent. At 183 m the traces show little evidence of turbulence. However, at 8 m some enhancement of turbulence occurs. Whilst this example may illustrate a rather unusual occurrence, it does emphasize the complex situations that may arise in stable conditions.

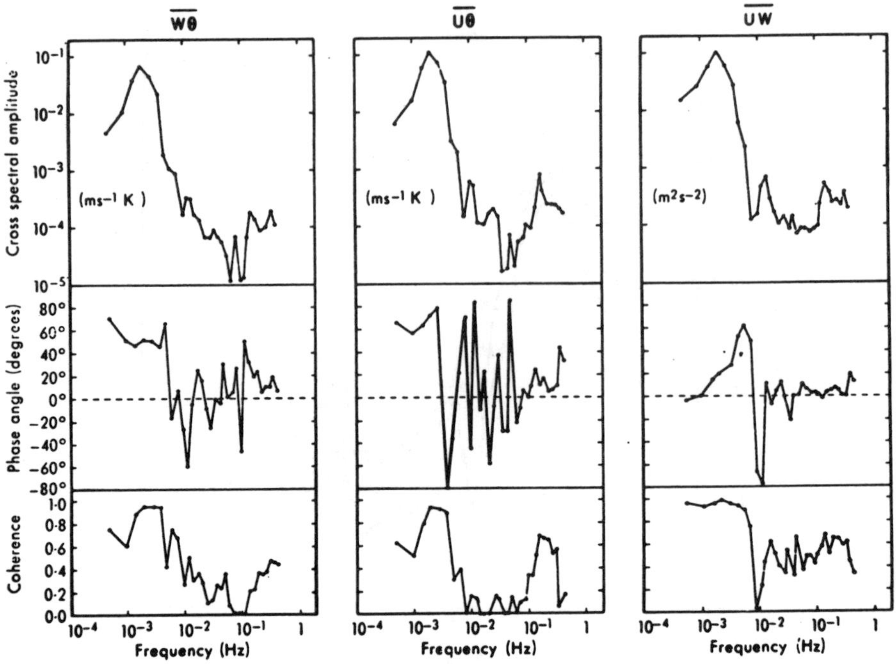

Figure 4.28. Plots of the (i) cross-spectral amplitudes, (ii) phase angles and (iii) coherences for the covariances $\overline{w\theta}$, $\overline{u\theta}$ and $\overline{uw}$ from the probe at 183 m (Caughey and Readings, 1975).

Since turbulence levels are very low at the 183 m level (Table 4.1) the wave characteristics are most clearly revealed here. Figure 4.27 shows the power spectra for u, w and T at this height and it is apparent that nearly all the spectral energy is contained in a single well-defined low-frequency peak. There is no evidence for an inertial subrange, the increase in spectral

Table 4.1. The variation of the standard deviation of vertical velocity, in the frequency range 10^{-2} to 10^{-1} Hz with height

Probe height (m)	σ_w (10^{-2} m s^{-1})
183	1
93	8
46	21
8	24

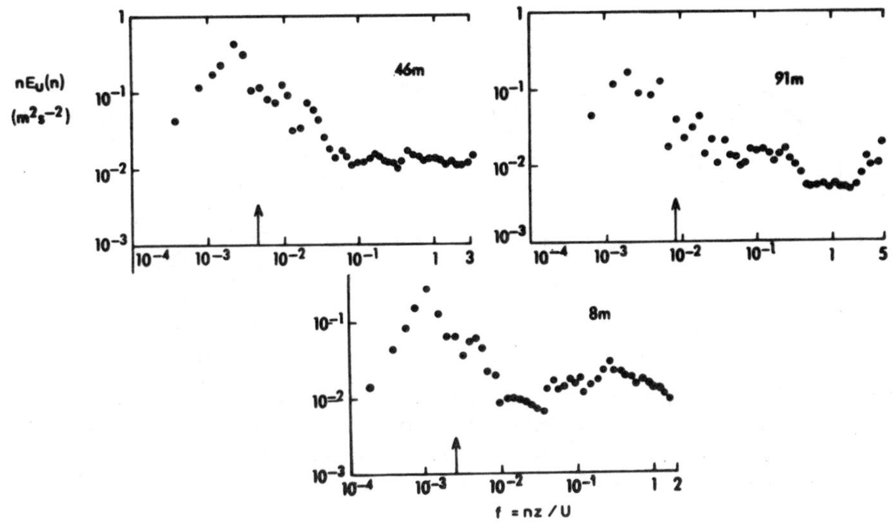

Figure 4.29. Power spectra of the horizontal wind speed over the period 2300-0415 8/9 November 1972 at heights of 8, 46 and 91 m. The arrows denote the positions of the Brunt-Vaisala frequencies (Caughey, 1977).

density at the high frequency end being due to noise. When a well defined wave is present, the cross-spectra should also exhibit peaks associated with high levels of coherence and stable phase angles (which are defined in Lumley and Panofsky (1964)). The observed data did indeed tend to exhibit this behavior (Figure 4.28). All the spectra have pronounced peaks in the frequency range $0.002 < n < 0.003$ Hz, which corresponds to a period of between 5-8 minutes. The coherences at these frequencies are much

Table 4.2: A comparison of the phase angles and coherences obtained in the present study with those from Axford's (1971) observations

Source of data	Component	Phase angle	Coherence
Present study (183 m)	$\overline{w\theta}$	48°	0.98
"	$\overline{u\theta}$	72°	0.96
"	$\overline{uw}$	22°	0.98
Axford (1971)	$\overline{w\theta}$	78°	0.90
"	$\overline{u\theta}$	160°	0.90
"	$\overline{uw}$	98°	0.85

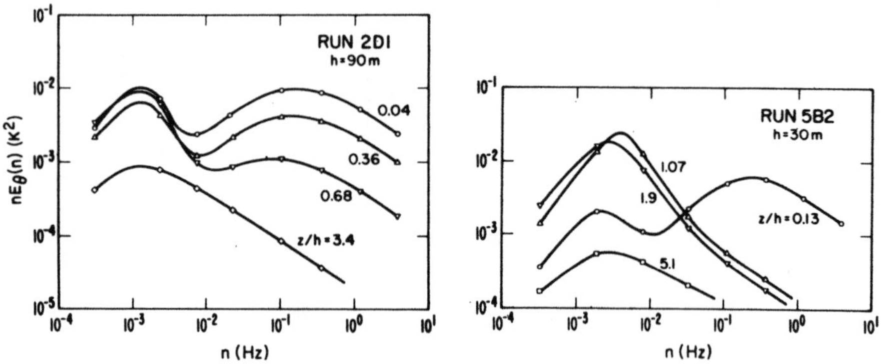

Figure 4.30. Power spectra of temperature from the Minnesota stable runs 2D1 and 5B2. These illustrate the presence of a spectral minimum at $\sim 10^{-2}$ Hz and the decrease in turbulence energy with increasing height in the SBL.

closer to the ideal value of unity than those reported by Axford (1971) (see Table 4.2). This Table also lists the phase angles between the various components and it can be seen that Axford's values are quite different from the present ones, although both form a self consistent set (i.e., $\phi_{w\theta} + \phi_{\theta u} + \phi_{uw} = 0$). For a stable temperature profile and reversed wind gradient, the following values would have been expected

$$\phi_{w\theta}(n) = 90^{\circ}, \quad \phi_{u\theta}(n) = 180^{\circ}, \quad \phi_{uw}(n) = 90^{\circ} \qquad (4.19)$$

(these values are also applicable to Axford's case). It is clear that though Axford's values are in fairly good agreement with the theoretical ones the boundary layer values are not. This was considered due to the proximity of the surface and non-linear wave effects (Caughey and Readings, 1975).

In general, waves occur with a variety of wavelengths and amplitudes and, hence, their spectral significance varies from one occasion to another. Caughey (1977) has shown that near the ground the "wave-subrange' occurs in the vicinity of the Brunt-Vaisala (B-V) frequency, defined by (4.12), and is generally separated by a spectral gap (or minimum) from the turbulence-subrange which occurs at higher frequencies (Figure 4.29). This gap is generally clearly discernible within the first 10 m or so of the SBL. At higher levels however, the turbulence region shifts to lower frequencies and the wave region to higher frequencies (as the B-V frequency increases with height) so that the net effect is that the gap closes and the spectrum assumes a single-peaked character. This means that extraction of the turbulence component (by

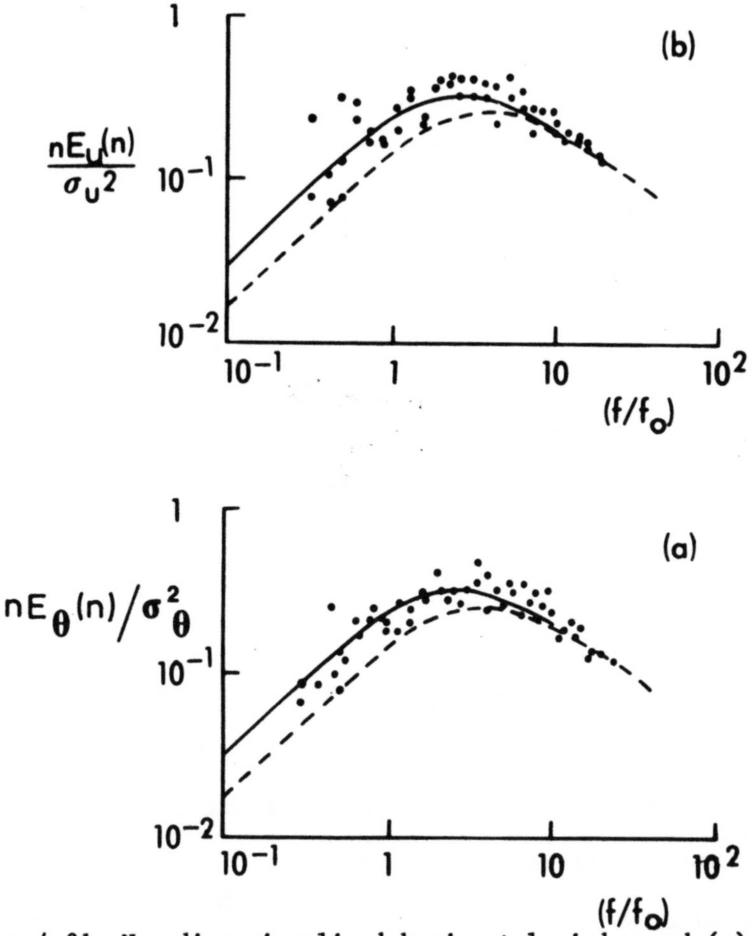

Figure 4.31. Non-dimensionalized horizontal wind speed (a) and temperature spectra (b) at 8m plotted versus f/f_0 compared with the universal curve given by Kaimal (1973); A = 0.3 is drawn as a solid line; A = 0.164 is drawn as a dashed line (Caughey, 1977).

integration of the spectrum at frequencies above the gap frequency) becomes impossible and, hence, one cannot distinguish between wave and turbulent fluctuations using this technique. However, the Minnesota data showed that for heights less than the SBL depth (h) a double-peaked character to all spectra and cospectra was apparent (Figure 4.30) so that the elimination of the wave contributions was indeed possible throughout the SBL depth.

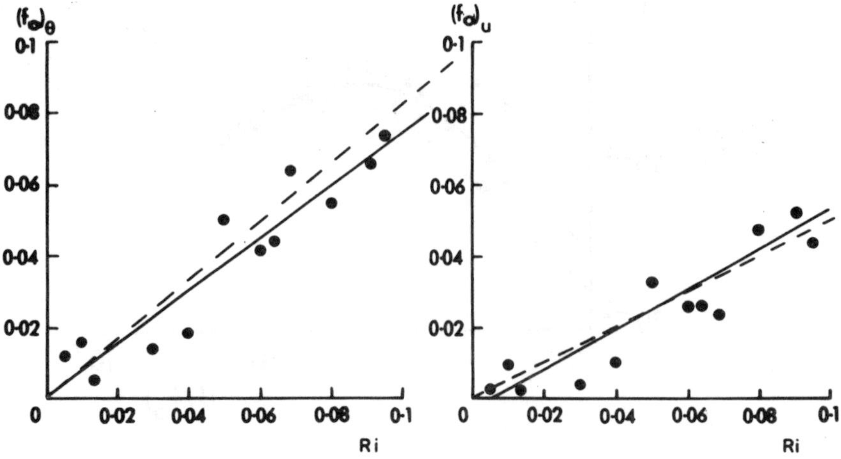

Figure 4.32. Relationships between the scaling parameter f_0 for horizontal wind and temperature and Richardson number Ri, over the range $0.005 < Ri < 0.1$. The solid lines represent linear regressions through the data whilst, the dashed lines are from Kaimal (1973) (Caughey, 1977).

Turbulence Spectra in the Stable Surface Layer

In his analysis of the stable Kansas data, Kaimal (1972, 1973) applied a low frequency cut-off to the data to remove the unwanted low-frequency region arising from waves/trends in the time series. Of course, the exact position of the gap depends on the various factors which determine the relative positions (on the frequency scale) of the turbulence subrange and the low frequency spectral contributions, so that the appropriate choice of cut-off filter frequency can only be made from an examination of the turbulence spectra. Kaimal (1973) achieved the collapse of stable-layer spectra into universal curves by introducing a new scaling parameter (f_0), which is the intercept on a reduced frequency scale ($f = nz/U$) at which the extrapolated inertial subrange slope meets the line $n\,E_\alpha(n)/\overline{\alpha^2} = 1$ (or $n\,C_{\alpha\beta}(n)/\overline{\alpha\beta} = 1$) for cospectra where $\overline{\alpha^2}$ and $\overline{\alpha\beta}$ are the variances and covariances associated with the turbulence subrange alone. This work was extended by Caughey (1977) to higher levels in the SBL and to a greater range of stabilities. The spectral results for temperature and horizontal wind speed at 8 m are shown in Figure 4.31. A good approximation to both spectral shapes was obtained with the relation

$$\frac{n\,E_\alpha(n)}{\sigma_\alpha^{\,2}} = \frac{A(f/f_0)}{1 + A\,(f/f_0)^{5/3}} \qquad (4.20)$$

where $\alpha = u$ or θ and $A = 0.3$.This agrees well with the line from Kaimal (1973) (i.e., A = .164). The difference between the A values is almost certainly due to the difference in the frequency bands over which the variances have been computed, i.e., the best fit A value will depend on the lower frequency limit (chosen rather subjectively) in the spectral gap region and the upper limit set by the sampling rate. For complete spectral prediction the parameter f_0 needs to be specified and this is done by relating it to the Richardson number (Ri). Typical scatter plots from Caughey (1977) are given in Figure 4.32 - the dashed lines indicate the variation obtained by Kaimal (1973). The best fit to the data are

$$\left.\begin{array}{l}(f_0)_\theta = 0.74\ Ri + 0.0013 \\ (f_o)_u = 0.53\ Ri - 0.0033\end{array}\right\} \quad \text{for } 0.005 < Ri < 0.1. \qquad (4.21)$$

These results indicate that, at 8 m, the characteristic scale (λ_m) for temperature fluctuations reduces from ~0.9 km at Ri = .01 to 0.1 km at Ri = 0.1; the corresponding figures for the longitudinal velocity component being ~4 km and ~0.2 km. For the variation of $(f_0)_w$ and $(f_0)_v$ Kaimal (1973) obtained the relations

$$(f_0)_w = 2.8\ Ri \text{ and } (f_0)_v = 1.5\ Ri \qquad (4.22)$$

The cospectra also showed the presence of spectral gaps, however, for horizontal heat flux (as noted by Kaimal (1973)) the slope in the inertial subrange region is better represented by a line of slope $n^{-3/2}$ than the n^{-2} predicted by Wyngaard and Coté (1971). The spectral gap or dip appeared quite broad and U-shaped, which renders the choice of a low-frequency limit subjective, however, for consistency with the power spectra, and because there are signs of an upward trend in spectral densities at lower frequencies, a limit for integration was set at 0.008 Hz (Caughey, 1977).

When plotted in non-dimensional form versus f/f_0 the cospectra for stress, horizontal and vertical heat fluxes at 8 m are well represented by the universal curves given by Kaimal (1973). The observed values for the f_0's were

$$(f_0)_{uw} = 0.22 \quad (f_0)_{w\theta} = 0.33 \quad (f_0)_{u\theta} = 0.14 \qquad (4.23)$$

and with Ri ~ 0.06 these are in good agreement with the values expected from Kaimal's relations

$$(f_0)_{uw} = 3.1\ Ri, \quad (f_0)_{w\theta} = 5.6\ Ri, \quad (f_0)_{u\theta} = 2.0\ Ri \quad (4.24)$$

being 0.19, 0.34 and 0.12 respectively. Non-dimensional cospectral forms were well fitted by the relations

$$\frac{n\ C\alpha\beta\ (n)}{\overline{\alpha\beta}} = \frac{0.88\ (f/f_o)}{1 + 1.5\ (f/f_o)^{2.1}}, \qquad \overline{\alpha\beta} = \overline{w\theta} \text{ or } \overline{uw}$$

$$\frac{n\ C\alpha\beta\ (n)}{\overline{\alpha\beta}} = \frac{0.85\ (f/f_o)}{1 + 1.7\ (f/f_o)^{2.2}}, \qquad \overline{\alpha\beta} = \overline{u\theta}. \tag{4.25}$$

In general, for more stable conditions (i.e., Ri's between 0.2 and 0.3) Caughey (1977) found that the turbulence intensities were low and spectral estimates fell to near-noise limits. However, some periods were observed in which the turbulence was sporadic (i.e., bursts lasting ½–4 minutes occurred) and Ri varied between ~0.2 and 0.7. For the latter case the high-frequency fall-off in the temperature spectrum was greater than −2/3 whilst in the former it was less. These conclusions compare well with those of Okamoto and Webb (1977) who found that for Ri ~ 0.2 some runs produced a near zero spectral slope, whilst others gave slopes between −1.2 and −2.5

In summary, therefore, the spectral distributions of turbulent fluctuations in the surface SBL can be obtained from a measurement of the local Richardson number. Caughey (1977) also provided evidence that these spectral shapes hold good at higher levels in the SBL and this question is explored further in the next section.

Turbulence Behavior through the SBL Depth

The Minnesota experiment provided a good opportunity to examine turbulence behavior through the entire SBL depth. Unfortunately, conditions were strongly evolving with time so that turbulence levels eventually fell to near-noise limits and the runs were terminated (Caughey et al., 1979). Shown in Figure 4.30 are the temperature spectra for runs 2D1 and 5B2 which had SBL depths of 90 m and 30 m respectively. These show the reduction in intensity and movement to larger scales of the turbulence subrange. As noted earlier, a spectral minimum is apparent for all heights less than the SBL depth and hence the wave fluctuation contributions can be eliminated. Accordingly, the turbulence statistics were obtained by integration over the band width 0.001 Hz < n < 5 Hz. Note also the significant change in spectral behavior above the SBL – the high frequency energy drops abruptly.

In analysing the Minnesota data, the approach used by Kaimal (1973) and Caughey (1977), which demonstrated the universal nature of spectra and cospectra in the lower layers of the SBL, was employed. This formulation forced the different spectra/cospectra to collapse into a single curve in the inertial subrange and since the overall shape varied little with height, the fit remained good

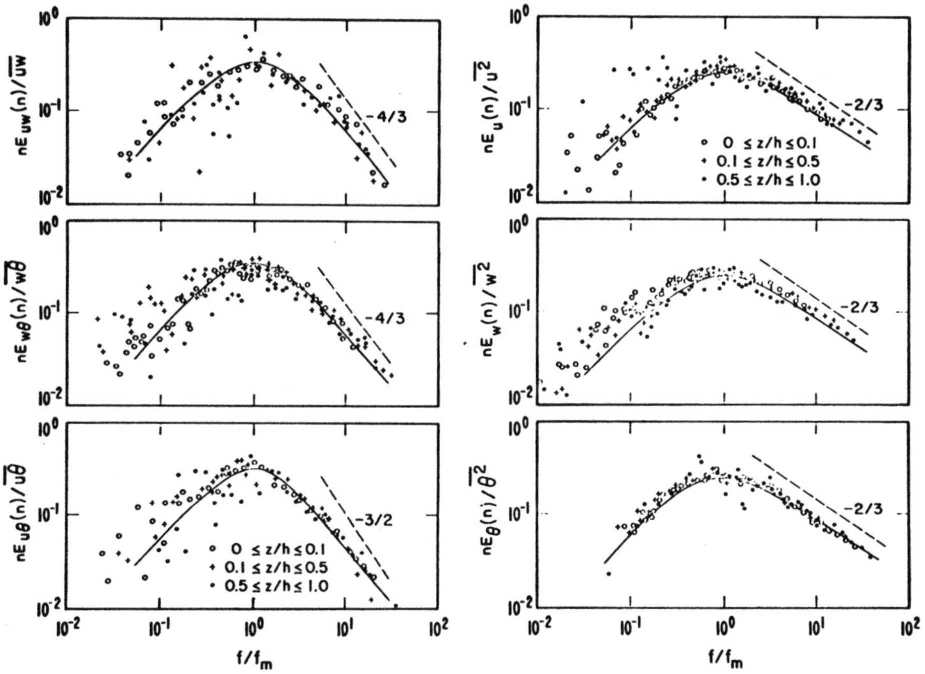

Figure 4.33. Normalised logarithmic spectra of u,v,w, $\overline{uw}$, $\overline{w\theta}$, $\overline{u\theta}$ plotted against f/f_m. (Caughey et al., 1979) Curves are from the Kansas data (Kaimal, 1973).

across the entire bandwidth. However, to consider behavior through the entire SBL depth, two departures were made from this scheme. The first is that f_m (reduced frequency at the spectral peak) is used instead of f_0, since the latter depends on knowledge of the variances/covariances and the integrations required to obtain these become more uncertain towards the SBL top. Secondly, Ri is replaced by z/h to explore the influence of boundary-layer depth on the spectral and cospectral scales.

The relationship between f_0 and f_m comes from the spectral and cospectral forms in Kaimal (1973), i.e.,

$$f_0 = \begin{cases} 0.26\, f_m & u,\ v,\ w,\ \theta \\ 1.27\, f_m & uw,\ w\theta \\ 1.39\, f_m & u\theta. \end{cases} \tag{4.26}$$

Substituting for f_0 in relations (4.20), (4.25) gives the curves shown in Figures 4.33. The secondary peaks which occur roughly two decades below the turbulence peaks, are not shown in the plots.

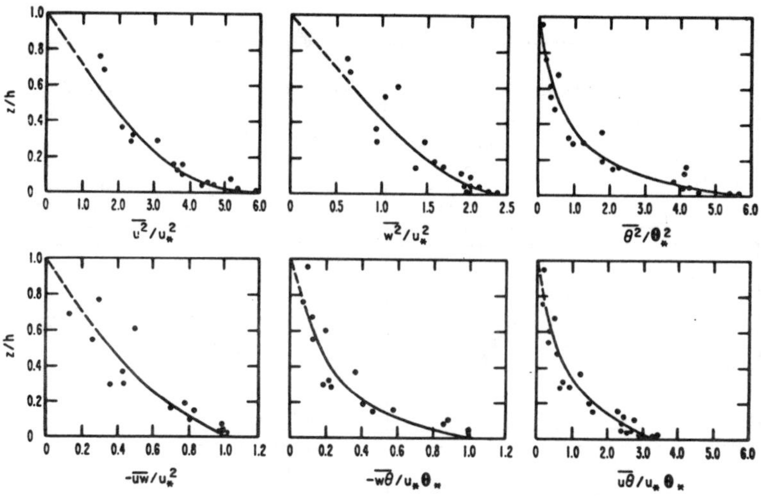

Figure 4.34. Verical profiles of the dimensionless variances and covariances. (Caughey et al., 1979)

The best agreement with the surface distributions is seen in the u and θ spectra (except for u above 0.5 h). The w data show more energy on the low frequency side compared to the Kansas data. A characteristic feature of the plots is the increase in scatter for $z/h > 0.1$. The normalized intensity at the peak is the same in all plots.

The dimensionless coordinates of Figures 4.33 mask any height dependence that exists in the spectral intensities and in the characteristic wavelength, λ_m. A consistent pattern emerges when these quantities are plotted versus z/h and appropriately normalized with surface-layer scales (Figures 4.34 and 4.35). All the average turbulence statistics show a quite rapid decrease with height becoming very small in the vicinity of the SBL top. The peak wavelength data collapse remarkably well considering the fact that the SBL depth was varying with time. Near the surface ($z < 0.1$ h), the wavelength is different for each variable; λ_m's are largest for u and θ, and smallest for w. Despite these difficulties, there exists a tendency for λ_m/h in all plots to approach unity at the SBL top, only θ and uθ indicate somewhat larger values.

In order to compare the Kansas behavior in the surface layer for the different λ_m's in the coordinates of Figures 4.35, one can use the Kansas relation between Ri and z/L (Businger et al., 1971), i.e.,

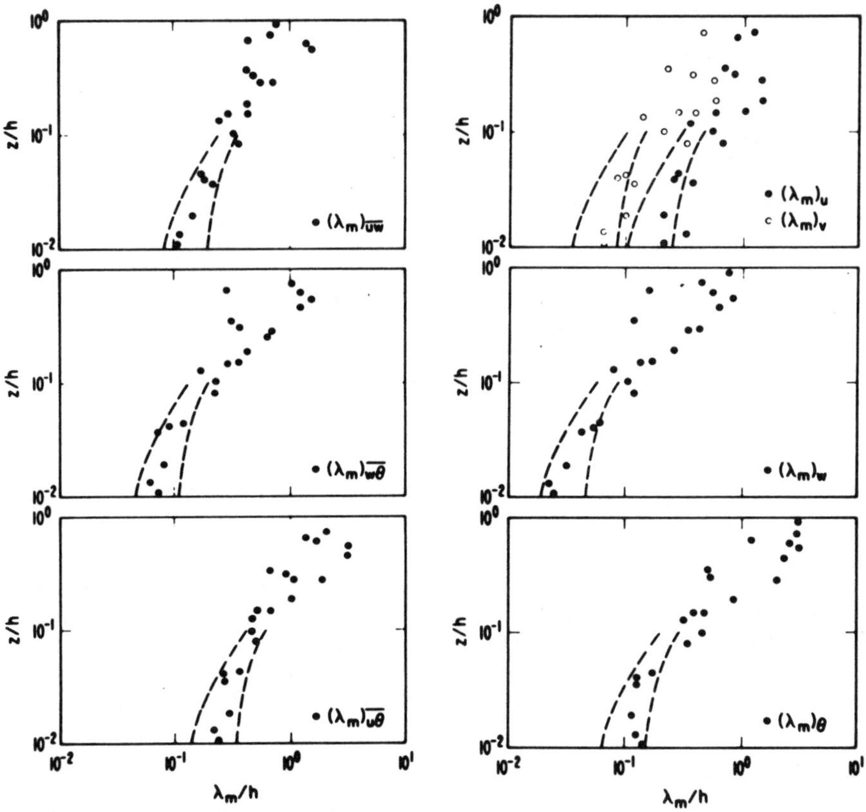

Figure 4.35. Vertical profiles of λ_m for u, v, w, θ, $\overline{uw}$, $\overline{w\theta}$, $\overline{u\theta}$. Curves are from the Kansas data for h/L = 3 and 8, the range of the present data. (Caughey at al. 1979)

$$Ri = \frac{[0.74 + 7.4 (z/h)(h/L)][(z/h)(h/L)]}{[1 + 4.7 (z/h)(h/L)]^2} \qquad 0 < \frac{z}{L} < 2 \quad (4.27)$$

A test of Equation (4.27) for the Minnesota data indicates good agreement with the Kansas behaviour for z/h < 0.1 (Caughey et al., 1979). The curves drawn in Figures 4.35 are predictions of Equation (4.27) for h/L = 3 and 8 since the Minnesota data fall between these extremes.

Dissipation rates ε and N, for the turbulence kinetic energy and half the temperature variance are plotted in dimensionless form in Figure 4.36. Both these quantities decrease with height rather gradually in the lower layers (0.01 h < z < 0.1 h) but more

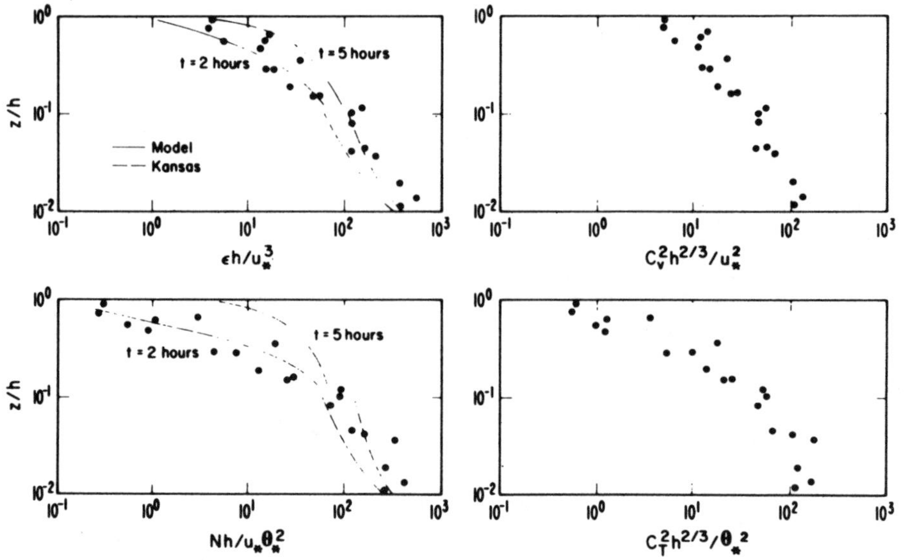

Figure 4.36. Vertical profiles of the dimensionless dissipation rates and structure parameters. Solid curves are from Brost and Wyngaard (1978) model calculations of ε and N for two times after transition, for a near-surface cooling rate of 4 K hr^{-1}. Dashed curves are from the Kansas data for h/L = 3 and 8 (Caughey et al., 1979).

abruptly above this. Figure (4.36) also includes dimensionless plots of the structure parameters $C_V{}^2$ and $C_T{}^2$, which are related to ε and N by

$$C_v^2 = 2\ \varepsilon^{2/3} \quad \text{and} \quad C_T^2 = 3.2\ N\ \varepsilon^{-1/3}. \tag{4.28}$$

The Kansas results (Wyngaard and Coté, 1971) show that the flux divergence terms in the budget of turbulent kinetic energy and temperature variance are negligible in the surface layer. It follows from (1.57) and (1.63), that for quasi-stationary and horizontally homogeneous conditions this implies

$$\begin{aligned} \varepsilon &= -\overline{uw}\ \partial U/\partial Z + (g/\Theta)\ \overline{w\theta}, \\ N &= -\overline{w\theta}\ \partial\Theta/\partial z. \end{aligned} \tag{4.29}$$

Non-dimensionalizing the above equation gives

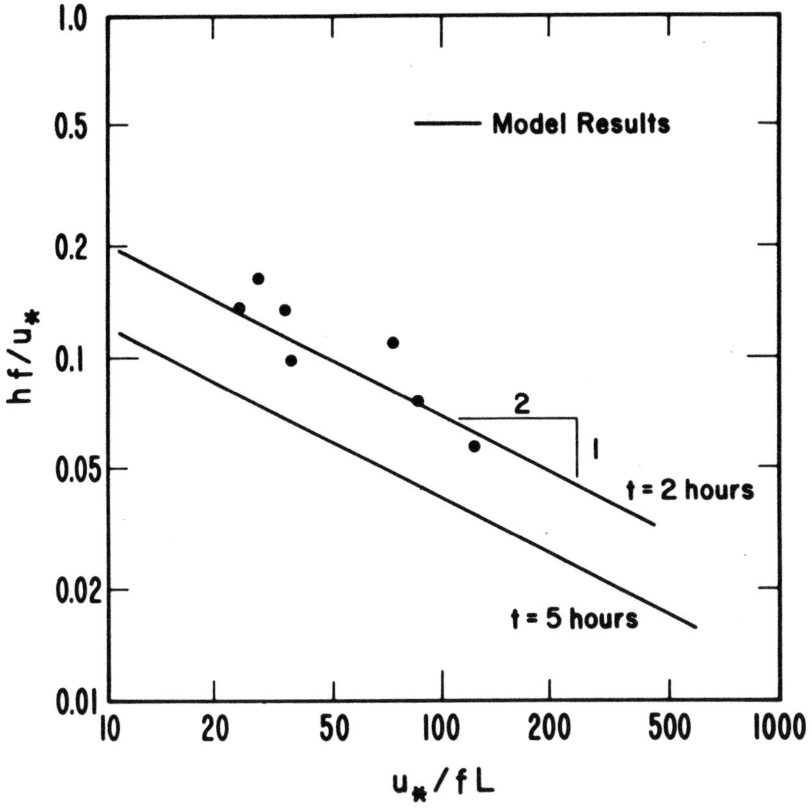

Figure 4.37. A test of the similarity prediction of Zilitinkevich for SBL depth h. Solid lines are from the Brost and Wyngaard (1978) model for a near surface cooling rate of 4 K hr^{-1} at 2 hr and 5 hr after transition (Caughey et al., 1979).

$$\frac{\varepsilon\, h}{u_*^3} = \frac{1}{k}\,\frac{[1 + 3.7\ (z/h)(h/L)]}{(z/h)}\ ,$$

$$\frac{Nh}{u_*\theta_*^2} = \frac{1}{k}\,\frac{[0.74 + 4.7\ (z/h)(h/L)]}{(z/h)}\ . \qquad (4.30)$$

These relations are shown as the dashed lines on Figure 4.36 for $k = 0.35$, $z/L = 3$ and 8 and $z < 0.1$ h. Although there is some scatter, these lines do form quite a good fit to the Minnesota surface layer data.

The solid curves in Figure 4.36 are based on the model calculations of Brost and Wyngaard (1978). The curves drawn are for a near-surface cooling rate of ~ 4 °C hr^{-1} which is close to the 3.6 °C hr^{-1} at z = 1 m observed in Minnesota for the first two hours after transition. The mid-points of the Minnesota runs occurred on average about 1 hr after transition, but because of the rapid transient response and possible sensitivity to the initial conditions the model results are only considered more than two hours after transition. Nevertheless the model results for ε and N at 2 hours agree quite well with the experimental data.

Depth of the SBL

Zilitinkevich (1972) predicted that the dimensionless depth, hf/u_*, of the idealized, steady-state, stably stratified atmospheric boundary layer behaves as

$$\frac{hf}{u_*} = c\left(\frac{u_*}{fL}\right)^{-\frac{1}{2}} \tag{4.31}$$

where c is a universal constant. Although the Minnesota h values decreased substantially with time Figure 4.37 shows the value for each run plotted in the form suggested by Equation (4.31). The results are seen to be fairly consistent with this behavior and also with the Brost-Wyngaard model at 2 hr; the full line in Figure 4.37 at 2 hr corresponds to c = 0.7. According to the model c decreases and approaches 0.4 as the SBL approaches steady state.

4.4. CONCLUDING REMARKS

In such a wide ranging subject it has been necessary to be highly selective in choosing material for presentation. The detailed results have been drawn from boundary-layer studies over fairly flat terrain, although the behavior observed in the marine CBL has similarities. For the convective case the conclusions may be summarized as follows:

(a) Turbulence structure is dominated by convective plumes which extend through a significant fraction of the boundary-layer depth. These interact with a capping inversion to produce entrainment effects which contribute to the heat and moisture budgets of the CBL and are therefore relevant to boundary-layer developement.

(b) Spectra of the velocity components in the mixed layer can be generalized within the framework of mixed layer similarity. The universal spectral forms hold good to ~0.8 h, where the influence of the upper boundary becomes important. At present there is no generally accepted method for non-dimensionalising interfacial data.

(c) The temperature spectra above 0.1 h cannot be generalized in the same manner as the velocity spectra, because of variability in the low frequency behavior. However, the inertial subrange is more predictable, decreasing as $(z/h)^{-4/3}$ up to ~0.5 h and increasing above this.

(d) Entrainment mechanisms range from small-scale Kelvin-Helmholtz breakdown of the interface to large-scale overturning of convectively induced hummocks.

(e) The characteristic wavelength, λ_m, for w and θ increases linearly with height in the range z = −L to 0.1 h but more slowly above this to reach an approximately constant value of ~ 1.5 h. The peak wavelength λ_m for u and v stays roughly constant (~ 1.5 h) throughout the CBL depth.

(f) Free-convection predictions for the variances of w and θ also appear valid to ~0.1 h, while C_T^2 fits the prediction to ~0.5 h.

With regard to the SBL, although the Minnesota site was unusually flat, the terrain slope was significant enough to influence the vertical structure of the mean winds. The Ekman spiral evolved 'abnormally' in three of the five evenings analyzed, evidently because of baroclinic and/or advective effects, but also due to the terrain slope. These complicating factors may well be common features of the SBL following transition. Nevertheles the SBL turbulence data show remarkable order when plotted in appropriate coordinates. The structure in properties of various parameters is in good agreement with the Kansas surface layer data. The extension of these properties through the remainder of the SBL, points to a steady drop in intensity of all parameters as their peak wavelengths approach to the layer depth, h. Reasonable as these trends seem, it is not possible to be sure that the observed structure is really that of an equilibrium SBL. Observations over polar sites or in warm-air advection (when SBL's can exist long enough to approach equilibrium) are needed to verify the Minnesota results. Similarly, there is a need for detailed studies of disturbed stable conditions.

The investigation of turbulence structure and the diffusive action of cloudy boundary layers in which radiative processes, can play a major role is still a topic requiring urgent attention. A few studies of stratocumulus have been reported (Coulman, 1978 and Caughey et al., 1981) and the contrast with the daytime CBL high lighted. The studies need to be extended to cover boundary layers with various types and amounts of low cloud. Other topics, such as the transport characteristics of water vapor and the influence of topography on boundary layer structure, also await detailed experimental study.

5. DIFFUSION IN THE CONVECTIVE BOUNDARY LAYER

R.G. Lamb

U.S.E.P.A., Research Triangle Park, N.C. 27711, U.S.A.

5.1. INTRODUCTION

In this chapter we will consider the behavior of foreign material released into fluids bounded below by a horizontal, uniform surface whose temperature is greater than that of the fluid. In this state, thermal energy acquired by fluid in contact with the surface results in a decrease in the density of this fluid; and therefore, in the presence of a gravitational field, potential energy is acquired which puts the fluid system as a whole in a state of unstable equilibrium (hence the common terminology unstably stratified fluid).

If the fluid is initially at rest and no forces are present that would subsequently induce large-scale horizontal motion, then under conditions first derived by Lord Raleigh (see, for example, the recent review of convection by Normand and Pomeau, 1977), small perturbations can initiate the transformation of the fluid's potential energy into kinetic energy. Such motions are called free convection. In the corresponding situation where the fluid is driven horizontally by some external force, the motions are called forced convection. The latter state is by far the most prevalent in the atmosphere; but as we shall see later, atmospheric convection is very often similar in its characteristics to free convection. This lecture will focus exclusively on those situations in which this similarity exists (see also Section 1.8).

Generally speaking, the forced convective state of the atmosphere is associated with strong winds and weak surface heating. Within this state there exists the well-known horizontal roll convection regime that is often identified by long, parallel bands of clouds at the top of the boundary layer. Studies of the Eulerian properties of motion in this state have been conducted, for example, by Lemone (1973), but neither observational nor numerical studies of diffusion have been performed.

We will also restrict attention to situations in which atmospheric convection is confined to shallow layers near the ground, i.e., 1-2 km deep, by the stabilizing action of synoptic scale subsidence, associated with anticyclones. When subsidence is weak or absent or when there is mean upward motion, deep convection can occur, resulting possibly in precipitation. This state is of concern in regional scale-dispersion studies but we will not consider it here.

Before going into details, let us look briefly at the philosophy and recent history of diffusion modeling.

Two of the basic premises underlying applied science are that physical systems obey physical laws; and that if proper initial and boundary descriptions of a given system are specified, the mathematical equations that describe those laws can be solved to predict the behavior of that system under given conditions. The objective is the utilization of knowledge of the physical world to solve problems of concern to man. The problem might be weather forecasting, designing supersonic aircraft, or mundane tasks like predicting the concentration of material that will result from the release of some pollutant into the atmosphere.

Before the advent of computers, advances in the applied sciences were greatly impeded by the inability to express the solutions of the governing equations in terms of known functions. Progress was made principally in only two areas: those problems in which the governing equations reduce to forms simple enough to solve analytically, and those problems in which similitude exists between the given system and a scale model that could be analyzed in the laboratory.

Examples of the latter category in the field of turbulent

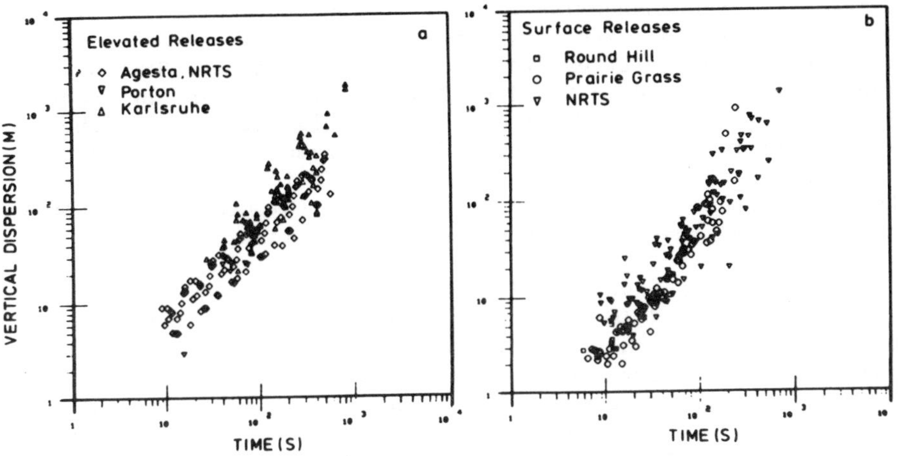

Figure 5.1. Estimates of vertical plume spread from atmospheric difussion experiments in unstable conditions. (a) Elevated sources. (b) Surface sources (Irwin, 1983).

diffusion are the wind tunnel and laboratory studies of material dispersion. In Chapter 6 some of these experiments are described and some of the laboratory studies of free-convective diffusion will be discussed later in this chapter. Also within the empirical solution category are the expressions drawn from many atmospheric diffusion experiments. Irwin (1983) has performed an exhaustive study of all of the currently available data and was kind enough to furnish me with the summaries presented in Figure 5.1. (Some data sets, such as that collected at Brookhaven and reported by Singer and Smith (1966), no longer exist.) These values are for unstable conditions only and were derived from the Gaussian plume formula using the measured values of wind speed, lateral plume spread σ_y and maximum ground-level concentration. Included in the surface release data shown in Figure 5.1(b) are measurements from the Round Hill, Prairie Grass and National Reactor Testing Station (NRTS) studies. The Prairie Grass data form the principal basis of the well known Pasquill-Gifford σ_y and σ_z expressions and in this role they have been the mainstay of atmospheric diffusion estimates for many years. The elevated release data shown in Figure 5.1(a) include the Agesta (release height = 50 m), Porton (150 m), NRTS (50 m) and Karlsruhe (100 m) measurements.

In the category of theoretical advances in turbulent diffusion, there are the many results drawn from the classical K-theory diffusion equation, whose analytic solution is known for a number of functional forms of the wind and diffusivity. Also included are the so-called similarity expressions for variables like the mean elevation of diffusing particles that have been derived for free-convective diffusion, for the constant-flux layer, and for other restricted regions of parameter space where solutions can be inferred from dimensional analyses. Some of the results are reviewed in Sections 7.4 and 7.5 but also by Pasquill (1974, 1976, 1978) and Weber (1976). I would rather treat these as a well established body of knowledge, representing the theoretical accomplishments of the precomputer era, and focus this lecture on the new methods that are emerging with the computer age and on the knowledge that these new techniques are providing.

As computers have come into being and their capabilities have grown, work devoted in analytical mathematics to solving differential equations has shifted to the problem of formulating analogs of these equations that a digital computer can solve directly. The progress made within the last 15 years in the development of techniques for solving non-linear partial differential equations of the type encountered in fluid mechanics has fostered renewed interest in turbulence and, in particular, intensified attacks on its one remaining mathematical obstacle: the closure problem, which has been extensively discussed in Chapters 1 and 3. In these chapters several closure relationships were developed.

It turns out, as one might expect, that the severity of the errors incurred by a poor choice of closure relationships is dependent upon the type of averaging that has been performed on the governing equations. The worst cases to handle are ensemble averages, which were discussed in Section 3.2. In practice, they consist of averages over the entire time axis in statistically-stationary systems, and averages over one or more space axes in homogeneous flows. The difficulty in these situations is that the closure relationships must embody the effects of the <u>entire</u> spectrum of velocity, temperature and concentration fluctuations. Since the largest-scale perturbations are affected by the 'geometry' of the system, a closure scheme that works well in one situation may produce deplorable results in another. Donaldson (1973) illustrates the results of an application of his model to the calculation of ensemble mean turbulence parameters in a stationary, horizontally homogeneous, unstable atmospheric boundary layer. While some of the predictions are quite good, e.g., $\sigma_w(z)$, others, particularly the vertical heat flux, bear almost no resemblance to the corresponding measurements.

In the early 60s when computers were still relatively slow, it was generally impractical to treat three- and four-dimensional problems. In those days there was little choice but to 'integrate out' time and/or space dimensions and to face the problems we have just cited in dealing with the accompanying closure problem. By the late 60s computer speeds had increased sufficiently that, for the first time, four-dimensional simulations were within the realm of feasibility. During this period a pioneering effort was begun by Deardorff to utilize numerical solutions of the Navier-Stokes equations to simulate high Reynolds-number fluid flows. In his first study, Deardorff (1970a) simulated turbulent, neutrally stratified channel flow. In this case the governing equations were averaged over a finite volume of space-time, surrounding each grid point in the simulated domain (this approach was introduced in Section 3.3 as volume-averaged models). In Deardorff's calculation the spatial dimensions of the averaging interval were of the order of 1/20 the channel depth h and the averaging period was 0.0033 h/u_*, where u_* is the friction velocity. With the equations averaged in this way, the dominant 'energy containing' scales of motion were handled explicitly in the equations, rather than implicitly in the closure relationships. The solutions derived from the resulting time-dependent three-dimensional system of equations exhibited the unsteady, chaotic motion characteristic of turbulent flows. The forcing functions were held constant, so ensemble means were approximated by averaging the solutions at selected spatial points over a sufficiently large number of time steps.

In subsequent studies Deardorff solved numerically the systems of equations that describe the dry, neutrally and unstable stratified planetary boundary layer. In one of these studies, reported in detail in 1972 and preliminarily in 1970, Deardorff made the important discovery that if the depth h of the convectively mixed layer is much larger than the magnitude of the Obukhov length L, which is a rough measure of the depth of fluid in which turbulence is being generated by shear stress, the turbulence observed above an altitude z of about 0.1h resembles that observed in free convection, where there is no mean wind at all. Specifically, Deardorff found that when $-h/L > 10$, turbulence in the upper 9/10 of the mixed layer scales with h and the convective velocity scale w_*, defined in (2.54). This type of scaling, which is called mixed-layer scaling was discussed in Section 2.6. Support for this conclusion was later presented by Deardorff and Willis (1973) from analyses of both laboratory and atmospheric measurements of turbulence; and support has continued to accumulate in the years since then (see Section 4.2). These particular confirmations of the similarity of free convection and atmospheric turbulence when $-h/L > 10$, apply only to Eulerian, or fixed point, measurements of velocity, temperature and the like. However, it would seem natural that similarity should also exist in the Lagrangian, or fluid particle reference frame, statistics manifest in the dispersion of passive, non-buoyant material released above the shear layer, i.e., $z^s > -L$, or from surface releases after a sufficient travel time. Proceeding under this assumption, Deardorff and Willis (1975) studied diffusion phenomena from a simulated ground-level cross-wind line source in a water tank model of the convective mixed layer. An important result of this study was that beginning at a travel time of about 0.5 h/w_*, the centerline of a ground-level plume begins to rise rather rapidly until by a travel time of the order of h/w_*, it has reached an elevation of about 0.75h.

This finding was so grossly out of line with the conventional Gaussian model of diffusion that speculation was raised, e.g., by Pasquill (1978), that inasmuch as the ascent of the centerline of the surface-source plume is a manifestation of free convection, it is inapplicable to atmospheric diffusion except under very low wind-speed conditions. An estimate of the limiting wind speed can be derived from the definition of w_* (2.54) and the criterion for similarity, $-h/L > 10$, given earlier. For the case $u_* = 0.05U$ we get

$$U < 6w_*. \tag{5.1}$$

Since in the atmosphere w_* is often larger than 1 m s^{-1}, one would expect from (5.1) that the Deardorff-Willis diffusion phenomenon is not a rarity. Indeed, the upturn in the σ_z data displayed in Figure 5.1(b) after a travel time of about 200 s is consistent with the 'lift-off' of the ground-level plume observed in the laboratory. (Recall that the σ_z data of Figure 5.1 are based on measurements of U, σ_y and ground-level concentration.) In Section 5.3 we will discuss observational evidence that support the validity of Deardorff's free-convective scaling hypothesis and also the relevance of the tank diffusion results to atmospheric dispersion.

The Deardorff-Willis laboratory study stimulated a great deal of interest in numerically modeling convective diffusion. Apparently the first effort along this line was that of Lewellen and Teske (1975) who used Donaldson's second-order closure model. In a simulation of free-convective diffusion of material from a continuous ground-level line source, their model predicted the 'lift-off' phenomenon but it was not nearly as pronounced as that observed by Deardorff and Willis in the laboratory. Since the Lewellen-Teske-Donaldson model predicts steady-state ensemble mean values directly, it is subject to the entire range of problems and uncertainties, mentioned earlier, that accompany the closure problem.

The only way to mitigate these difficulties is to apply the governing equations to local averages over 'volumes' of space-time much smaller than the dominant scales of motion, as discussed above. Wengle (1979) has applied this approach to the mass continuity equation of an inert nonbuoyant chemical species, i.e.,

$$\frac{\partial \underset{\sim}{c}}{\partial t} + \frac{\partial(\underset{\sim}{u}_i \underset{\sim}{c})}{\partial x_i} = S, \tag{5.2}$$

to predict local space-time averaged concentration $\{C(x_i,t)\}$, defined by equation (3.7) including an integration over a time interval, i.e.,

$$\frac{\partial\{C\}}{\partial t} + \{U_i\}\frac{\partial\{C\}}{\partial x_i} + \frac{\partial}{\partial x_i}\{u_i'c'\} = S, \qquad (5.3)$$

where primed variables denote deviation from the local, space-time average. By contrast, corresponding to (5.3), Lewellen and Teske treated the equation

$$U_i\frac{\partial C}{\partial x_i} + \frac{\partial}{\partial x_i}\overline{u_i c} = S \; . \qquad (5.4)$$

Here, U_i, C and the overbar denote ensemble averages. The variables u_i and c represent deviations from the local ensemble mean, e.g., $c = \underset{\sim}{c} - C$, (while $c' = \underset{\sim}{c} - \{C\}$).

Wengle chose an identical averaging volume to that used by Deardorff (1974) in his model so that he could use archived velocity output of that model as the source of $\{U_i\}$ (Equation (5.3)). Lewellen and Teske obtained U_i by solving a form of the ensemble averaged momentum equation. Wengle approximated the product mean values $\{u'_i c'\}$ that appear in (5.3) by a simple gradient transfer relationship (see Equation (3.96)) and solved the resulting Equation (5.3) using a pseudospectral technique. Lewellen and Teske used Donaldson's second-order closure method to approximate $\overline{u_i c}$, which results in a set of 18 equations, and solved the system using conventional finite-difference techniques.

The point we made earlier regarding the relationship between the severity of the closure problem and the nature of the averaging performed on the equations, can be illustrated here. Wengle used the simple gradient transfer hypothesis to 'close' (5.3) and it appears that he got good results. However, the same hypothesis applied to (5.4) in the context of convective diffusion would lead to solutions C that are unacceptably poor replicas of the true mean concentration field. Later we will see why this failure occurs. In fact, as we shall also see, there is reason to believe that even the higher-order closure approximation that Lewellen and Teske used in (5.4), is itself deficient in applications to convective diffusion. An excellent discussion of the problems and uncertainties associated with the ensemble averaged and volume averaged models was presented in Section 3.3.

Diffusion models that start from the mass continuity equation (5.2) are called Eulerian models. The closure problem that arises when this equation is averaged, requires the imposition of approximations that affect the velocity and concentration fields jointly, as in the term $\{u'_i c'\}$ in (5.3). Consequently, some closure schemes that work reasonably well for area source emissions have been found to break down altogether when applied near a point source. A simple example of this is that when the gradient transfer approximation is applied to $\{u'_i c'\}$ near a point source, the diffusivity parameter must be a function of travel time from the source rather than a local property of space. In principle, we can circumvent the constraints that closure methods place on the concentration field, and at the same time avoid the operational problems associated with the numerical solution of differential equations, by working with the Lagrangian counterpart of (5.2). This is the approach that the present author has taken for the last several years. In the next section we will outline this and then, in Section 5.3, apply it to convective diffusion.

5.2. FORMULATION OF A LAGRANGIAN DIFFUSION MODEL

Starting from the definition of concentration

$$\underset{\sim}{c}(x_i,t) \equiv m/\delta v, \tag{5.5}$$

where m is the number of material particles in the sample volume δv centered at x_i at time t, it can be shown (Lamb, 1975, 1980) that the ensemble mean concentration is given by

$$C(x_i,t) = \int \int_0^t p(x_i,t;x_i',t')S(x_i',t')\, dt'dx_i' \tag{5.6}$$

and the concentration variance

$$\overline{c(x_i,t)c(x_i,t')} = \iint \int_0^t \int_0^{t'} p_2(x_i,t,x_i,t';x_i'',t'',x_i''',t''') S(x_i'',t'')\, S(x_i''',t''')\, dt''dt'''dx_i''dx_i''' \tag{5.7}$$

where S is the source-strength function. In order to understand the meanings of p and p_2, it is necessary first to understand what is meant by an ensemble.

Physical laws such as the equations of motion (momentum, entropy and mass conservation), delineate regions of function space that contain all functions that describe a particular physical system. Let us denote the set of all functions that satisfy the equations that govern fluid velocity $\underset{\sim}{u}_i(x_i,t)$ by G. Specifying the physical properties of a fluid (density, viscosity, etc.) along with its velocity distribution $\underset{\sim}{u}_i(x_i,t_0)$ at some

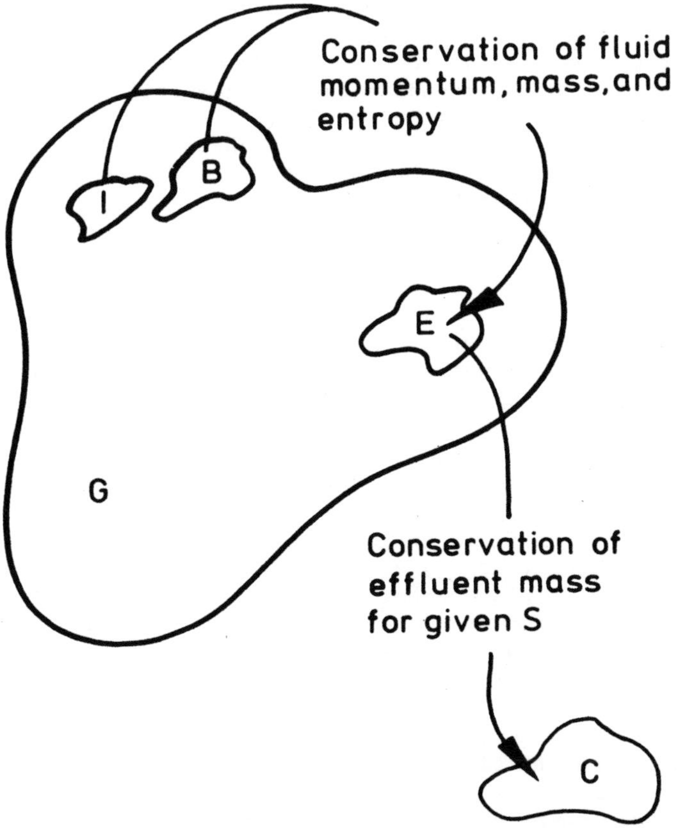

Figure 5.2. Schematic illustration of the function sets that correspond to a discrete specification of the initial state of a fluid, I; the boundary conditions on the fluid, B; and the set E of velocity functions $\underset{\sim}{u}_i(x_i,t)$ in to which the points in (I,B) map through the equations representing physical law. Set C represents the mapping of points in E into concentration space for a given distribution $S(x_i,t)$ of material sources.

instant t_0, as well as the locations and characteristics of its boundaries, and so on, effects the selection from G of a single function $\underset{\sim}{u}_i(x_i,t)$ that describes the velocity field in that fluid under the conditions specified. Since our ability to quantify the state of a natural system is finite (we can distribute only a finite number of probes throughout it) we can never specify precisely the state of a system such as the atmosphere. As a result, for any description that we give based on a discrete set of observations there exists an innumerable set of continuous functions each, of which is consistent with that finite set of observations. Let I denote the set of such functions that correspond to a particular (discrete) observation of the initial velocity of a given fluid and let B represent the corresponding set associated with the boundary condition specifications. Then, for each combination of a function in I, i.e., $\underset{\sim}{u}_i(x_i,t_o)$, and a function in B, there is a one-to-one correspondence through the equations of motion to a function in G. This subset of functions in G represents the ensemble of flow fields $\underset{\sim}{u}_i(x_i,t)$ associated with the initial and boundary condition specifications and we will refer to it as E. Figure 5.2 illustrates the relationships among G, I, B and E in schematic form.

Perhaps an example will help establish the concept a little better. Suppose that a scientist in Laboratory A and a scientist in Laboratory B each have 'identical' closed chambers in which they plan to study convection. (Quote marks here and below signify that the description is true within the available power of observation.) Each scientist places a velocity probe at the same relative location $(x_i)_o$, say in his chamber and then waits for several days for the fluid to come to a 'state of rest'. Subsequently, he switches on a powerful lamp to illuminate the blackened bottom of his chamber and velocity recording is begun with time reckoned from the instant the lamp is lighted. After an hour or so of recording, each experiment is terminated. Later the scientists get together to compare records.

Even if we ignore possible differences in the lamps used in the studies, our intuition tells us that the velocity records produced in these experiments would differ in a 'random' way for all times after some initial period. Since it is axiomatically assumed that fluid motion obeys certain physical laws - whether those laws are expressed precisely by the Navier-Stokes equations is another matter - and that the motions are unique functions of the initial and boundary conditions, then the only way that the velocity records in our two hypothetical experiments can differ is for the initial states of the fluids in the two chambers to have differed, due to the presence of unresolvable perturbations, and/or for the boundary characteristics to have differed, due to microscopic differences in the chamber surfaces, dimensions, etc. If each scientist repeats the experiment, even if the boundary state in each chamber is unchanged, the initial states will differ from the original one and two entirely different velocity records will result. In fact, no matter how many times the experiments are performed, we can expect that the velocity records obtained will differ from one experiment to another and from one chamber to another, because the function sets I and B are infinite. Thus, it is pointless to discuss the individual velocity fields $\underset{\sim}{u}_i$ within E. Only averaged properties of this ensemble of velocities are meaningful. For example, the ensemble mean velocity $U_i((x_i)_o,t_o)$

is the vector obtained by evaluating all members of E at the point $((x_i)_0,t_0)$ and averaging the results. Similarly, the joint moment $\overline{\underset{\sim}{u}_{i1}\ \underset{\sim}{u}_{i2}}$ (the indices 1 and 2 denote $(x_i,t)_1$ and $(x_i,t)_2$) is obtained by the following process: take a function $\underset{\sim}{u}_i$ from E; evaluate it at $(x_i,t)_1$ and $(x_i,t)_2$; form the dot product of the two vectors; repeat the process for all $\underset{\sim}{u}_i$ in E and average the dot products obtained each time.

The essential point here is that we can specify the state of a fluid system only to the extent that we can delineate a region of function space whose members include the precise state itself. The boundaries of this set of functions are dependent on the type and quantity of conditions that we impose as constraints and only averaged properties of the functions in the set are useful.

With these points established we can now define the functions p and p_2 that enter in (5.6) and (5.7), respectively. The former is defined for chemically inert material by

$$p(x_i,t;x_i',t') = \overline{\phi(x_i,t;x_i',t')}/\delta v, \qquad (5.8)$$

where δv denotes the (small) sample volume defined in (5.5) and $\phi = 1$ for any $\underset{\sim}{u}_i$ in E if $\underset{\sim}{u}_i$ is such that a particle released at (x_i',t') is in δv centered at x_i at time t and $\phi = 0$ otherwise. A similar defintion relates p_2 and ϕ_2. Since each $\underset{\sim}{u}_i$ in E maps onto a point in concentration space through the mass continuity equation (5.2), for the case of zero initial concentration and no material absorption at the boundaries, both being specifiable with virtual certainty, there exists an ensemble C, of concentration distributions corresponding to E for each emissions function S. This is also illustrated in Figure 5.2. Equations (5.6) and (5.7) express the mean values over C, i.e., ensemble concentration statistics, in terms of properties of E and the specified emission distribution S. Implicit in our use here of the term 'ensemble average' is the assumption that each point in I and B is equally probable. Indeed, as we have just pointed out, it is hard to imagine precisely the same velocity field $\underset{\sim}{u}_i(x_i,t)$ occurring twice, regardless of the number of experiments performed.

In this lecture we are concerned with the unstable planetary boundary layer. In this case the ensemble E is the set of velocities associated with the inital and boundary condition sets

I = 'fluid at rest'; (5.9a)

B = horizontally 'infinite, uniform' surface; constant kinematic heat flux at the surface; subsidence inversion that maintains a constant mixed-layer depth h in the steady-state. (5.9b)

The forcing function is a horizontal pressure gradient that, in the steady-state, maintains an average flow speed U in the mixed layer and a negative mean vertical velocity W adequate to hold the mixed-layer depth h constant against the erosion produced by convective turbulence. In practice, we shall often assume that these solutions also apply to the quasi-stationary, convective boundary layer, when the time change of the boundary-layer height, dh/dt, is much smaller than the convective velocity scale w_*.

We want to use the Lagrangian Equations (5.6) and (5.7) to determine the concentration ensemble properties of this situation. For this purpose we first need a description of the velocity ensemble E.

It turns out that with horizontally uniform boundary conditions, as specified above in the definition of B, and a uniform horizontal pressure field, as we assume in the governing equations, the information that we need about the mean properties of the set E of velocity fields can be extracted (approximately) from a single member $\underset{\sim}{u}_i(x_i,t)$ of the set. The properties of E that can be derived from any one of its members and the precise conditions under which the equivalence of properties holds, are set forth in the ergodic theory of statistics, which we will not discuss here.

Our immediate goal is to exploit this relationship to derive an estimate of the probability density p, defined by (5.8), from a single record $\underset{\sim}{u}_i(x_i,t)$ generated by Deardorff's (1974) model under the conditions specified in (5.9) and with a steady, uniform horizontal pressure field. In other words, we assume that this model-generated velocity field, after modifications that we will discuss later, is a member of the set E defined above.

Due to steadiness and the horizontal homogeneity of (5.9) and the pressure field, the probability density function p must have the form

$$p(x_i,t;x_i',t') = f(x - x',y - y',z,z',t - t'). \qquad (5.10)$$

Let us approximate the right-hand side of this expression in terms of a single-velocity record, given at M points in each horizontal plane by:

$$f(\xi,\eta,z,z',\tau) \simeq$$

$$\frac{1}{M\delta v} \sum_{m=1}^{M} \hat{\phi}(x_m+ \xi,y_m+ \eta,z,\tau;x_m,y_m,z',0) \equiv \frac{1}{M\delta v} \Sigma\, \hat{\phi}_m \qquad (5.11)$$

where $\hat{\phi}_m$, in analogy with (5.8), equals 1 if in the given velocity field a particle released at (x_m, y_m, z') at time $t = 0$ is in δv centered at $(x_m + \xi, y_m + \eta, z)$ at time $t = \tau$, and equals 0 otherwise. Combining (5.10) and (5.11) and comparing the result with (5.8), we see that (5.11) is equivalent to

$$\overline{\phi} \simeq \frac{1}{M} \sum_{m=1}^{M} \hat{\phi}_m \tag{5.12}$$

where the $\hat{\phi}_m$, $m = 1,2,\ldots,M$ are defined for a <u>single</u> velocity field. We need a quantitative measure of how the accuracy of (5.12) depends on the locations (x_m, y_m, z') of the particle release points that define $\hat{\phi}_m$ and on the total number of M of functions $\hat{\phi}$ on which the estimate of $\overline{\phi}$ is based.

Let

$$\varepsilon = \overline{\phi} - \frac{1}{M} \sum_{m=1}^{M} \hat{\phi}_m . \tag{5.13}$$

Keeping in mind that a set $\hat{\phi}_m$, $m = 1,2,\ldots,M$ exists for each velocity field in E and that the overbar denotes averaging over E, we obtain from (5.13)

$$\overline{\varepsilon} = \overline{\phi} - \frac{1}{M} \sum_{m=1}^{M} \overline{\hat{\phi}}_m . \tag{5.14}$$

Now $\overline{\hat{\phi}}_m = \overline{\hat{\phi}(x_m + \xi, y_m + \eta, z, \tau; x_m, y_m, z', 0)}$; but due to the assumed horizontal homogeneity, the right-hand side of this expression must be independent of (x_m, y_m). Thus,

$$\overline{\hat{\phi}}_m = \overline{\phi}\ [= f(\xi, \eta, z, z', \tau)\ \delta v] \tag{5.15}$$

and hence

$$\overline{\varepsilon} = 0 . \tag{5.16}$$

This result shows that the estimate (5.12) is 'unbiased', or that the 'expected' error ε is zero.

Squaring (5.13) and averaging we find the expected mean square error:

$$\overline{\varepsilon^2} = \frac{1}{M^2} \sum_{j=1}^{M} \sum_{k=1}^{M} \overline{\hat{\phi}_j \hat{\phi}_k} - \overline{\phi}^2 . \tag{5.17}$$

Writing the covariance in the form $\overline{\hat{\phi}_j\hat{\phi}_{j+m}}$, we have by virtue of horizontal homogeneity

$$\overline{\hat{\phi}_j\hat{\phi}_{j+m}} = R_m \tag{5.18}$$

which does not depend on j. Making use of this relationship in (5.17), we obtain after some manipulations

$$\overline{\varepsilon^2} = \frac{\overline{\phi^2}}{M} + \frac{2}{M^2}\sum_{m=1}^{M-1}(M - m)R_m - \overline{\phi}^2. \tag{5.19}$$

In implementing (5.12) it is convenient to let the particle release points form a square grid on the release plane z = z'. Suppose that these M points are spread over a square, horizontal region of width L. The area of each cell of the grid of release points is therefore

$$a = L^2/M. \tag{5.20}$$

In order to estimate R_m in (5.19), let us assume that there is perfect correlation in quantities measured at two points on a given plane z' if the separation distance of those points is less than the Eulerian integral space scale Λ at the level z'. Expressing this assumption in terms of (5.20), we have

$$\begin{aligned} R_m &= \overline{\phi^2}, && \text{if } m < \left(\frac{\Lambda}{L}\right)^2 M \\ R_m &= \overline{\phi}^2, && \text{otherwise.} \end{aligned} \tag{5.21}$$

Substituting this approximation into (5.19) and dividing the result by $\overline{\phi}^2$ to obtain the mean square fractional error, we get

$$\frac{\overline{\varepsilon^2}}{\overline{\phi}^2} = \frac{\overline{\phi^2}}{M\,\overline{\phi}^2}\left[\mu\left(2 - \frac{\mu + 1}{M}\right)+1\right] + \left[\frac{(M - \mu - 1)}{M}\left(1 - \frac{\mu}{M}\right) - 1\right] \tag{5.22}$$

where

$$\mu \equiv (\Lambda/L)^2 M . \tag{5.23}$$

Kaimal et al. (1976) found from observations that, in a convective boundary layer of depth h, the integral length scale $\Lambda \sim \lambda_m \simeq 1.5h$ for elevations higher than about z = 0.1h. The velocity record from Deardorff's model that we will use to compute $\overline{\phi}$ covers a spatial domain L ≃ 5h. Since we are using a 'top hat' approximation of the correlation function in (5.21), an effective value for Λ for use in computing μ might be $\Lambda \sim h$. The eddies that are responsible for the correlation of the $\hat{\phi}_m$ initially dissipate in time so that as travel time t increases, the effective value of Λ, and hence also μ, decreases to zero. Based on evidence that we present later, we believe that the time scale of the large eddy decay is about $2h/w_*$. Consequently, we have

$$\begin{aligned} \mu &\simeq M/25 \ , \quad t = 0 \ ; \\ \mu &= 0 \qquad\ \ , \quad t > 2h/w_* \end{aligned} \tag{5.24}$$

In order to estimate $\overline{\phi^2}/\overline{\phi}^{-2}$, let us assume that for any travel time t a particle is equally likely to be found anywhere inside a volume V given by

$$V = \sigma^2 \sigma_z \tag{5.25}$$

where σ_z and $\sigma = \sigma_x = \sigma_y$ are the root mean square particle displacements at time t. We want the dimensions of the sample volume δv to be small fractions of the corresponding dimensions of V so that we can resolve the structure of plumes. In particular, we want

$$\delta v = \xi^3 \sigma^2 \sigma_z \tag{5.26}$$

where, preferably, $\xi < 1$. It now follows from the definition (5.8) of ϕ that

$$\overline{\phi^2}/\overline{\phi}^{-2} = \xi^{-3} \qquad \text{(3-D)} \tag{5.27}$$

$$\overline{\phi^2}/\overline{\phi}^{-2} = \xi^{-2} \qquad \text{(2-D)} \tag{5.28}$$

Most of our interest later will be with cross-wind integrated or two-dimensional plumes, for which (5.28) applies.

Finally, on combining (5.22), (5.23), (5.24) and (5.28) we obtain the following limiting error estimates with M = 1225 and $\xi = 0.2$:

$$\frac{(\overline{\varepsilon^2})^{\frac{1}{2}}}{\overline{\phi}} \simeq \begin{cases} 140\%, & t = 0; \\ 15\%, & t > 2h/w_*. \end{cases} \tag{5.29}$$

The large initial error is not troublesome partly because (5.29) is only an estimate and partly because small travel times correspond to points immediately near the source where generally we have no interest. Keep in mind that these error estimates pertain to cross-wind integrated values of $\overline{\phi}$. If we attempt to resolve the three-dimensional plume, (5.27) applies and the errors in the results are much larger.

The calculation of the $\hat{\phi}_m$ requires only a description of the trajectory of each of the M particles released. Let $X_i^m(t)$ denote the vector position at time t of the particle released at (x_m, y_m, z'), i.e.,

$$X_i^m(t) = X_i(x_m, y_m, z', t). \tag{5.30}$$

By definition

$$\frac{d}{dt} X_i^m = \hat{u}_i(X_i^m(t), t) + u_i'(X_i^m(t), t), \tag{5.31}$$

where $\hat{u}_i$ represents the velocity distribution given by Deardorff's model and u'_i represents the velocity perturbation due to all scales of motion (or eddies) smaller than that resolvable by Deardorff's model.

The process by which we determine u'_i is described in detail in Lamb (1981a) so I will only outline it here. The basic idea is to delineate a set of functions, in a manner identical to that discussed earlier in connection with Figure 5.2, whose set mean values are consistent with the known mean values of u'_i and then to select members from this set at random for each of the M release points. For example, Deardorff's model provides a value of the local mean square of u'_i. We know also that u'_i is approximately isotropic, that it is comprised of velocity fluctuations in the inertial subrange, and we have also a rough idea of the form of its temporal autocorrelation and integral time scale. All of this information does not specify u'_i uniquely but it must be preserved by functional forms that we use to represent u'_i in the particle trajectory calculations. This is accomplished in the computer modeling by using for u'_i functions generated by the algorithm

$$u_i'(X_i, t) = \alpha u_i'(t - 2\Delta t) + \beta u_i'(t - \Delta t) + \gamma \rho_i \tag{5.32}$$

where ρ_i is a computer-generated isotropic random vector with zero mean and variance

$$\overline{\rho_i^2} = \frac{2}{3} E, \qquad i = x, y, \text{ or } z \tag{5.33}$$

where $E = \frac{1}{2}\overline{(u'^2 + v'^2 + w'^2)}^{\frac{1}{2}}$ is the mean subgrid scale kinetic energy given by Deardorff's model at the particle's position x_i at time t. The pseudo-constants α, β and γ in (5.32) must satisfy the following two constraints:

$$\alpha^2 + \beta^2 + \gamma^2 + \frac{2\alpha\beta^2}{1 - \alpha} = 1 \ , \qquad (5.34)$$

$$T_L'/\Delta t = \frac{(1 - \alpha + \alpha\beta)}{(1 - \alpha)(1 - \alpha - \beta)} \ , \qquad (5.35)$$

where T_L' is the integral time scale of u'_i. Since u'_i has a bandwidth limited spectrum, we would expect (but are not certain) that its integral time scale is zero. The smallest value of T_L' that can be achieved with the algorithm (5.32) is $\Delta t/2$. Thus, one criterion for selecting the (α, β) pair is that it satisfies (5.35) with $T_L' = \Delta t/2$. Another constraint is that the values chosen for α and β yield an auto-correlation function $\overline{u_\alpha'(t)u_\alpha'(t + t')}$ that is consistent with the bandwidth limited spectrum of u_i'. A final condition is that (α,β) satisfy (5.34) for real values of γ. (See Lamb (1981) for further discussion of these points.)

The results that we show in the next section were computed using an early version of the u'_i generator in which α is effectively zero. The value of β was chosen assuming that $T_L \simeq \Delta/E^{\frac{1}{2}}$, where $\Delta = (\Delta x \ \Delta y \ \Delta z)^{1/3} \simeq 90$ m is the scale of the grid cells used in Deardorff's model. I believe now that the proper value of T_L' for the subgrid scale turbulence should be close to zero, for reasons cited above, and thus the plume simulations should be redone. The largest changes would probably occur near the ground, where the subgrid turbulence plays a dominant role in the diffusion process.

Having outlined the procedure by which the density function p required in Equation (5.6) for the ensemble mean concentration C can be derived from a single velocity record, we turn in the next section to a description of the results we have obtained for diffusion from point sources, $S = Q(t)\delta(x_i - x_i^s)$ in Equation (5.6), in the convective boundary layer described qualitatively by (5.9).

5.3. NUMERICAL SIMULATIONS OF NON-BUOYANT MATERIAL DIFFUSION AND COMPARISONS WITH OBSERVATIONS

We wish to emphasize at the outset that the results presented in this section pertain to non-buoyant material only. In other words, they represent the Lagrangian statistics of turbulence in the convective planetary boundary layer. The addition of buoyancy to the dispersing material results in radically-different mean concentration distributions, as we shall discuss later.

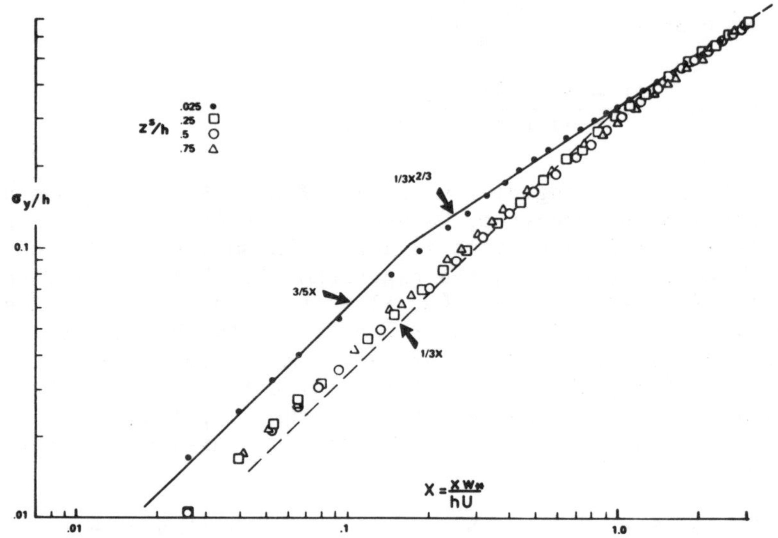

Figure 5.3. Calculated root-mean-square cross-wind spread of particles given by Equation (5.36) from sources of heights $z^s = 0.025$ h (solid circles), 0.25 h (squares), 0.5 h (circles) and 0.75 h (triangles) as a function of non-dimensional travel time, $X = xw_*/(hU)$, where x/U is the travel time, t. The solid and dashed lines are functional approximations of the data discussed in Section 5.5 (Lamb, 1979).

We want to stress too that we will consider here only the quasi-free convective boundary layer, $-h/L > 10$, and source heights $z^s > \|L\|$.

Let us look first at the results of the simulations before comparing them with available data.

Results

A great deal of information about characteristics of diffusion is revealed by the travel-time dependence of the mean square vertical and lateral spreads of particles released from an instantaneous point source. These variables are defined by

$$\sigma_z^2(t;z^s) = \int_0^\infty \int_{-\infty}^\infty \int_{-\infty}^\infty (z - z^s)^2 \, p(x,y,z,t;0\ 0,z^s,0) \, dxdydz$$

$$(5.36)$$

$$\sigma_y^2(t;z^s) = \int_0^\infty \int_{-\infty}^\infty \int_{-\infty}^\infty y^2 \, p(x,y,z,t;0,0,z^s,0) \, dxdydz$$

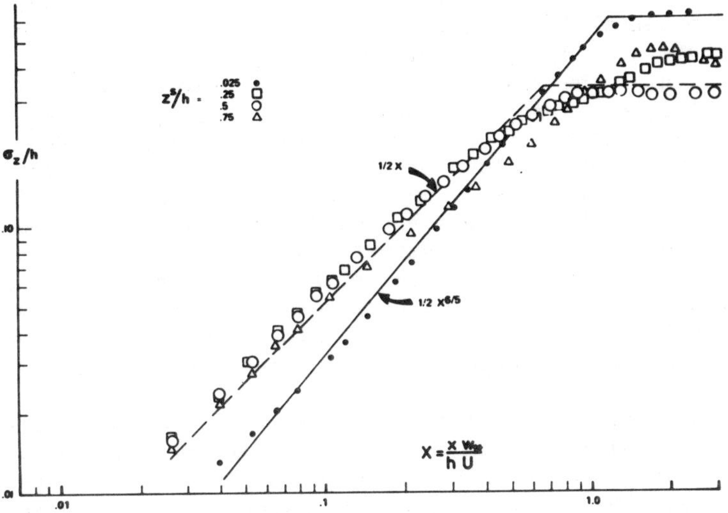

Figure 5.4: Calculated root-mean-square vertical particle spread given by Equation (5.36) as a fuction of downstream distance. The notation is the same as in Figure 5.3 (Lamb, 1979).

where the release point $x_1{}^s = (0,0,z^s)$ and the release time $t_0 = 0$. These equations assume implicitly that the mean wind is directed along the x-axis and they also utilize the assumed stationarity and horizontal homogeneity of the velocity fields (cf, 5.10).

Figure 5.3 shows the simulated values of the lateral spread σ_y; the corresponding values of σ_z are presented in Figure 5.4. In each figure data for the near-surface release ($z^s = 0.025h$) are depicted by closed circles, and results for the elevated releases $z^s = 0.25, 0.5, 0.75h$ are represented by open squares, circles, and triangles, respectively. Also, in both figures all variables have been nondimensionalized using the mixed layer depth h as a length scale and h/w_* as a time scale. As we noted earlier, Deardorff (1970, 1972) showed that in boundary layers in which $-h/L > 10$ the statistical properties of turbulence above $z \simeq 0.1h$ acquire universal forms when scaled by h and w_* (see also chapters 2 and 4). Further evidence of this will be found in the next section when we compare the numerical results presented here with measurements made by Willis and Deardorff in the laboratory.

The profiles σ_y shown in Figure 5.3 reveal the existence of two dispersion regimes, one for surface sources and one for elevated releases. We have not determined what source height z^s marks the transition between these two regimes but it probably lies near $z^2 = 0.1$ h. The model results show that the lateral spread of material is completely independent of source height for sources above $z^s \simeq 0.2h$, i.e., $\sigma_y^2(t;z^s) = \sigma_y^2(t)$, for $z^s > 0.2h$. The results show that for travel times t less than 0.1 h/w_*, $\sigma_y^2 \sim t^2$, as expected, but that σ_y^2 is larger for a surface release than for an elevated one. The latter result is consistent with the findings of a numerical turbulence simulation reported by Deardorff (1973) and a laboratory study of Willis and Deardorff (1974), both of which showed that the mean-square-lateral-turbulent velocity fluctuations decrease with height near the ground. By contrast, Panofsky et al. (1977) presented data that show no systematic change of lateral turbulence energy with height (see Equation (4.15)). Further investigations are required to understand the discrepancies among these experiments.

The two dispersion regimes revealed by the σ_y^2 data given in Figure 5.3 are also evident in the σ_z^2 profiles displayed in Figure 5.4. The model results indicate that for small travel times ($t < 0.1$ h/w_*) the vertical spread of material from an elevated source is considerably larger than that of particles released near the ground. This is a direct consequence of the known increase of the mean square vertical turbulent velocity fluctuations σ_w with height in an unstable boundary layer (See, for example, Figure 4.18(a) and Equation (4.14)). The data in Figure 5.4 also show that at travel time t of about $2/3(h/w_*)$, σ_z^2 for a surface release, surpasses that of material from an elevated source. This is a manifestation of the 'sweep-out' phenomenon first observed by Willis and Deardorff (1976). We will discuss this phenomenon later.

The σ_z^2 profiles shown in Figure 5.4 for the elevated releases do not exhibit the high degree of z^s invariance that the σ_y values possess. This is undoubtedly due to the effects of the ground and the elevated inversion on the vertical structure of turbulence in the mixed layer. For example, we see in Figure 5.4 that at a travel time of approximately 1/5 h/w_*, the rate of increase of σ_z for material released at height $z^s = 0.75$ h becomes appreciably smaller than that of particles released at heights in the range $0.25 < z^s/h < 0.5$. This occurs because after a travel time of 1/5 h/w_*, a significant number of particles released from a height $z^s = 0.75$ h have reached the inversion base that caps the mixed layer and their upward motions have been stopped or reversed. From this time onward the rate of growth of σ_z is dominated by the rate at which particles are descending in downdrafts. If the vertical structure of turbulence were symmetrical about the midpoint of the mixed layer, one would expect the travel time dependence of σ_z to be the same for particles released from 0.75 h as for particles released from 0.25 h.The fact that σ_z increases more slowly for particles released at $z^s = 0.75$ h than from $z^s = 0.25$ h suggests an asymmetry

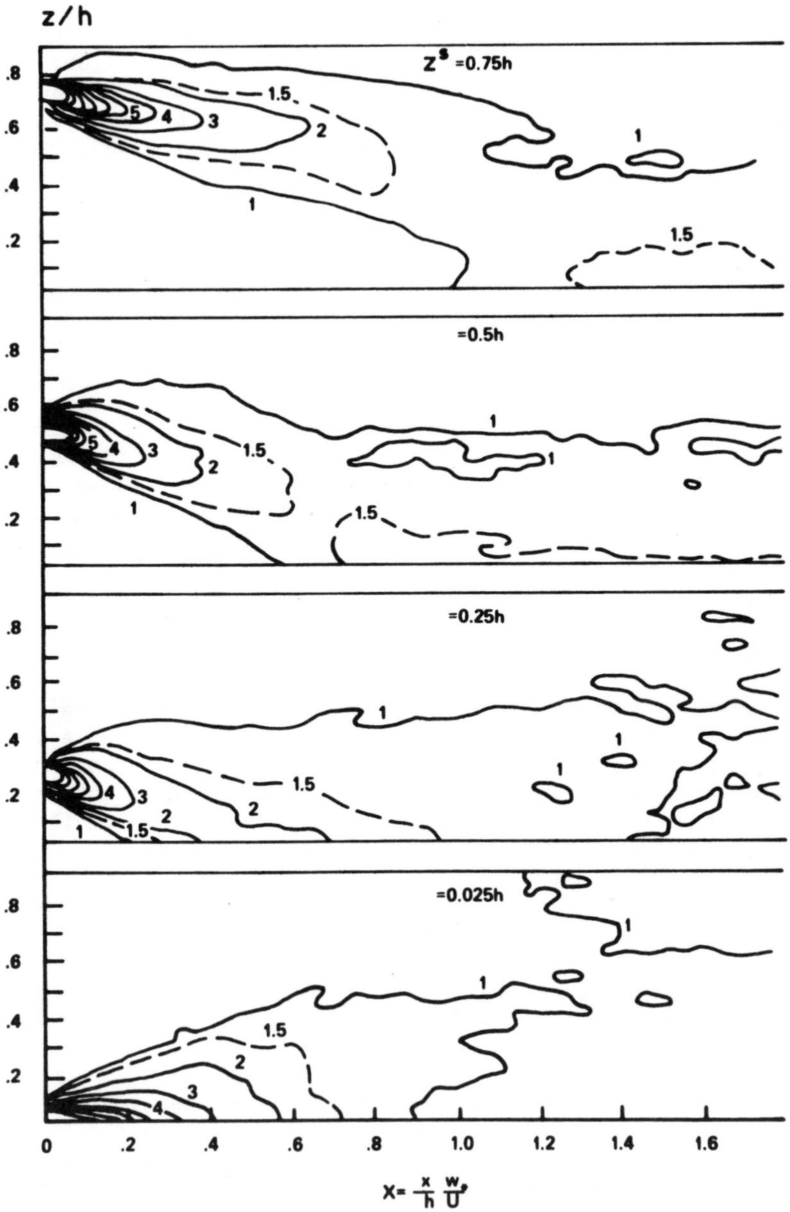

Figure 5.5. Calculated non-dimensional cross-wind integrated concentration $C^y(x,z)hU/S$ for point sources of heights z^s = 0.75 h, 0.5 h, 0.25 h and 0.025 h (Lamb, 1979).

of the vertical turbulence structure, with descending fluid moving more slowly on the average than ascending fluid. This is in fact the case as we shall discuss later.

The two dispersion regimes discussed above are quite evident in the distributions of mean material concentration. Figure 5.5 shows the calculated mean cross-wind integrated concentration C^y of non-buoyant material from continuous point sources of strength S particles per unit time of heights z^s = 0.025, 0.25, 0.5 and 0.75 h. The corresponding mean ground-level concentrations C(x,y,0) are shown in Figure 5.6. We should note here that the irregularity at large distances from the source, are the result of statistical fluctuations in the concentration estimates that arise from the finite size of the ensemble of particle trajectories on which our estimate of the kernel p is based. The magnitude of the variance of these fluctuations was estimated earlier (Equation (5.29)).

A characteristic of the elevated source diffusion regime revealed by the C^y profiles of Figure 5.5 is that beginning near the source, the plume centerline, as defined by the locus of maximum concentration, descends until it intercepts the ground. This phenomenon was first pointed out in Lamb (1978). (Incidentally this is an example in which the predictions of a theoretical model were later confirmed by observations.) In that paper it was argued that the descent of the plume centerline is due to the size, long life, and organized nature of downdrafts in a convective mixed layer. Except for shallow layers near the ground and the inversion base, downdrafts occupy more than half of the horizontal plane at each level of the mixed layer. (We will discuss this further in Section 5.4.) This phenomenon and the constraint of mass continuity results in upward fluid motions being faster on the average than sinking motions - hence the difference in the σ_z profiles for sources of heights z^s = 0.75 h and 0.25h described earlier. Because of the greater areal coverage of downdrafts, a majority of the material emitted by an elevated source descends immediately after release; and because of the long lives of the convective circulations, descending material descends for a relatively long period. The net result is that the greatest concentration of material is found along a descending path from the source.

By contrast, material released by a surface source can only ascend or move horizontally. Particles released into the base of an updraft begin their ascent immediately while those emitted into a downdraft move approximately horizontally until eventually they too are swept into updrafts and are carried upward. On the average, the lifetimes of the updraft circulations are long enough that particles that begin to ascend immediately after release are still rising when those that have lingered for a time in the

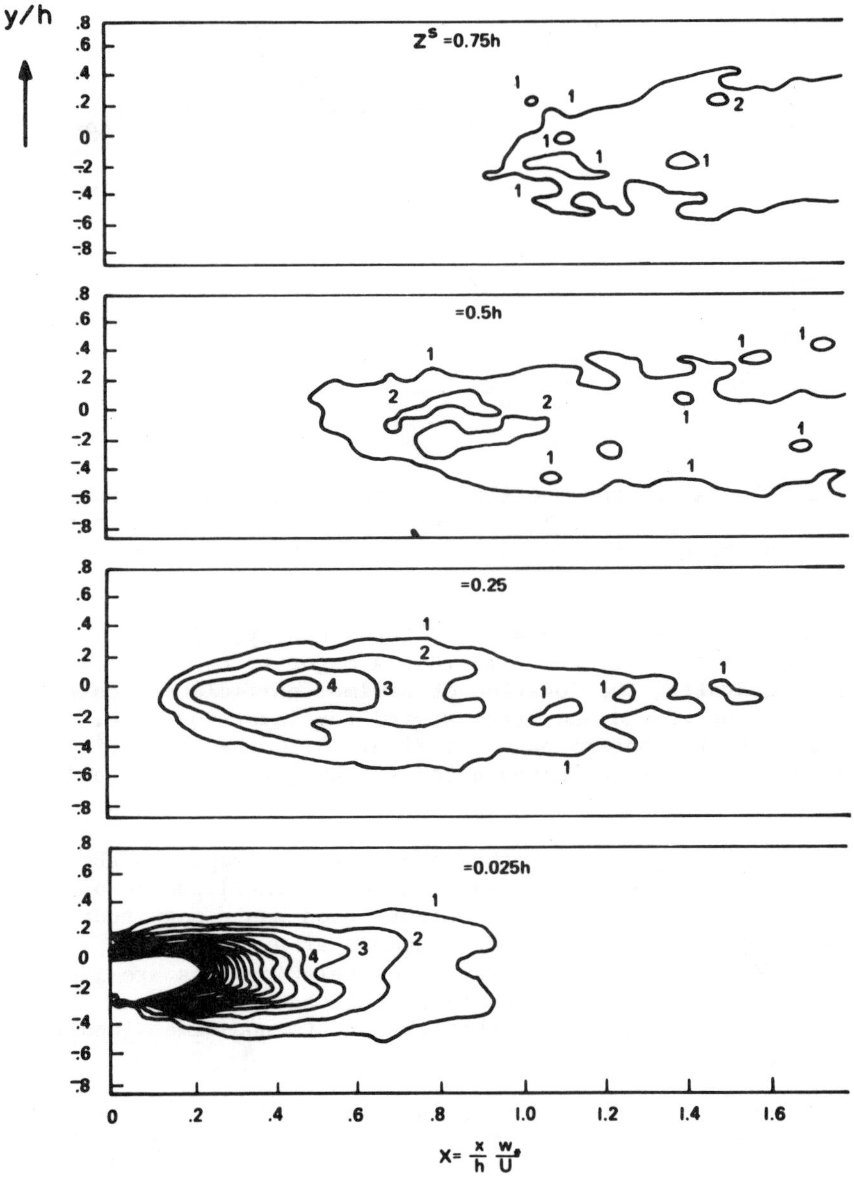

Figure 5.6. Calculated non-dimensional ground-level concentration $C(x,y,0)h^2U/S$ for point sources of heights z^s = 0.75 h, 0.5 h, 0.25 h and 0.025h (Lamb, 1979).

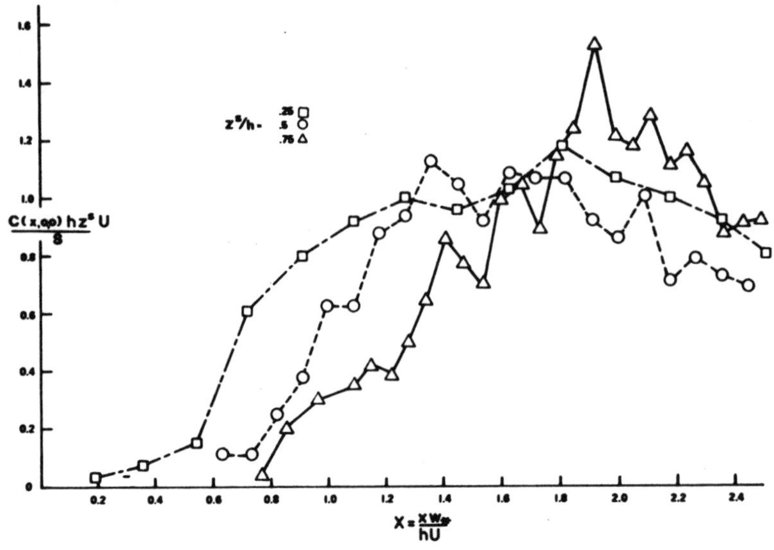

Figure 5.7. Calculated profiles of non-dimensional ground-level concentration $C(x,0,0)hz^sU/S$ along the plume centerline for sources of heights z^s = 0.75 h, 0.5 h and 0.25 h (Lamb, 1979).

surface layer begin their ascent. Consequently, after a sufficiently large travel time that a majority of particles have entered updrafts, the location of maximum particle concentration lifts off the ground and rises toward the inversion base. According to the bottom panel of Figure 5.5, the lift-off of the maximum concentration begins after a travel time of the order of h/w_*.

A close inspection of the concentration isopleths shown in Figures 5.5 and 5.6 reveals that for an elevated source of height z^s, the maximum ground-level concentration has a magnitude $C_{max}(x,0,0) \simeq 1.2S/(z^shU)$ and it occurs at a distance $x_{max} \simeq 2z^sU/w_*$ from the source. These relationships are implied by Figure 5.7 where we have used z^s rather than h to non-dimensionalize x, and z^sh rather than h^2 to non-dimensionalize $C_{max}(x,0,0)$.

It is noteworthy that C_{max} depends on z^s and h jointly. Using the Gaussian plume formula, Pasquill (1974, page 274) we obtain

$$C_{max} \sim \frac{S}{(z^s)^2\, U} \left(\frac{\sigma_z}{\sigma_y}\right) x_{max} . \tag{5.37}$$

This result and ours, are consistent if $\sigma_z/\sigma_y \simeq z^s/h$ at x_{max}. The following is a heuristic argument that this is in fact the case.

The maximum ground-level concentration occurs at the point x_{max} where material released into downdrafts first reaches the ground.The equation for x_{max} implies that the effective mean fluid speed within downdrafts is $-w_*/2$ (we corroborate this with additional data later). Thus, the travel time to the point of maximum ground-level concentration is $\sim 2z^s/w_*$. Since for small travel times $\sigma_z^2 \sim \sigma_w^2 t^2$ and since within downdrafts $\sigma_w \sim w_*$, we conclude that $\sigma_z \simeq z^s$ at x_{max}.

Now the lateral dimensions of the downdrafts that carry material to the surface are about 1 - 1.5 h (Kaimal et al., 1976; Willis and Deardorff, 1979). Since on the average fluid at the edges of downdrafts, near groundlevel, is entering updrafts rather than continuing to move horizontally, lateral spread σ_y during the time that material is confined to a single downdraft should be limited by the lateral dimensions of the downdraft and thereby be sensitive to h - hence $\sigma_y \simeq h$ at x_{max}.

The above arguments are not meant to be taken literally. Rather they are intended to provide only a rough physical explanation for the above expression of the maximum concentration.

The dependence of C_{max} on U rather than w_* is a result of the implicit condition $U > w_*$ in our simulations (see (5.38)). Obviously, in the free-convective limit $U \to 0$, C_{max} must become dependent on w_*. This case occurs only infrequently in the atmosphere and therefore we shall not consider it here.

I want to caution the reader most strongly about applying the above results to sources of buoyant material. As we shall see in Section 5.6 there is evidence that the addition of even a small buoyancy flux to effluents causes a radical change in the character of dispersion in convective conditions.

Comparison of the Model Results with Observations

Measurements of the three-dimensional distribution of mean concentration within a plume are so difficult and expensive to make in the atmosphere that none of the field studies cited in Section 5.1 that have been performed to date, contain all of the data needed to test the numerical model predictions thoroughly. The Prairie Grass data contain measurements that provide a partial check on the predictions of dispersion from a ground-level source. Nieuwstadt (1980) analyzed these data, using the free convective scaling principles, and compared them with the computer-model predictions discussed above for the low-level release ($z^s = 0.25$ h) and with the laboratory model results presented by

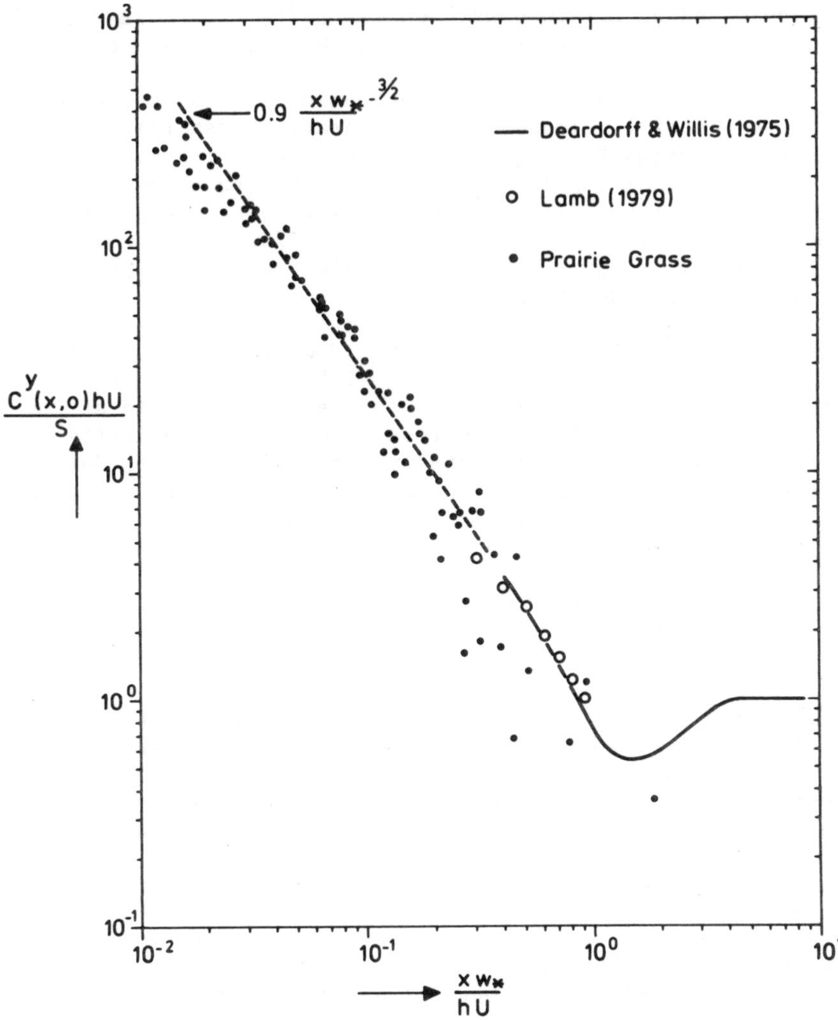

Figure 5.8. The dimensionless cross-wind integrated concentration at ground level as a function of the mixed-layer scaling variable xw_*/hU. The circles indicate predictions of the computer model described here for a source height of 0.025 h (Nieuwstadt, 1980).

Deardorff and Willis (1975) for a surface source. Nieuwstadt's findings are presented in Figures 5.8 and 5.9.

The first of these shows a departure at small travel distances of the measured cross-wind integrated concentrations from those predicted by the 3/2 power law of free-convective

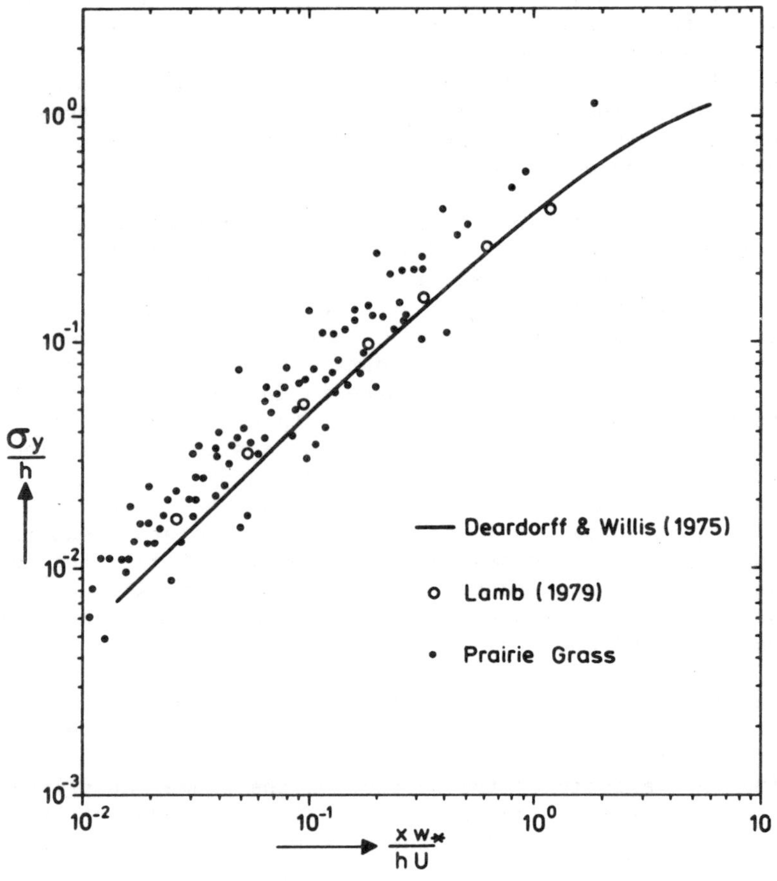

Figure 5.9. The dimensionless lateral dispersion coefficient as a function of the mixed-layer scaling variable xw_*/hU. Circles represent the numerical model predictions shown earlier in Figure 5.3 for a release height of 0.025 h (Nieuwstadt, 1980).

similarity theory. This discrepancy is not unexpected because the source in these experiments was at ground level and, therefore, the material released is initially under the control of shear, rather than buoyancy, generated turbulence. Figure 5.8 shows that the measured concentrations are in fairly good agreement with both the laboratory and the numerical model results. However, at travel times greater than about 0.3 h/w_*, the atmospheric ground-level concentrations appear to decrease slightly faster than those predicted by either the laboratory or numerical models. Briggs and McDonald (1978) noted the same predictions. The faster rate of decline in the measured concentrations could be caused by a loss of the tracer material in the field experiments due to surface uptake or other causes. Unfortunately, attempts to estimate the actual removal rate have not been successful (see Barad and Haugen, 1959). Nieuwstadt's comparison of the lateral plume spread profiles $\sigma_y(x)$, shown in Figure 5.9 reveal quite good agreement between the Prairie Grass data and the predictions of both the laboratory and numerical models.

The fact that the physical characteristics of plumes within two systems of such vastly different scales as the atmosphere and a laboratory chamber, should agree quantitatively when non-dimensionalized by h and w_* is proof of the correctness of the free convective scaling hypothesis. The supportive evidence that we have cited above, plus recent data analyses by Briggs (1980), leave little doubt that this hypothesis is a sound basis for atmospheric diffusion studies. An important consequence for computer modeling is that the results of a single simulation can be applied to a wide range of mixed-layer depths and surface heat flux conditions. Another important consequence is that one can determine quite easily and accurately from laboratory experiments certain dispersion information that is very difficult and costly to measure in the atmosphere. Indeed, laboratory data are currently the only basis upon which the elevated source model results presented above can be checked.

The first laboratory study of elevated-source diffusion in convective conditions was performed by Willis and Deardorff (1978) (referred to herein as WD) and it was motivated by interest in checking the validity of the descent of the plume centerline predicted by the numerical model. Recall that these authors had earlier found that the centerline of a ground-level source plume ascends.

In WD the mean, cross-wind integrated concentration C^y was determined using a method that is accurate only if the effects of streamwise diffusion on the concentration distribution is negligible. Deardorff and Willis (1976) have prevously estimated that insofar as point or line source plumes are concerned, streamwise diffusion effects are insignificant provided that

$$U/w_* > 1.2 \ . \qquad (5.38)$$

This criterion is barely satisfied in the problem simulated above numerically ($U/w_* \simeq 1.3$) and consequently the C^y field obtained in the laboratory simulation may not agree well with the numerical results presented in Figure 5.5. If the limit given by condition (5.37) is incorrect and should, in fact, be larger than 1.2 or if there are significant deficiencies in the numerical or laboratory models, then the C^y fields will disagree accordingly. In order to examine both of these possiblities, we have calculated the

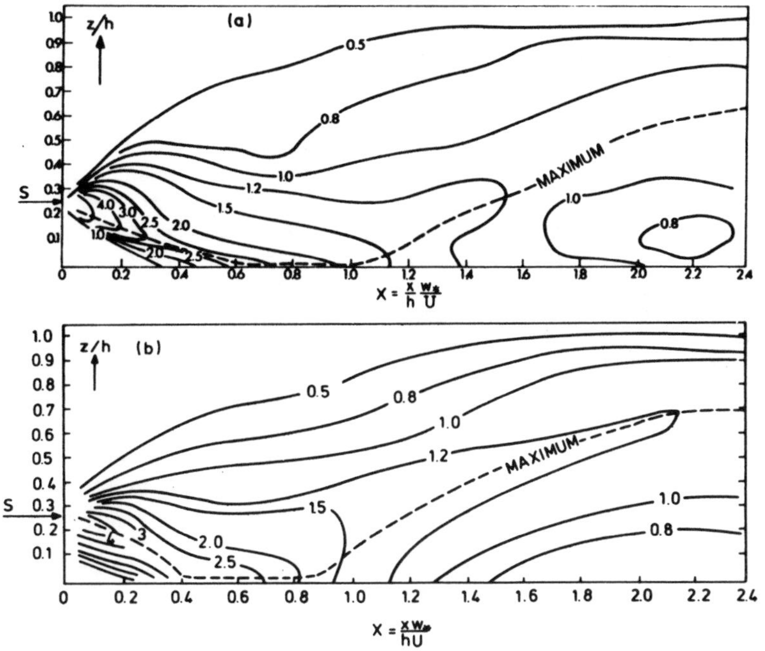

Figure 5.10: Comparison of concentrations distribution $C_z{}^y$ (see Equation 5.39)) predicted by the numerical model (a) (Lamb, 1978) with the corresponding field measured in the laboratory (b) (Willis and Deardorff, 1978).

quantity

$$C_z^y(X,Z) = \int_{-\infty}^{\infty} \int_{-\infty}^{\infty} p(x,y,Zh,Xh/w_*;0,0,z^s,0)\ dxdy \qquad (5.39)$$

where $X = xw_*/hU$ (see Section 2.6) and $Z = z/h$. This is the numerical counterpart of the function used by Willis and Deardorff to represent C^y. Thus, by comparing the function given by WD with the calculated $C_z{}^y$ field, we can obtain a measure of the inherent differences between the computer and laboratory models. And by comparing $C_z{}^y$ with the numerical C^y profile shown in Figure 5.5, we can test the validity of criterion (5.38).

The calculated values of $C_z{}^y$ are shown in Figure 5.10(a). On comparing these results with the C^y values given by WD, shown in Fig. 5.10(b), we find that overall the agreement is very good. The maximum ground-level values of C^y and $C_z{}^y$ are nearly the same, $\simeq$ 2.5, and they occur at approximately the same distance from the source, $x \simeq 0.6\ h/w_*$. However, the tongue of $C^y = 1.2$ extends to a higher altitude in the laboratory than in the numerical results and there is a tendency in the numerical simulation, but not in the laboratory model, for a secondary concentration maximum to occur at ground-level after the locus of the principal maximum has ascended. We believe that this last discrepancy is caused by the scheme we have used in the numerical model to parameterize the effects of subgrid scale velocity fluctuations. Recall from Section 5.2, see Equation (5.32), that we represent the subgrid velocity field by a random velocity component u'_i whose statistical properties are determined by the local subgrid scale turbulent energy content of the fluid. At ground-level the subgrid scale energy constitutes a large fraction of the total velocity variance (see Deardorff (1974)) and therefore the nature of the parameterization scheme is more critical in this region. Any one of several deficiencies in the scheme could cause a fictitious ground-level maximum to occur. One possibility is that the Lagrangian integral time scale behaves differently near a reflective boundary than we have assumed. Another possible cause is that the subgrid and resolvable scale velocity fields are coupled in a manner that the parameterization scheme does not adequately describe. Nevertheless, overall the numerical and laboratory models are in acceptably good agreement.

Comparing the C_z^y distribution of Figure 5.10(a) with the numerical simulated C^y profile shown in Figure 5.5 for the release height $z^s = 0.5h$, we can see evidence in at least three places of the effects of streamwise diffusion. The first is in the maximum ground-level concentration. The function C_z^y, which neglects streamwise diffusion, has a maximum ground-level value about 20% larger than that of C^y, shown in Figure 5.5. This result is not surprising because diffusion acts to destroy concentration gradients and thereby to lessen the maximum concentration values. A second effect of streamwise diffusion is apparent in the rate at which material first reaches ground-level. For example, the isopleth $C^y = 1$ intercepts the ground near $x = 0.25hU/w_*$ whereas the corresponding isopleth for C_z^y, in which streamwise diffusion is absent, reaches ground level farther downstream, near $x = 0.32hU/w_*$. A third area in which the effects of streamwise diffusion are apparent is in the downstream portion of the plume, where the elevated maximum of the C_z^y distribution occurs. The elevated maximum is quite distinct in C_z^y (Figure 5.10(a)) but not in C^y (Figure 5.5). Again, given the nature of the diffusion process this result is to be expected, especially since the elevated maximum in C_z^y is only slightly larger in amplitude than the cross-sectional mean value of C_z^y. Whether in reality the elevated maximum is entirely absent in the C^y field is uncertain, due to errors in the numerical simulation discussed earlier. (We should note here that the C^y field could not be calculated to as great a distance from the source as C_z^y because due to the effects

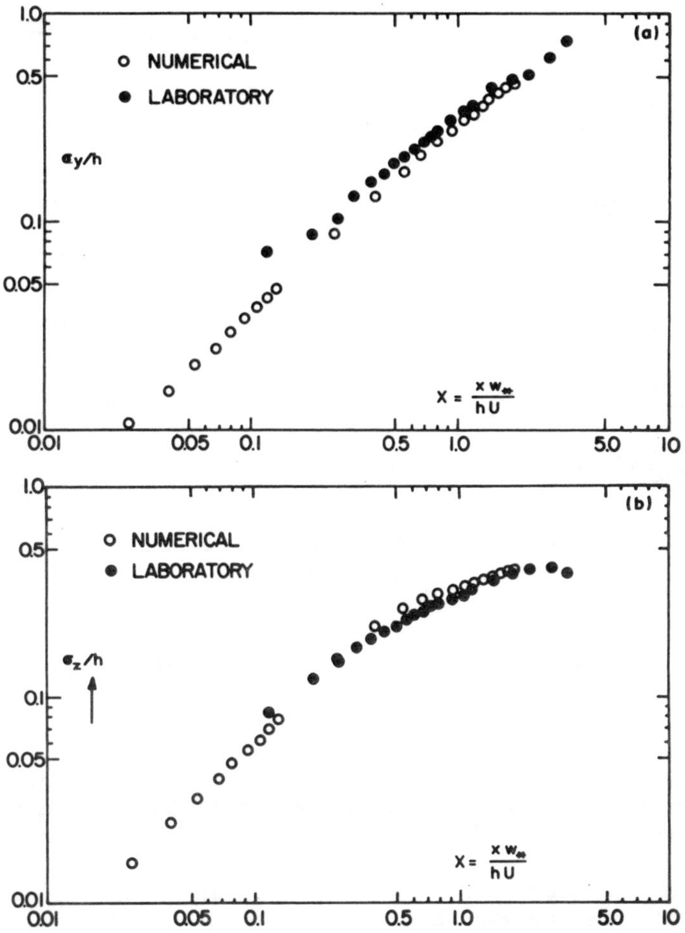

Figure 5.11. Comparison of the root-mean-square lateral (a) and vertical (b) displacement of particles from a source of elevation 0.25h as predicted by the numerical model and the laboratory model (Lamb, 1978).

of streamwise diffusion and the limited simulated particle travel time (~1450 s), steady-state values of C^y were unachievable at distances much beyond $x = 1.5\ hU/w_*$).

None of the discrepancies between C_z^y and C^y cited above can be considered large and we conclude from this that the results of the numerical simulation tend to confirm the validity of the Deardorff-Willis criterion, Equation (5.38), for neglecting streamwise diffusion.

The computed mean square lateral and vertical particle

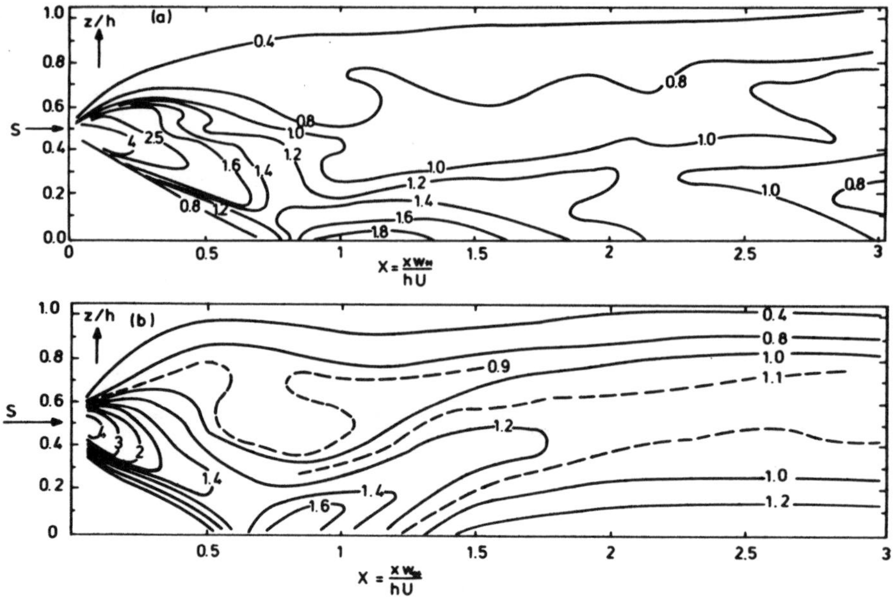

Figure 5.12. Distribution of C_z^y (see Equation (5.39)) predicted by the computer model (a) and simulated by the laboratory model (b) (Willis and Deardorff, 1981).

displacements for a source height $z^s = 0.25$ h are compared with the corresponding laboratory measurements of WD in Figure 5.11. The agreement in these dispersion parameters is excellent.

Willis and Deardorff (1981) recently investigated diffusion from a source in the middle ($z^s = 0.49$ h) of the mixed layer using their tank model. The distribution of cross-wind integrated concentration that they found, neglecting streamwise diffusion, is reproduced in Figure 5.12(b). The analogous field, namely $C_z{}^y$ given by Equation (5.39), predicted by the model is shown in Figure 5.12(a). The laboratory results in this case agree much better with the numerically generated $C_z{}^y$ field, which is essentially the cross-wind integrated plume concentration with streamwise diffusion 'turned off', than they do with the simulated field shown in Figure 5.5 in which streamwise diffusion is retained. On comparing the numerical results shown in Figure 5.12(a) with the C^y distribution shown in Figure 5.5 for $z^s = 0.5$ h, we see that both the intensity of the surface maximum in $C_z{}^y$ at the point of touchdown of the plume centerline and the subsequent 'rebound' of the plume centerline are apparently artifacts of the neglect of streamwise diffusion. Willis and Deardorff (1981) noted that the pronounced rebound of the plume centerline that they observed was partly due to the small positive buoyancy that their tracer droplets acquire with time, but they found no evidence that this was the sole cause of the rebound nor that the rebound was due entirely to the neglect of streamwise diffusion. Perhaps future studies will clarify the true cause of this phenomenon.

The secondary maximum at ground-level in the cross-wind integrated concentration fields $C_z{}^y$ shown in Figure 5.12 is a consequence of an important feature of the fluid motion in a convective boundary layer. Note first that the occurrence of the secondary maximum requires the existence of counter-gradient material fluxes, or equivalently, a negative vertical eddy diffusivity (which physically is not plausible). Apparently, the mechanism responsible for this phenomenon is the following.

As plume material from an elevated source descends in a downdraft, it begins to move horizontally as it approaches the ground rather than 'rebounding' as K-theory assumes. In particular, once material descends below an elevation of about $z = 0.1$ h, it moves horizontally on the average until it finds its way into an updraft (see Lamb, 1981b). The time material spends in the layer $z < 0.1$ h is long enough (about h/w_*) that it accumulates in this layer as a result of the steady influx of new material from above; hence the formation of the secondary maximum. Further insights into the characteristics of fluid motion in convective mixed layers can be gained from the laboratory study of Willis and Deardorff (1979).

Taking into consideration the enhancement of the plume centerline rebound attributable to the tracer buoyancy in the laboratory study, we conclude from the data presented in Figure 5.12 (and that presented earlier as well) that the numerical and laboratory models are in remarkably good agreement. This justifies the use of the numerical model to explore the characteristics of convective turbulence that are responsible for the diffusion phenomena we have found. We pursue this analysis in the next section.

5.4. THE STRUCTURE OF TURBULENCE IN THE CONVECTIVE BOUNDARY LAYER

On first inspection the plume structures exhibited in Figure 5.5 are quite mystifying. They indicate, for example, that at a distance of approximately hU/w_* downwind of two continuous point sources, one at an elevation of $z^s = 0.75$ h and the other directly beneath it at ground-level (viz $z^s = 0.025$ h), the axes of the averaged plumes from these two sources would cross forming an 'X' shaped pattern. This bizarre result is a manifestation of two important features of convective turbulence: the long Lagrangian integral time scale of the vertical velocity, and differences in

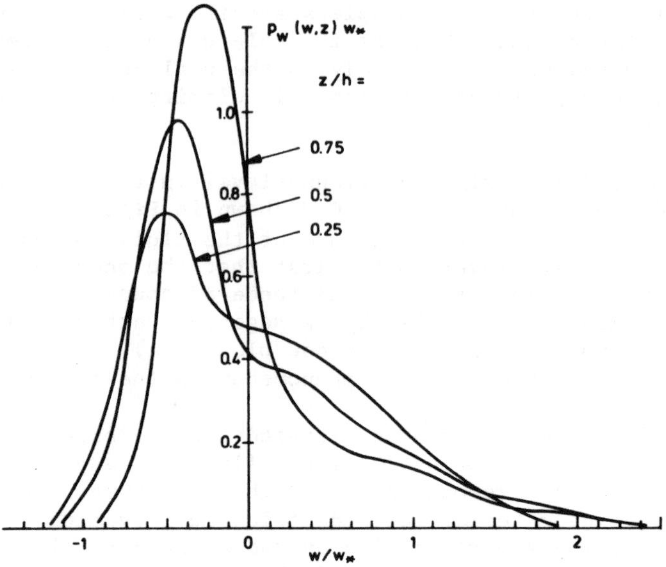

Figure 5.13. Probability density of vertical velocity at three levels of a convective mixed layer.

the vertical velocity's mean and mode (the most frequent value of the velocity). In neutral and stable boundary layers with horizontal flow that is non-divergent, the mean vertical velocity is zero and the mode is approximately the same. Under these conditions the axes of non-buoyant plumes are horizontal, as we are accustomed to envisioning them. (To be precise, the axes of plumes in these conditions are tilted downward slightly by the effects of wind shear, as dicussed in Chapter 6. Note, however, that in convective conditions there is virtually no shear above an elevation $z \simeq \|L\|$ and, therefore, the pronounced downward slope of the elevated plumes shown in Figure 5.5 is not attributable to wind shear.) In a convective boundary layer with non-divergent horizontal flow, the vertical velocity has a zero mean value but a strongly negative mode. This is evident in Figure 5.13 which was derived from data generated by Deardorff's numerical turbulence model.

One aspect of the probability density profiles displayed in Figure 5.13 that is immediately obvious, is that they are not Gaussian. At an elevation of 0.25 h the most probable vertical velocity (or mode) is $-w_*/2$, rather than zero; but as one ascends to higher elevations, the magnitude of the mode decreases. It has a value of approximately $-w_*/4$ at $z = 0.75$ h and at levels well inside the capping inversion, it becomes zero and there the density $p_W(w)$ acquires the Gaussian form.

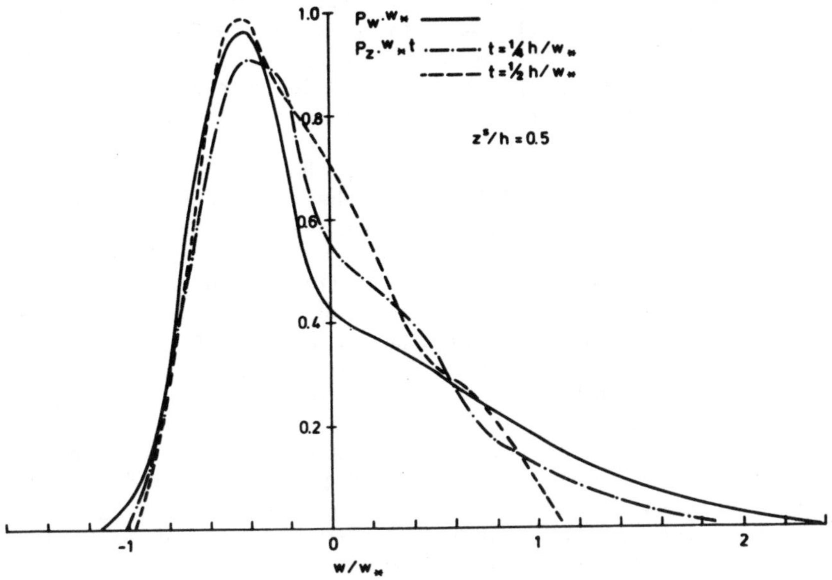

Figure 5.14. Plots of the probability density of the vertical velocity fluctuations at z = 0.5 h (from Figure 5.13) superposed on the vertical displacement density functions p_z of particles released at the same elevation.

Note also that the area under the $p_W(w)$ profile is larger in the negative half of the w axis than in the positive half. This indicates that within the convective boundary layer (except in shallow layers near the ground and the capping inversion base), sinking fluid occupies more area than rising fluid. The constraints of mass continuity requires, therefore, that ascending fluid has larger speeds on the average than sinking fluid. This is evident in Figure 5.13 in the positive skewness of the probability density profiles.

If non-buoyant particles are released into a turbulent fluid, the probability density of their initial speed is identical to that of the fluid velocity at the point of release. It follows that the probability density of vertical particle displacement for small travel time t satisfies

$$p_z(z',t;z^s) = p_w(\frac{z'}{t};z^s)\ t^{-1}\ , \quad \text{small } t \tag{5.40}$$

Here z' denotes displacement from the release level $z = z^s$.

One indicator of the magnitude of the Lagrangian integral time scale of w is the length of the interval of travel times over which (5.40) is satisfied. In Figure 5.14 we have copied from Figure 5.13 the probability density p_w at $z = 0.50$ h and superposed on it plots of the probability density of vertical particle displacement p_z given by the model for a $z^s = 0.5$ h release for two travel times: $t = 0.25$ and 0.50 h/w_*.

We see that at both travel times, the modes of p_z are nearly identical to that of p_w and that the shapes of the three density functions corresponding to vertical velocities more negative than the mode, i.e. $w < -0.4\ w_*$, are practically the same. These similarities indicate that the time scales of the dominate downward velocity fluctuations, or downdrafts, are larger than 0.5 h/w_*. In fact, the coherence of the centerline of the plume from the $z^s = 0.75$ h release shown in Figure 5.5 indicates that the time scale is at least 1.5 h/w_*, the travel time at which the plume centerline intercepts the ground. Incidentally, since the centerline is defined as the locus of maximum concentration, it is approximately the locus of the mode of p_z and, as we discussed earlier, reflects the mode values of the vertical velocity. Indeed, the slope of the centerline elevation $z_{c\ell}$ of each of the elevated source plumes shown in Figure 5.5 is roughly $dz_{c\ell}/dt \simeq -0.5\ w_*$, a value near the mode of p_w.

Deardorff (1970b) and Lamb (1981b) give evidence that the mean life times of updrafts and downdrafts is larger than 2 h/w_*. Since most particles released anywhere within the mixed layer will have either struck the ground or the inversion base within this time, for many practical purposes one can consider the integral time scale of vertical motion in convective turbulence to be infinite. Weil and Furth (1981) have adopted this assumption to develop a simple Monte Carlo model of diffusion in the mixed layer. The results of their calculations are remarkably close to those presented in the previous section.

In contrast to the persistence of downward motion, Figure 5.14 indicates that particles released from $z^s = 0.5$ h experience strong upward motions for only a short time. The differences in the p_z and p_w profiles suggest that on the average, particles that start upward with speeds in excess of 0.6 w_* are decelerating so that within a period of about 0.25 h/w_* their speeds have declined to values in the range $-$ 0.4 w_* to 0.6 w_*. The strongest upward velocities are associated with the core regions of updrafts which are constantly loosing kinetic energy to the ambient fluid. In addition, the fast-moving fluid in the interior of updrafts that is not detrained, encounters the inversion base within a short time period and, subsequently, its motion is reversed.

The long life times of updrafts and downdrafts make it very frequently possible for diffusing particles to traverse the entire depth of the mixed layer in one steady motion. This is

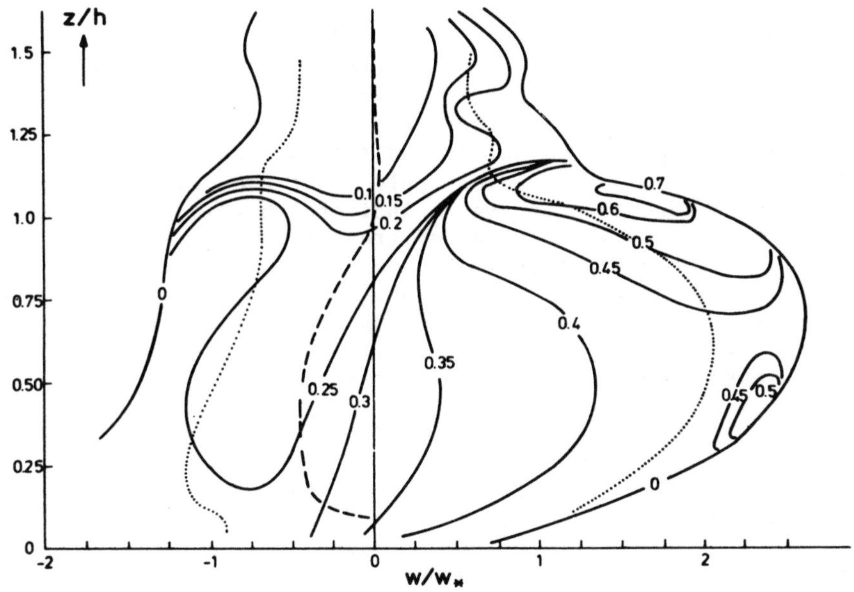

Figure 5.15. Average rms kinetic energy contained in all fluctuations smaller in scale than $\Delta = 0.08h$ as a function of vertical velocity (due to all perturbations larger than Δ) and elevation z. Dotted lines mark the one percentile w values and the dashed line is the locus of the mode of w. In the simulation from which these results were taken $w_* = 1.95 \text{ m s}^{-1}$.

incompatible with the tenets of Eulerian K-theory, in which turbulent diffusion is envisaged as a Brownian motion or random walk type process in which particles can traverse large distances only after suffering many upward and downward displacements. Lamb and Durran (1978) demonstrated this inherent discrepancy by showing, using the concentration fields presented in the previous Section, that the effective eddy diffusivity in convective conditions is strongly-dependent on the source height, rather than being a function solely of elevation as K-theory assumes. For this reason, dispersion parameters derived for convective conditions from K-theory principles are unreliable.

Even second order closure models are likely to yield erroneous predictions of concentration in the mixed layer, particularly those that assume a form for third order velocity moments, as discussed in Section 3.2, because as we showed earlier the skewness of the w distribution, manifest in $\overline{w^3}$, has a profound effect on the concentration distribution.

Figure 5.15 provides additional information on the structure of convective turbulence. Shown there is the combined rms kinetic energy contained in eddies smaller than about $\Delta = 90$ m, the grid scale of Deardorff's model, as a function of elevation z and the local vertical velocity w (due to all eddies larger than Δ). To assist the interpretation of these data, we have plotted the locus of the mode of p_W (dashed curve) and the one percentile points of p_W (dotted curves). At any level, only 1% of the fluid has a vertical speed less than the value indicated by the dotted curve on the left and only 1% has a speed greater than the dotted curve at the right. The data shown in Figure 5.15 were derived from Deardorff's model and, although quantitiatively, they are quite sensitive to the particular closure approximations that were used in the calculations, they provide qualitiative information that is useful in assessing the fate of dispersing material.

We note first that at all elevations, upward moving fluid contains more energy in small-scale fluctuations than sinking fluid. Part of this energy is undoubtedly generated by shear in the updraft jets, but a considerable portion is probably due to the upward transport of turbulence produced by shear in the surface layer, adjacent to the ground. Figure 5.15 shows a pronounced peak in small-scale energy in updrafts at the inversion base. This is largely attributable to turbulence generated by Kelvin-Helmholtz instability along the surfaces of updrafts where they have penetrated into the capping inversion and are spreading laterally (see also Section 4.2 on entrainment). The turbulent environment of the updrafts at the inversion base coupled with the rapid lateral spread of the updrafts themselves probably plays an important role in preventing buoyant material emissions from penetrating into the inversion above the mixed layer.

According the Figure 5.15 small-scale turbulence levels within downdrafts are low. This is expected, considering the stabilizing effects on the lapse rate induced by subsidence. The weak turbulence in downdrafts suggests that non-buoyant emissions will not mix rapidly with environmental fluid, and this can have an important consequence on ground-level concentrations of chemically reactive materials.

I have attempted in this Section to sketch the features of convective turbulence that are chiefly responsible for the dispersion phenomena described in Section 5.3. These features include the highly-skewed distribution of vertical velocity in convective turbulence, shown in Figure 5.13; the long mean life times of the updraft and downdraft circulations that constitute the bulk of convective motion - this is evident in the partial equivalence of p_W and p_Z shown in Figure 5.14; and the differences in the intensities of small-scale turbulence within updrafts and downdrafts that are effective in mixing the plume material with the ambient fluid. Additional descriptions of the convective boundary layer that are relevant to dispersion studies can be found in Lenschow and Stephens (1980, 1981), Lamb (1981b), and various works of Deardorff and Willis cited in the references.

5.5. FORMULAS FOR APPLICATIONS

In order to put the results of the model simulations presented earlier into a form that can be used easily in applied studies, we shall attempt in this section to develop a formula that will yield concentration estimates that agree as closely as possible with the values derived from the model simulations. I want to stress again that the formulas that will be presented here apply to non-buoyant emissions only. In Section 5.6 we will consider the modifications that are necessary to treat buoyant material.

In view of the popularity and wide spread use of the Gaussian formula in present day air pollution engineering work, it seems appropriate to incorporate our model simulation results into this formula rather than into one of an altogether different form. The standard Gaussian formula for the concentration distribution C is given by Equation (7.31). In this equation σ_y and σ_z are empirical 'dispersion coefficients' which are functions of atmospheric stability and distance x from the source. The ensemble average C is under steady conditions equal to the time average, when the averaging time is much greater than h/w_*. For smaller averaging times statistical fluctuations occur, which can be estimated with (5.7). This concentration variability is further discussed in Section 7.6.

By integrating (7.31) with respect to y over the entire y axis, we obtain the Gaussian expression for the cross-wind integrated concentration, namely

$$C^y(x,z) = \frac{Q}{\sqrt{2\pi}\,\sigma_z U}\left\{ \exp\left[-\frac{(z-H)^2}{2\sigma_z^2}\right] + \exp\left[-\frac{(z+H)^2}{2\sigma_z^2}\right]\right\} \qquad (5.41)$$

where H is the (effective) emission height, previously z^s. In seeking a suitable modification of Equation (5.41) we are guided by the single, most evident discrepancy between the concentration field given by this equation and that produced by the numerical simulations. Equation (5.41) predicts that the maximum concentration occurs at the source height H at all distances from the source (except at sufficient distances downstream where appreciable quantities of material have been reflected from the ground), whereas the model simulations show that the maximum concentration occurs along a descending path from elevated sources and an ascending path from surface releases.

In fact, the descent of the plume centerline is so pronounced that for sources higher than about 0.3h, no value of σ_z exists that makes the ground-level C^y values given by the Gaussian formula (5.41) agree with those predicted by the model. We are guided, too, by the finding that the mean square lateral particle spread and, to a lesser extent, the vertical spread (5.36) are invariant with respect to source height z^s when $z^s > 0.1h$. We should point out here that in stationary, homogeneous turbulence we would have that σ_y and σ_z are given by (5.36). Based on these obervations, we propose the following modifications of the plume formula, Equation (7.31):

(1) Replace σ_y^2 and σ_z^2 by the results following from (5.36). In particular

$H > 0.1\ h$

$$\frac{\sigma_y}{h} = \frac{1}{3} X \quad \text{for} \quad X < 1 \quad \text{and} \quad \frac{\sigma_y}{h} = \frac{1}{3} X^{2/3} \quad \text{for} \quad 1 < X < 3$$

$$\frac{\sigma_z}{h} = \frac{1}{2} X \quad \text{for} \quad X < \frac{2}{3} \quad \text{and} \quad \frac{\sigma_z}{h} = \frac{1}{3} \quad \text{for} \quad X > \frac{2}{3}$$

$H < 0.1\ h$ (5.42)

$$\frac{\sigma_y}{h} = \frac{3}{5} X \quad \text{for} \quad X < \frac{1}{6} \quad \text{and} \quad \frac{\sigma_y}{h} = \frac{1}{3} X^{2/3} \quad \text{for} \quad \frac{1}{6} < X < 3$$

$$\frac{\sigma_z}{H} = \frac{1}{2} X^{6/5} \quad \text{for} \quad X < \frac{6}{5} \quad \text{and} \quad \frac{\sigma_z}{h} = \frac{3}{5} \quad \text{for} \quad X > \frac{6}{5}$$

where $X = (xw_*)/(hU)$, as discussed in Section 2.6. The relationships given in Equations (5.42) are plotted in Figures 5.3 and 5.4, respectively. It is clear from Figure 5.4 that a better approximation exists for some of the elevated source σ_z data than the one we have chosen. In fact, for all source heights H except 1/3 h and 2/3 h the limiting value $\sigma_z = 1/3\ h$ given by (5.42) is inconsistent with the true value,

$$\lim_{x\to\infty} \sigma_z^2 = h^2/3 - hH + H^2.$$

Our rationale for choosing (5.42) is that it is the simplest expression that fits all the data reasonably well. Furthermore, one of our goals in modifying the plume formula is to keep the modifications as simple as possible so that the final formula will be easily usable in performing routine dispersion calculations.

(2) Our second proposed change to the plume formula is to replace the source height H by a virtual source height $\zeta(x;H)$ defined by

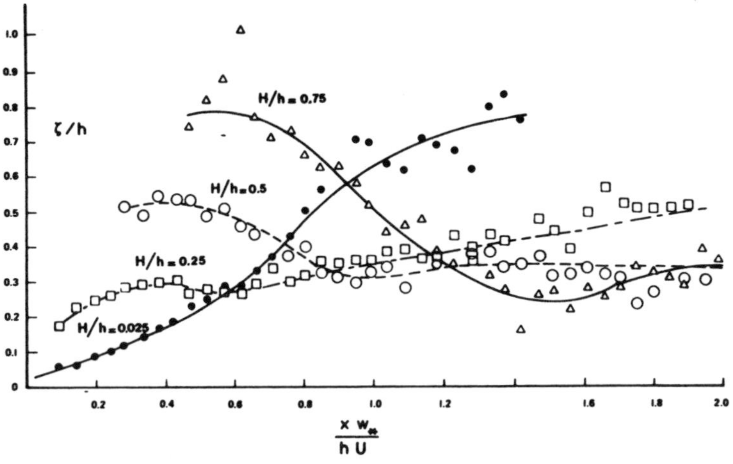

Figure 5.16. Virtual source height profiles $\zeta(x;H)$ determined from Equation (5.43) using the calculated plume centerline concentrations $C(x,0,0)$ and σ_y and σ_z formulations given by Equations (5.42).

$$\zeta(x;H) \equiv [-2\sigma_z^2 \ \ell n\{\pi\sigma_y\sigma_z C(x,o,o)U/Q\}]^{\frac{1}{2}} \qquad (5.43)$$

where σ_y and σ_z are obtained from (5.42) for the appropriate release height H. Note that the virtual source height ζ is just the value of H that an observer at the ground-level centerline of a point source plume would estimate from (7.31) using a local measurement of the mean concentration and σ values obtained from (5.42).

In the plume formula (5.41), the effective reflection of material at the inversion height is accounted for in the form (5.42) of σ_z/h.

Before considering how well the predictions with the Gaussian plume formula incorporating these modifications agree with the results of the numerical simulations, let us examine the characteristics of the virtual source height ζ. This parameter provides a means of comparing characteristics of the convective diffusion process with the classical Gaussian model.

The virtual source height profiles $\zeta(x;H)$ derived from the model results are presented in Figure 5.16 for each of the four release heights H treated earlier. One interesting feature of these profiles is that at the distance x where the plume first reaches ground level, the virtual source height ζ is approximately

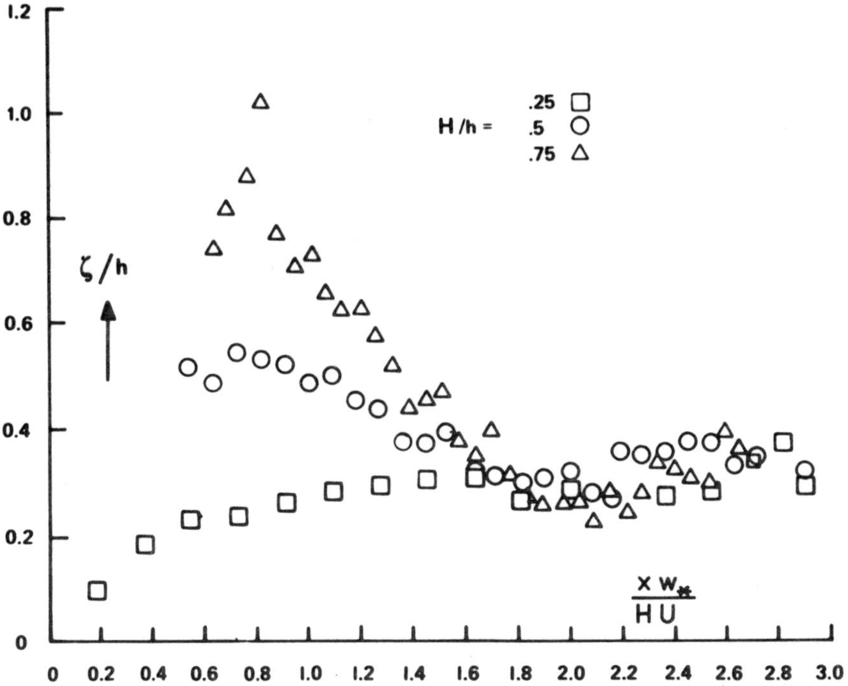

Figure 5.17. Virtual source height profiles $\zeta(x;H)$ for elevated sources ($H > 0.1h$) plotted as in Figure 5.16 but with distance x scaled by H rather than h (Lamb, 1979).

equal to the release height H. However, there is a tendency in all the elevated releases for ζ to be slightly smaller than H at the extreme upwind, ground-level edge of the plume. This indicates that material is not spread out as much at this point as it would be if it were distributed in a Gaussian manner. Particles at the ground-level upwind edge of the plume consist almost entirely of those that were emitted directly into the core regions of downdrafts where sinking motions are strongest. The relatively high concentrations of material in the upwind edge of the plume may reflect the low level of small-scale turbulence in the core areas of downdrafts indicated in Figure 5.15.

Just downstream of the point, where the plume first touches ground, the virtual source height attains a maximum value that is larger than H, indicating that material is spread over a larger area than in a Gaussian distribution. This point is so close to the source that the vast majority of the particles observed at ground level have spent their entire journey from the source in downdrafts. Thus, a possible explanation of the enhanced plume

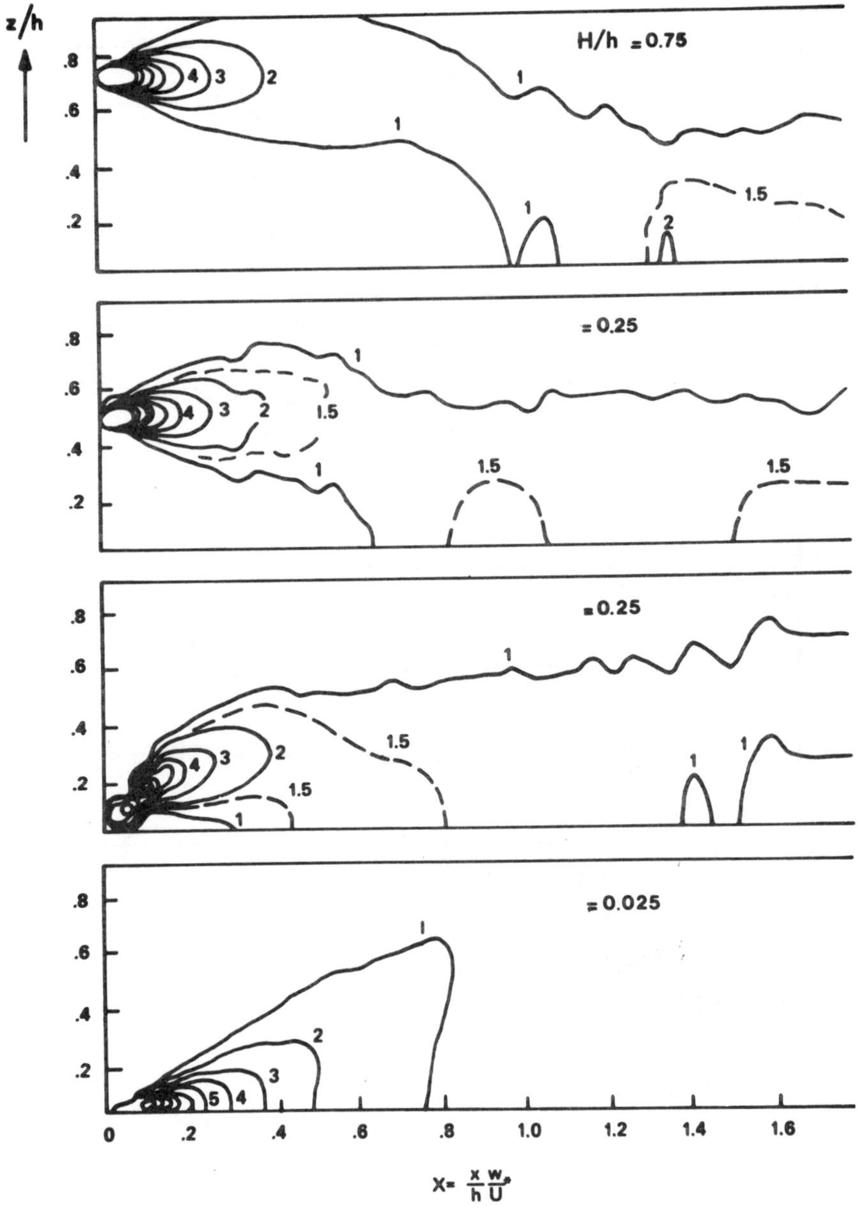

Figure 5.18. Nondimensional cross-wind integrated concentration $C^y(x,z)hU/Q$ given by the modified Gaussian plume formula for source heights $H(\equiv z^s)$ indicated. This figure can be compared with figure 5.5 to assess the accuracy of the modified Gaussian model (Lamb, 1979).

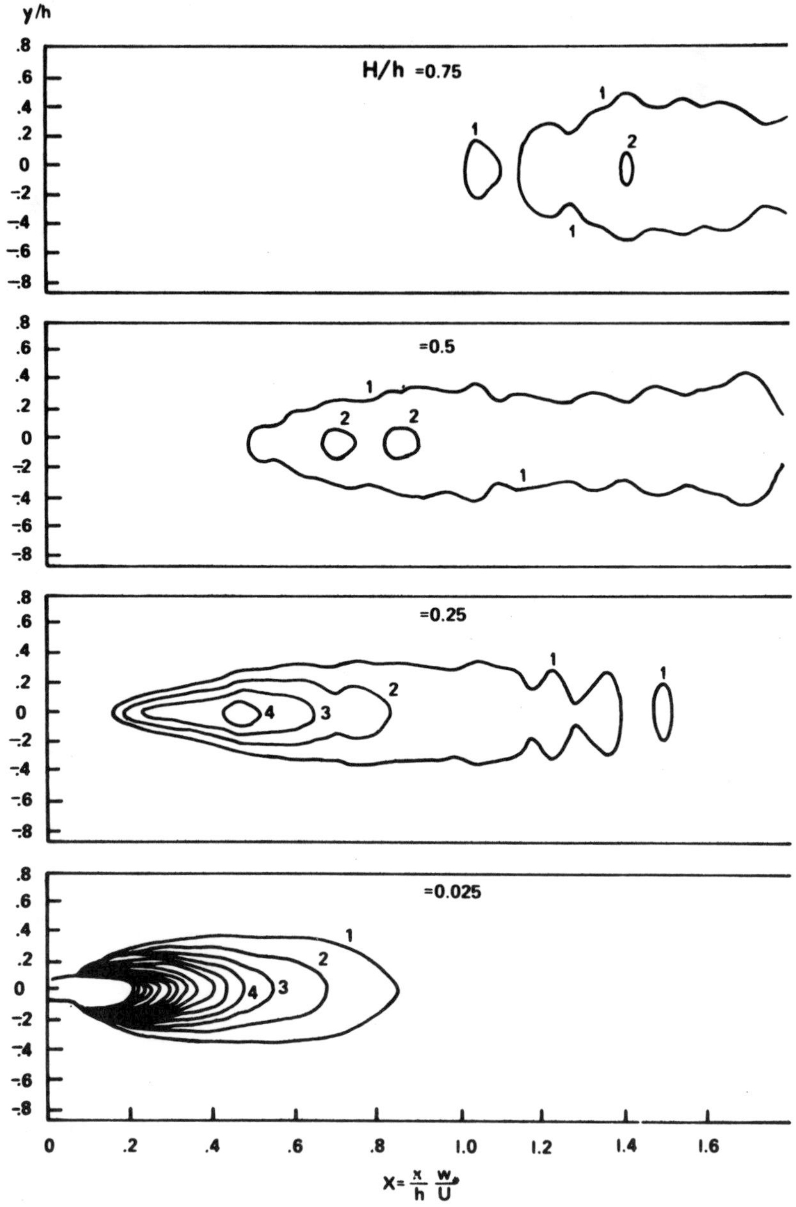

Figure 5.19. Non-dimensional ground-level concentrations $C(x,y,0)h^2U/Q$ given by the modified Gaussian formula for source heights indicated. This figure can be compared with Figure 5.6 to assess the accuracy of the modified Gaussian model for receptors near ground-level.

spreading is the conversion of the vertical kinetic energy of downdrafts into horizontal kinetic energy as the sinking fluid strikes the ground.

Downstream of the point, where it achieves a maximum value, ζ decreases steadily to a minimum near the point $x = 2\ HU/w_*$, where the ground-level concentration is largest. This decrease in ζ is a direct result of the descent of the plume centerline dicussed earlier.

An interesting feature of the ζ profiles for elevated sources ($H > 0.1\ h$) is that they become approximately independent of H at distances x greater than $2\ HU/w_*$. This can be seen in Figure 5.17. It is also apparent from this figure that at $x \simeq 2\ HU/w_*$ where the ground-level concentration is highest, the virtual source height ζ has a value of about h/4 for all values of H greater than 0.1 h.

Our definition of the virtual source height, Equation (5.43), was motivated primarily by the desire to have the predictions of the modified Gaussian formula agree with the predictions of the numerical model along the ground-level centerline of the plume. To achieve this capability with the Gaussian formula, we sacrificed some accuracy at all other locations in the plume, because, as we pointed out earlier, the material distributions predicted by the numerical model are not Gaussian. The magnitude of the discrepancies between the model simulations and the predictions of the modified Gaussian formulas can be seen in Figures 5.18-5.21. The first of these shows the mean, cross-wind integrated concentrations given for each of the four release heights treated earlier. The corresponding ground-level concentration estimates are displayed in Figure 5.19. Irregularities in the concentration isopleths shown in Figures 5.18 and 5.19 are the result of statistical fluctuations in the ζ values we used.

Comparing Figures 5.5 and 5.18, we see that there is fairly good qualitative agreement between the model predictions of C^y and the values of the modified Gauss model. The latter reproduces the descending centerlines of elevated source plumes at H = 0.5 and 0.75 h rather well. However, it completely misrepresents the shape of the plume very close to the source at H = 0.25 h and it overestimates the width of the ascended portion of the plume from a surface source. Comparing Figures 5.6 and 5.19, we see that, in general, the plume widths and impact areas at ground level are in quite good agreement with the model simulations.

A quantitative measure of the discrepancies between the simulated concentration profiles and those given by the modified Gauss formula can be obtained from Figures 5.20 and 5.21. The former compares the predicted and simulated cross-wind integrated concentration profiles at specified distances x downstream of

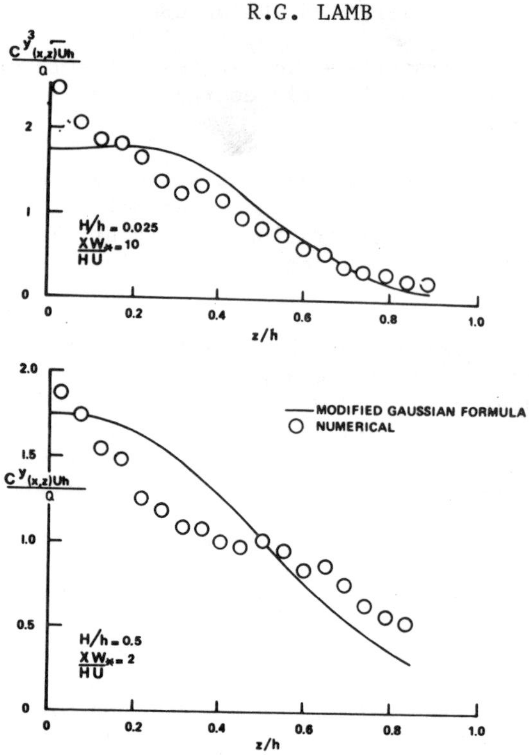

Figure 5.20. Comparison of the simulated mean cross-wind integrated concentration (circles) with the values given by the modified Gaussian model (solid lines) for source heights $z^s(\equiv H) = 0.5$ h and 0.025 h. The concentration profiles are evaluated at $x = 2HU/w_*$ and $10HU/w_*$, respectively.

sources of heights H = 0.025 and 0.5 h. Comparisons of the corresponding ground-level mean concentrations are given in Figure 5.21 (The plots given for H = 0.5 h are nearly identical to those that we found for source heights H = 0.25 h and 0.75 h.)

Looking first at Figure 5.20, we see that the Gaussian formula underestimates the total quantity of material at ground-level from both surface and elevated sources. The error is about 20% in the case of elevated sources and about 30% for surface releases. Over the entire depth of the mixed layer, the accuracy of the Gaussian model is remarkably good considering the fact that the parameters it contains are defined independently of C^y.

Figure 5.21 reveals that the simulated values of C(x,y,0) possess an asymmetry about the plume centerline y = 0 that may be due to statistical fluctuations in the concentration estimates. Further studies will be required to determine whether this is the case.

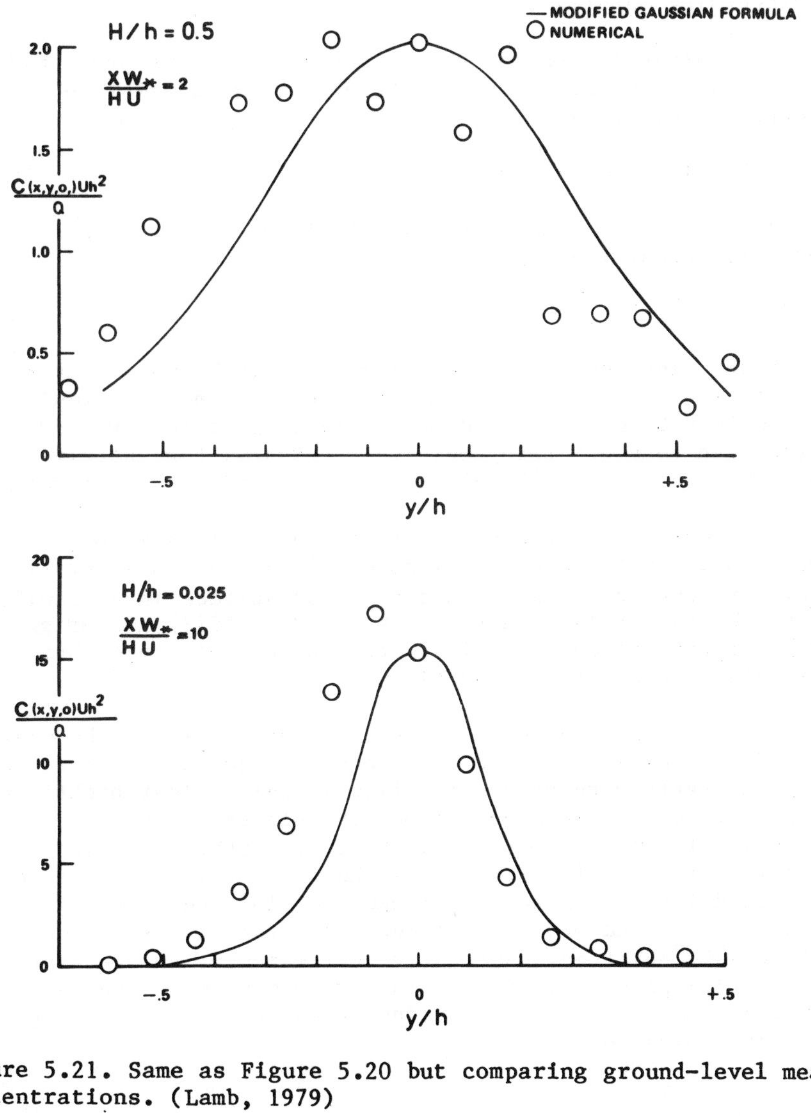

Figure 5.21. Same as Figure 5.20 but comparing ground-level mean concentrations. (Lamb, 1979)

We conclude that the effects of source height on material dispersion are described rather well by the Gaussian plume formula provided that the source elevation H is replaced by the virtual source height defined in (5.43) and given in Figure 5.17, and the dispersion coefficients are replaced by the particle displacement variances given by (5.42). Our goal is to determine quantitatively how ζ, σ_y and σ_z vary with the buoyancy of the material emissions. The hope is that a generalized set of parameters can be found that will permit the application of the modified Gaussian formula to any dispersing material.

The limitations on the use of formulas we have presented here should be kept in mind: non-buoyant effluents; quasi-free convective conditions, i.e.,

$$-h/L > 10 \quad \text{and} \quad z^s(\equiv H) > -L. \tag{5.44}$$

and sufficient mean wind that Taylor's 'frozen' turbulence hypothesis holds, i.e.,

$$U > 1.2\, w_* \, . \tag{5.45}$$

Without the last condition, the simple relationship $x = Ut$ between travel distances and travel time that we used in the formulas above would not be valid, and much more complicated expressions for the mean concentration would be necessary. Fortunately, it is only infrequently that (5.45) is not satisfied in the atmosphere.

For ground-level sources, more refined estimates of dispersion at short range (i.e. $\sigma_z \lesssim 0.1$ h) than those provided by the expressions above can be derived from surface layer similarity theory. The reader is referred to Van Ulden (1978) and Briggs (1982) for theoretical/empirical formulas of this type that are applicable in convective conditions.

Our analyses have been limited to sources within the mixed layer, $H < h$. However, it is not uncommon along shorelines and during the early morning hours following calm, clear nights to find $H > h$. Under these conditions the dispersion of material that enters the mixed-layer through its top (the so-called fumigation process) is affected by the of entrainment of inversion layer fluid into the mixed layer and the wind shear at $z \simeq h$ as well as by h, U and w_*. In a recent laboratory study Deardorff and Willis (1982) obtained quantitative estimates of the relationships of fumigating plumes. Their findings refute some of the basic assumptions employed by earlier investigators in formulating models of fumigation.

5.6. DISPERSION OF BUOYANT EMISSIONS IN A CONVECTIVE BOUNDARY LAYER

When passive material is released into a boundary layer, its subsequent motion is determined almost exclusively by the velocity of the ambient fluid (molecular diffusivity of the material plays a small role). By contrast, buoyant emissions generate an internal vorticity field which propels the emissions and the vorticity itself through the ambient fluid.

When the environment is non-turbulent, the qualitative effects of buoyancy are rather simple. Material released from an elevation z^s into a neutrally- or stably-stratified airstream ascends along a curve of decreasing slope and spreads out laterally. In a recent review, Briggs (1975) estimated that there exist between 50 and 100 formulas for predicting the rise Δz and spread of plumes under these conditions as functions of wind speed, stability, distance from the source, and certain source parameters (see also Section 7.3 on plume rise). The basic assumption underlying the use of these formulas is that beyond a certain distance from the source, a plume of buoyant material released from an elevation z^s looks essentially the same as a plume of non-buoyant material released from an upwind virtual source of elevation $z^s + \Delta z$. Under this assumption, formulas for estimating the mean concentration of passive material can be applied to buoyant emissions as well.

Shortly after the model results presented in Section 5.3 were first reported, it was brought to my attention by several people with access to pollutant monitoring data around large power plants, that if the model results were applied to buoyant plumes in the conventional way, the plume rise Δz needed to get the predicted concentrations to agree with the observations was absurdly large. These reports motivated an attempt to modify the model, described in Section 5.2, to obtain at least a 'first order' estimate of the effects of buoyancy.

For this purpose a component u_i'' representing the buoyancy generated particle velocity was added to the basic trajectory Equation (5.31). Values for u_i'' were obtained by solving the simplified, spatially averaged, time-dependent heat, momentum and mass conservations equations suggested by Briggs (1975). The results for the case of a source with a small buoyancy flux, $F = 13\ m^4 s^{-3}$ (see Equation 7.35 for the definition of F), and release height $z^s = 0.25\ h$ are compared in Figure 5.22 with the corresponding simulation of a nonbuoyant plume (from Figure 5.5). Although quantitative features of the buoyant plume simulation are not reliable, for reasons that I will give later, the results provide some important qualitative information. We see on comparing the two simulated plumes shown in Figure 5.22 that buoyancy has two basic effects: a slight upward displacement of the effective release point - this is probably due mostly to the momentum of the emissions, and an upward rotation of the plume axis. The magnitude of the displacement and the angle of rotation are both directly proportional to F. In the conventional applications of passive material dispersion formulas to buoyant emissions, an upward displacement of the plume axis is allowed but a rotation is not (the axis is assumed to be horizontal); and it is undoubtedly the neglect of the rotation effect that is responsible for the need cited earlier for unreasonably large

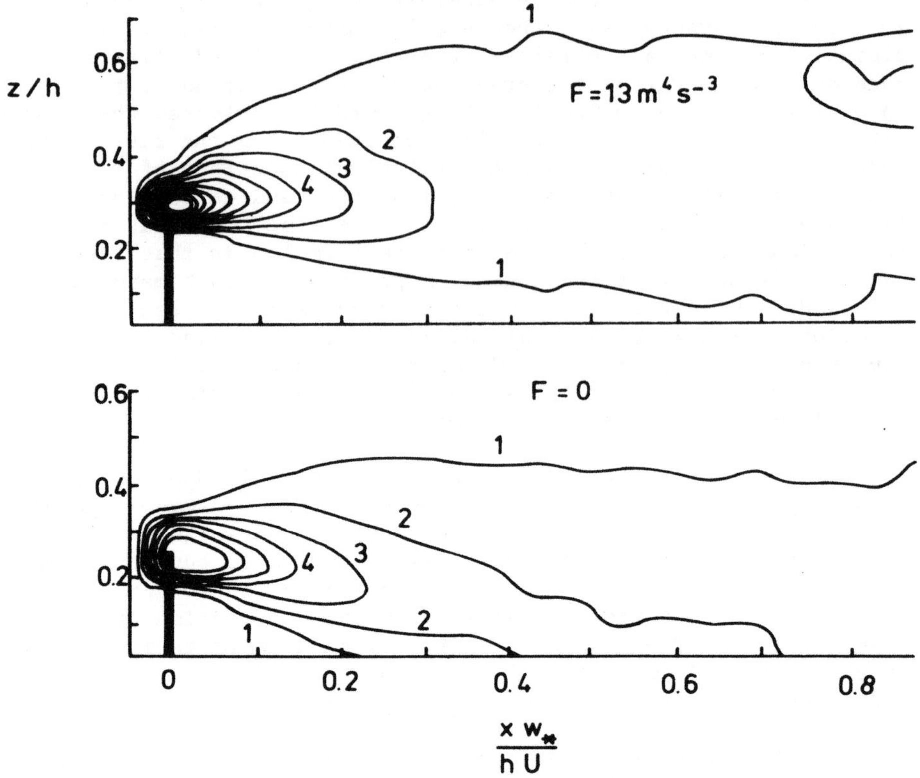

Figure 5.22. Results of a 'first order' simulation of a plume with a buoyancy flux $F = 13\ m^4\ s^{-3}$ in the convective mixed layer (top panel) compared with the corresponding simulation of a passive material plume (see Figure 5.5).

plume rises to make the passive plume results agree with the ground-level concentrations observed in buoyant plumes. In retrospect, the rotation phenomenon seems obvious, because if we add to the vertical turbulence velocities w a constant value w_B, representing the buoyancy speed, the mode of the probability density p_w (see Figure 5.13 and Equation (5.40)) will increase from approximately $-w_*/2$ to $(-w_*/2 + w_B)$; and hence the slope of the plume centerline will increase from $-w_*/(2U)$ to $(-w_*/2 + w_B)/U$.

One of the aspects of the buoyant plume simulations that led to doubts about the quantitative accuracy of the results was the ease with which only moderately buoyant emissions were able to penetrate the capping inversion. In the model, the rate of expansion of the plume was controlled by buoyancy until the estimated rate of turbulent energy dissipation inside the plume was equal to that in the local, ambient fluid. Beyond that point plume expansion was controlled solely by the ambient fluid. In our simulations plumes emitted as low as $z^s = 0.25\ h$ and with a buoyancy flux greater than about 200 $m^4\ s^{-3}$ arrive at the inversion base, after transport in updrafts, with sufficient speed and density deficit to penetrate a considerable distance into the capping inversion. The observational data that I have seen (principally time-lapse movies of power plant plumes), indicate that typical buoyant emissions are not able to escape the mixed layer under typical midday conditions. It also seems unrealistic that a buoyancy flux of only 13 $m^4\ s^{-3}$ should have as dramatic an effect on dispersion as the simulation shown in Figure 5.22 indicates. These results suggest that the assumption that a buoyant plume is immune to the forces of ambient turbulence up until a certain time in the plume's decay is incorrect, and that actually ambient turbulence continually affects the rise of buoyant emissions relative to the ambient fluid.

The question of how to treat turbulence effects has been a matter of dispute among plume-rise modelers for many years. In fact, under turbulent conditions, the concept of plume rise is itself not well defined because plumes do not follow simple, ascending trajectories as they do when environmental turbulence is absent. These problems are made especially critical in convective conditions by the deep, well organized vertical circulations associated with updrafts and downdrafts.

Within the past few years, several investigators have proposed simple models for estimating the 'effective' plume rise Δz_e. The effective rise is defined as the source height, $H = z^s + \Delta z_e$ that one would need in the conventional Gaussian model to predict accurately the maximum ground-level concentration of buoyant emissions. Apparently all of the formulas for Δz_e proposed to date make use of the '2/3 law' of plume rise in neutral, turbulent free conditions given by (7.34).

One of the models of effective plume rise Δz_e in convective conditions was developed by Briggs (1975). He assumed that in a downdraft of mean speed w_d the rise of a buoyant plume relative to the ambient air is the same as the rise of a plume in a neutral, turbulent free fluid. (On the average, the air within downdrafts is very slightly stably stratified.) In this case the bottom of a plume, which is assumed to have an elevation $z^s + 0.5\Delta z$, will strike the ground after a travel time t_d given by

$$w_d t_d = z^s + 0.5\Delta z. \tag{5.46}$$

Substituting (7.34) with $t = x/U$ into this expression for Δz, we can solve for the touchdown time t_d; and then we can put the value of t_d into (7.34) to obtain the effective plume rise Δz_e at touchdown, which is assumed to be the location of maximum surface concentration.

Weil (1979) suggested a Gaussian model for estimating ground-level concentrations in which the dispersion coefficients are expressions (5.42) modified to allow for the dispersion enhancement caused by buoyancy generated turbulence inside the plume; and the effective source height $z^s + \Delta z_e$ is the minimum value given by the '2/3 law', Briggs formula (5.46), or the height of a trapped plume. Weil found the predictions of this model to be in good (factor of two) agreement with observations.

Venkatram (1980) combined Briggs' model with available descriptions of convective turbulence statistics to estimate the probability that a plume will hit the ground within a given distance x of the source. He used this estimate of the impingement probability distribution as the basis of a simple model of ground-level concentration beneath the plume's centerline. Weil (1980) compared the predictions of this model with observational data and found that the model underpredicted ground-level concentrations in 9 out of 12 cases considered. This finding suggests that the model overestimates the ability of buoyant emissions to rise through turbulent fluid, just as our 'first order' model (Figure 5.22) seems to do. Since neither model allows ambient turbulence to affect the relative plume rise prior to some point in the final period of decay, it would seem that this treatment of ambient turbulence effects is perhaps the cause of the poor performance.

In the remainder of this section I would like to formulate a '1½ order' modification of the Lagrangian model developed in Section 5.2 that will permit continual influence of the ambient flow on the plume motion. The principal effect that we will attempt to include is that of the large-scale straining motion of the ambient turbulence. To visualize this mechanism, imagine two buoyant particles A and B emitted at a time interval Δt_0, say of one minute apart. If the ambient fluid is turbulent free, the plume material emitted between the release of A and B will lie generally along a smooth arc of length L_0. However, if the ambient flow is turbulent, the locus of the plume material connecting A and B will be erratic and have an arc length L that generally is greater than L_0. In effect, the plume is stretched by the ambient flow and this serves both to accelerate the rate of dilution of the plume material with ambient fluid and to alter the inertial motions of the plume, as we will discuss later.

As in the 'first order' model discussed at the beginning of this section (Figure 5.22), to generate the ensemble of buoyant particle trajectories needed to estimate the displacement probability densities p and p_2, we need in place of (5.31)

$$\frac{d}{dt} X_i^m = \hat{u}_i(X_i^m(t),t) + u_i'(X_i^m(t),t) + u_i''(X_i^m(t),t)$$

$$m = 1, \ldots\ldots M \qquad (5.47)$$

where X_i^m and u_i' have their same meanings as before, (see 5.30 and 5.31), and u_i'' is the component of velocity produced by the particle's buoyancy. We turn now to the formulation of a '1½ order' expression for u_i''.

For this purpose we will work with the vorticity equation. By taking the curl of the Navier-Stokes equations (1.16) in the Boussinesq approximation we find

$$\frac{\partial \tilde{\omega}_i}{\partial t} + \tilde{u}_j \frac{\partial \tilde{\omega}_i}{\partial x_j} = \tilde{\omega}_j \frac{\partial \tilde{u}_i}{\partial x_j} + \nu \frac{\partial^2 \tilde{\omega}_i}{\partial x_j^2} - \frac{\partial B}{\partial x_j} \varepsilon_{ij3} + \Omega_i \qquad (5.48)$$

where ν is the kinematic viscosity; $\tilde{\omega}_i$ is the i-th component of vorticity defined by

$$\tilde{\omega}_i = \varepsilon_{ijk} \, \partial \tilde{u}_j / \partial x_k . \qquad (5.49)$$

(Editorial note: to avoid confusion one should observe here that Lamb has digressed from the usual definition of vorticity given by $\tilde{\omega}_i = \varepsilon_{ijk} \partial \tilde{u}_k / \partial x_j = -\varepsilon_{ijk} \partial \tilde{u}_j / \partial x_k$.) The buoyancy term B is given by

$$B = \frac{g}{T} (T - T_0) \qquad (5.50)$$

where T is temperature and T_0 the temperature of the base state of the fluid. When deviations in temperature $T - T_0$ are small, the dominator can be approximated by T_0 giving (5.50) its more common form. The term Ω_i in (5.48) represents sources of vorticity associated with the momentum and volume fluxes produced by material sources.

The heat equation coupled with (5.48) through the buoyancy B is

$$\frac{\partial \tilde{T}}{\partial t} + \tilde{u}_i \frac{\partial \tilde{T}}{\partial x_i} = \kappa \frac{\partial^2 \tilde{T}}{\partial x_i^2} - \tilde{u}_i \delta_{i3} \left(\frac{\partial T_0}{\partial x_i} + \frac{g}{c_p} \right) + Q . \qquad (5.51)$$

Here κ is the molecular thermal diffusivity and Q represents heat emissions associated with material sources.

Let $\tilde{\omega}_i$ and $\tilde{T}$ represent the vorticity and temperature fields that describe the combined fluid system 'boundary layer plus material emmision'. Since the data from Deardorff's model represent the boundary layer fluid alone, we decompose $\underset{\sim}{\omega}_i$ and $\underset{\sim}{T}$, disregarding the subgrid contribution, as

$$\underset{\sim}{\omega}_i = \hat{\omega}_i + \omega_i'' , \quad \underset{\sim}{T} = \hat{T} + T'' + T_0, \quad \underset{\sim}{u}_i = \hat{u}_i + u_i'' \tag{5.52}$$

where $\hat{\omega}_i = \varepsilon_{ijk}\,\partial\hat{u}_j/\partial x_k$ etc. If the variables with the caret (^) denote conditions in the absence of heat and material sources, then

$$\frac{\partial\hat{\omega}_i}{\partial t} + \hat{u}_j \frac{\partial\hat{\omega}_i}{\partial x_j} = \hat{\omega}_j \frac{\partial\hat{u}_i}{\partial x_j} + \nu \frac{\partial^2\hat{\omega}_i}{\partial x_j^2} - \frac{\partial}{\partial x_j}\left(\frac{g}{T}\hat{T}\right)\varepsilon_{ij3} \tag{5.53}$$

and

$$\frac{\partial\hat{T}}{\partial t} + \hat{u}_i \frac{\partial\hat{T}}{\partial x_i} = \kappa \frac{\partial^2\hat{T}}{\partial x_i^2} - \hat{u}_i\delta_{i3}\left(\frac{\partial T_0}{\partial x_i} + \frac{g}{c_p}\right) \tag{5.54}$$

These are the equations satisfied by the data from Deardorff's model. Substituting (5.52) into (5.48) and (5.51) subtracting (5.53) and (5.54) from the results, and assuming that $1/T \simeq 1/T_0$ which is valid everywhere except near the source, we obtain the equations governing the components ω_i'' and T'' associated with the material sources

$$\frac{\partial\omega_i''}{\partial t} + (\hat{u}_j + u_j'')\frac{\partial\omega_i''}{\partial x_j} + u_j''\frac{\partial\hat{\omega}_i}{\partial x_j} \tag{5.55}$$

$$= \hat{\omega}_j \frac{\partial u_i''}{\partial x_j} + \omega_j''\frac{\partial}{\partial x_j}(\hat{u}_i + u_i'') + \nu\frac{\partial^2\omega_i''}{\partial x_j^2} - \frac{g}{T_0}\frac{\partial T''}{\partial x_j}\varepsilon_{ij3} + \Omega_i$$

$$\frac{\partial T''}{\partial t} + (\hat{u}_i + u_i'')\frac{\partial T''}{\partial x_i} + u_i''\frac{\partial\hat{T}}{\partial x_i} = \kappa\frac{\partial^2 T''}{\partial x_i^2} - u_i''\delta_{i3}\left(\frac{\partial T_0}{\partial x_i} + \frac{g}{c_p}\right) + Q \; .$$

Having solved (5.55) for ω_i'' we can obtain u_i'' from the general expression

$$u_i''(x_i) = -\frac{1}{4\pi}\int \frac{\varepsilon_{ijk}\, s_j\, \omega_k''(x_i')}{s^3}\, dx_i' \tag{5.56}$$

where $s_i = x_i - x_i'$ and $s = \| s_i \|$. (Note that dx_i stands for $dx_1 dx_2 dx_3$). On combining u_i'' with the field $\hat{u}_i$ from Deardorff's model, we have the combined motion of material from buoyant sources in the convective boundary layer.

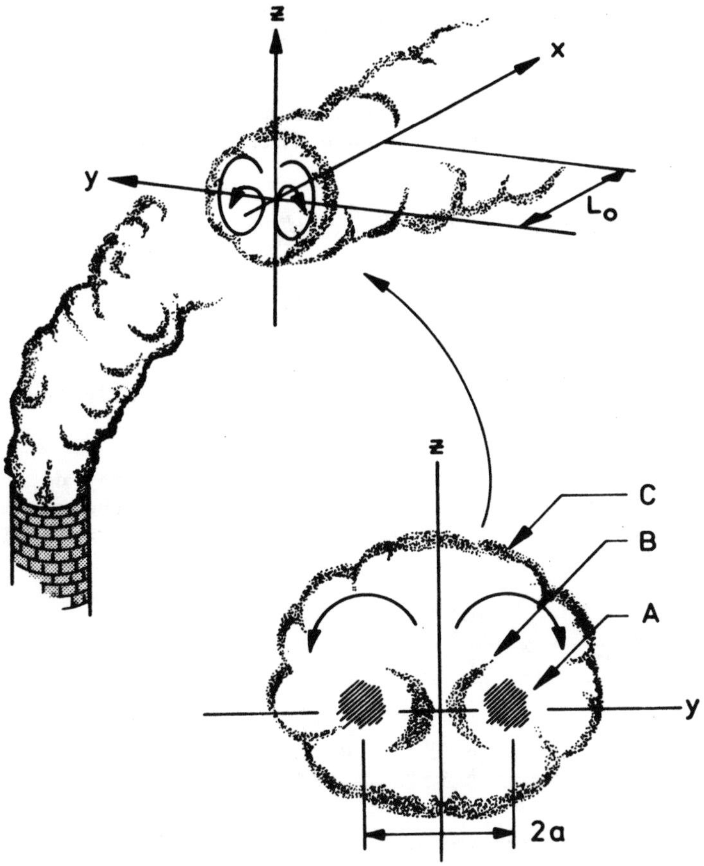

Figure 5.23. Schematic illustration of the idealized vorticity distribution within a bent-over buoyant plume.

In order to simplify the problem of solving (5.55), we will make use of the analogy pointed out by Scorer (1959) between a bent-over buoyant plume and a two-dimensional line thermal. This is depicted in Figure 5.23. The bent-over plume is envisaged as a pair of approximately parallel vortex tubes. The vorticity is created initially about the z axis ($\omega_z \neq 0$) by the ambient air flowing around the edges of the vertical column of fluid emerging from the source. As the wind bends the plume over, the vorticity vector rotates toward a horizontal position ($\omega_x \neq 0$) where it is then subjected to amplification by the buoyancy force B.

Consider a small longitudinal section of the bent-over plume of initial length L_0 as shown in Figure 5.23, and let the coordinate axis be aligned so that initially

$$\omega''_y \simeq \omega''_z = 0 \simeq u'' \quad \text{and} \quad \frac{\partial v''}{\partial x} \simeq \frac{\partial w''}{\partial x} = 0 \simeq \frac{\partial T''}{\partial x} \,. \tag{5.57}$$

If the origin of the coordinate system moves with a velocity $\hat{u}_i$, then to first-order accuracy we have initially (from (5.55))

$$\frac{\partial \omega''_x}{\partial t} + v'' \frac{\partial}{\partial y}(\omega''_x + \hat{\omega}_x) + w'' \frac{\partial}{\partial z}(\omega''_x + \hat{\omega}_x)$$
$$= \omega''_x \frac{\partial \hat{u}}{\partial x} + \nu\left(\frac{\partial^2 \omega''_x}{\partial y^2} + \frac{\partial^2 \omega''_x}{\partial z^2}\right) - \frac{g}{T_o}\frac{\partial T''}{\partial y} + \Omega_x \tag{5.58}$$

where $\hat{u}_i = (\hat{u}, \hat{v}, \hat{w})$, $\hat{\omega}_i = (\hat{\omega}_x, \hat{\omega}_y, \hat{\omega}_z)$ and similarly for u_i'' and ω_i''.

Although ω_y'' and ω_z'' may initially be zero, they do not retain this value. We find from (5.55) that under the assumptions (5.57) and with $\Omega_y = 0$, the evolution of ω''_y is governed initially by

$$\frac{\partial \omega''_y}{\partial t} + v'' \frac{\partial \omega''_y}{\partial y} + w'' \frac{\partial \omega''_z}{\partial z} =$$
$$\hat{\omega}_y \frac{\partial v''}{\partial y} + \hat{\omega}_z \frac{\partial v''}{\partial z} + \omega''_x \frac{\partial \hat{v}}{\partial x} - \left[v'' \frac{\partial \hat{\omega}_y}{\partial y} + w'' \frac{\partial \hat{\omega}_y}{\partial z}\right] \,. \tag{5.59}$$

A similar expression governs ω''_z. We see from this equation that ω''_y is created by four processes: stretching of $\hat{\omega}_y$ by the u_i'' field (first term on the right-hand side of (5.59)); rotation of $\hat{\omega}_z$ by shear in u_i'' (second term); rotation of ω''_x by shear in $\hat{u}_i''$ (third term); and advection of $\hat{\omega}_y$ (fourth term). All four of these processes represent interactions between the ω'' and $\hat{\omega}$ fields. To obtain a '1½ order' approximation of u_i'' we will neglect all interaction terms except the advection of the ω''_i field by u_i'' - this will be taken into account by the translation of the coordinate system - and the amplification (dimunition) of ω''_i by local stretching (contraction) of $\hat{u}_i$ - this will be handled by aligning the x-axis parallel to $\hat{u}_i$ and retaining the first term on the righthand side of (5.58). Within the scope of these restrictions conditions (5.57) hold for all travel time and (5.58) reduces to

$$\frac{\partial \omega''}{\partial t} + v'' \frac{\partial \omega''}{\partial y} + w'' \frac{\partial \omega''}{\partial z} = \omega'' \frac{\partial \hat{u}}{\partial x} + \nu\left(\frac{\partial^2 \omega''}{\partial y^2} + \frac{\partial^2 \omega''}{\partial z^2}\right) - \frac{g}{T}\frac{\partial T''}{\partial y} + \Omega_x \,. \tag{5.60}$$

Here we have dropped the subscript on ω_x'' for notational convenience. The corresponding temperature equation is found from (5.55) to be

$$\frac{\partial T''}{\partial t} + v'' \frac{\partial T''}{\partial y} + w'' \frac{\partial T''}{\partial z} = \kappa\left(\frac{\partial^2 T''}{\partial y^2} + \frac{\partial^2 T''}{\partial z^2}\right) - w''\left(\frac{\partial T_o}{\partial z} + \frac{g}{c_p}\right) + Q. \quad (5.61)$$

Equations (5.60) and (5.61) describe the vorticity and temperature distributions within the line thermal depicted in Figure 5.23. The vorticity is distributed throughout the thermal with $\omega'' > 0$ in the right half and $\omega'' < 0$ in the left portion. Outside the thermal $\omega'' = 0$ everywhere by virtue of our treatment of the ambient fluid vorticity in the component $\hat{\omega}_i$.

If we assume that vorticity is concentrated within the two small hatched regions labeled A in Figure 5.23,and if the section L_0 of the plume that we simulate is long enough to appear quasi-infinite at a point in its center, then we find from (5.56) that the plume, or thermal, moves bodily upward relative to the ambient fluid with a speed

$$\overline{w''} \simeq K/4\pi a \quad (5.62)$$

where 2a is the separation of the vortex tubes that comprise the thermal (see Figure 5.23), and K is the circulation around any closed path Γ in the y-z plane that passes through the center of the thermal. According to Stokes-formula

$$K = \int_S \omega_i'' n_i \, dS \quad (5.63)$$

where S is the area bounded by Γ, and n_i is the normal on S. We will use $\overline{w''}$ as the vertical component of the buoyancy velocity term u_i'' needed in the particle trajectory expression (5.47). Our next task, therefore, is to express the circulation K and thermal radius a in terms of travel time.

Consider K. Let the path Γ be an infinite rectangle with one side the z axis. The area S is then the half-plane $y \geq 0$. If we assume that the vortex tubes that comprise our idealized thermal each have a radius r_0, then (5.63) reduces to

$$K = \int_o^{2\pi} \int_o^{r_o} \omega'' r \, dr d\theta \ . \quad (5.64)$$

Due to the divergence of $(\hat{v}, \hat{w})$, i.e. $\partial\hat{u}/\partial x \neq 0$, the radius r_0 of the vortex tubes will change with time. Allowing for this we obtain upon differentiating (5.64)

$$\frac{\partial K}{\partial t} = 2\pi r_o \omega_o \frac{dr_o}{dt} + \int_o^{2\pi} \int_o^{r_o} \frac{\partial \omega''}{\partial t} r \, dr d\theta \tag{5.65}$$

where ω_o represents the value of ω'' at the edge of the vortex tube.

Integrating (5.60) over S, combining the result with (5.65), and making use of the divergence theorem we obtain

$$\frac{\partial K}{\partial t} = -\nu \int_{-\infty}^{\infty} \left(\frac{\partial \omega''}{\partial y}\right)_{y=0} dz + \frac{g}{T_o} \int_{-\infty}^{\infty} (T'')_{y=0} \, dz. \tag{5.66}$$

To arrive at this expression we assumed that within the vortex tubes, vorticity is uniformly distributed with value ω_o.

Note first that stretching of the vortex tubes does not alter the circulation K - changes in K require the exertion of a torque about the x axis.

The first term on the right-hand side of (5.66) represents the dissipation of circulation by viscosity. The form of the term implies the nature of the physical process involved. Namely, as vorticity crosses the plume centerline (y = 0) it is annihilated by molecular scale mixing with vorticity of the opposite sign. The loss of vorticity results not only in a dimunition of the circulation, see (5.63), but also in a reduction in the plume's overall velocity field u''_1 (see 5.56). Consequently, the viscous term in (5.66) represents also the transformation of the plume's kinetic energy into heat.

The last term on the right-hand side of (5.66) represents the generation of circulation by buoyancy. Note that this process is active only when there is warm fluid on the centerline (z axis) of the plume. The entrainment of cool fluid, in which $T'' = 0$, into the core of the plume diminishes the rate of generation of circulation. In fact, if entrainment is intense enough to split the plume into seperate, parallel columns of warm fluid, there is no generation of circulation at all. Later we will show that in the early stages of rise, vorticity is generated much more rapidly by buoyancy than it is destroyed by viscosity. For this reason we will neglect the viscous term in (5.66) in the formulation of our buoyant plume model.

To model the buoyancy generation term we will assume that T'' is uniform inside the thermal plume. With this assumption and the neglect of the molecular dissipation terms, (5.66) is modeled by

$$\frac{\partial K}{\partial t} \sim \frac{g}{T_o} a \gamma_1 \overline{T''} \tag{5.67}$$

where $\overline{T''}$ is the mean value of T'' in the plume and γ_1 is a constant.

An expression for $\overline{T''}$ can be derived from the heat equation, (5.61). Integrating (5.61) over all space and keeping in mind that we are treating a section of the plume of finite length (initially L_0) along the x-axis and that $T'' \neq 0$ only within the plume fluid, we obtain

$$\frac{\partial}{\partial t} (\overline{T''} V) = -\gamma_2 \overline{w''} V \left[\frac{\partial T_o}{\partial z} + \frac{g}{c_p}\right] \tag{5.68}$$

where V is the total volume of the plume segment and T_o is evaluated at the elevation of the plume at time t. The 'constant' γ_2 in (5.68) is used to compensate for the fact that since motion is induced in the ambient fluid that surrounds the plume, w'' is nonzero within a larger volume of fluid than that, V, in which $T'' \neq 0$. Integrating (5.68) we get

$$\overline{T''}(t) = \overline{T''_o} \frac{V_o}{V(t)} - \frac{\gamma_2}{V(t)} \int_0^t \overline{w''} V \left[\frac{\partial T_o}{\partial z} + \frac{g}{c_p}\right] dt \tag{5.69}$$

where $\overline{T''_o}$ and V_o denote initial values. We will approximate the plume volume V(t) by

$$V(t) = \gamma_3 \pi a^2 L \tag{5.70}$$

where L is the length of the plume segment at time t; a is the plume 'radius', shown in Figure 5.23; and γ_3 is a third 'constant'. Equations (5.67), (5.69) and (5.70) provide an approximation of the circulation K that we need in (5.62) to estimate the buoyancy-induced speed $\overline{w''}$. We turn next to the tasks of approximating the plume length L and radius a.

There are three processes that cause changes in the plume dimensions: stretching/contraction by the ambient fluid ($\partial\hat{u}/\partial x \neq 0$), entrainment due to small scale ambient turbulence, and expansions induced by the plume's internal vorticity field ω_1.

It is an observed fact that a line thermal in a quiescent fluid expands as it rises. Richards (1963) found in laboratory studies that

$$da/dt = \eta \overline{w''} \quad \text{(static ambient fluid)} \tag{5.71}$$

where η is a constant of order 0.5 and $\overline{w''}$ represents the rate of rise of the thermal. In the idealized model of the line thermal in which vorticity is confined to two thin tubes, there exists no mechanism to cause the thermal to expand. This is evident from (5.56) - neither vortex tube induces a lateral motion v'' at the location of the other. Thus, the observed expansion of thermals indicates that the internal distribution of vorticity is more complicated. One possibility is that the negative (positive) vorticity generated at the edges of the fluid entrained into the right (left) side of the thermal in the regions marked B in Figure 5.23 is effectively centered above the x-y plane. In this position it induces an outward velocity on each of the main vorticity centers (designated A in Figure 5.23). In this regard it is important to note that when the thermal is moving in stratified fluid in which $\partial T_0/\partial z > -g/c_p$, once the temperature deviation $\overline{T''}$ in the thermal becomes negative, vorticity is generated along the entire edges of the plume, marked C in Figure 5.23. This new vorticity will have a sign opposite to that of the vorticity in the corresponding central core A and consequently it will not only act to decelerate the upward motion of the thermal but also to enhance its rate of expansion.

Another vorticity distribution that can cause the thermal to expand laterally is that in which vorticity diffuses from the edges of the thermal and is swept by the ambient flow into positions underneath the thermal. Maxworthy (1972) found evidence of vorticity in the wake of non-buoyant vortex rings. We must conclude that the expansion of thermals in a static ambient fluid is the result of several processes that are too complex to be treated in a more rigorous manner than the semi-empirical formula (5.71).

Since the fluid is incompressible, changes in the plume radius are caused by stretching and contraction in the ambient fluid must accompany changes in the length L of the plume segment that result in no net change in the volume V. If $\partial \hat{u}/\partial x$ is approximately constant over V, then we have $dL/dt = L\, \partial \hat{u}/\partial x$ and from the divergence theorem $da/dt = -(a/2)\, \partial \hat{u}/\partial x$. In the simulations we will use the more accurate expressions

$$dL/dt = \Delta\hat{u} = \| \hat{u}_i(-L/2) - \hat{u}_i(L/2) \| \tag{5.72}$$

where $\Delta\hat{u}$ is the magnitude of the difference in the ambient fluid velocity $\hat{u}_i$ at each end of the plume segment; and

$$\frac{da}{dt} = -\frac{a}{2L}\frac{dL}{dt} = -\frac{a}{2L}\Delta\hat{u} \ . \tag{5.73}$$

We pointed out earlier in connection with Equation (5.60) that ω'' is amplified by stretching motions in the ambient fluid ($\partial\hat{u}/\partial x$ or $\Delta\hat{u}$ greater than 0). By this process the stability of the vortex tubes that comprise the plume is diminished. Consequently, if stretching continues long enough, small perturbations may initiate a rapid restructuring of the vorticity distribution into a more stable configuration, such as a series of vortex rings. Perhaps this is why plumes rising into turbulence-free shear layers often exhibit a lumpy character as though the plume is composed of a succession of puffs. Moore (1974) has argued on the basis of photographic evidence of the puffy nature of plumes that plume rise is a three-dimensional phenomenon rather than a two-dimensional one as the '2/3 law' Equation (7.34) assumes. The model that we are developing here possesses both two- and three-dimensional features. The former are exhibited when $\partial\hat{u}/\partial x = 0$ and three-dimensional characteristics are displayed otherwise (e.g., in a laminar shear flow, or in turbulence fluid).

As a result of the instability created by stretching, the plume cannot be squeezed to an arbitrarily small radius. Accordingly, we will assume that instability enhances the vorticity induced plume spread in such a manner that this mechanism of plume growth just balances the squeezing action caused by the ambient fluid. We will incorporate this assumption into the form we adopt for the parameter η, introduced in (5.71)

$$\begin{aligned} \eta &= \eta_o , & \partial\hat{u}/\partial x < 0 \\ \eta &= \eta_o + \tfrac{1}{2}\,(a/L)\,(\Delta\hat{u}/\overline{w''}), & \partial\hat{u}/\partial x > 0 \end{aligned} \qquad (5.74)$$

when $\overline{T''} > 0$. To account for the enhancement in the lateral growth rate of the plume mentioned earlier that we suspect to occur when $\overline{T''}$ becomes negative, we will assume somewhat arbitrarily that η_o is twice the nominal value when $\overline{T''} < 0$. In our mixed-layer studies this assumption will affect only the ability of the plume material to penetrate the capping inversion layer. (It turns out that increasing η_0 by a factor 2 or 3 after $\overline{T''} < 0$ has little effect on the equilibrium height of the plume in stably-stratified fluid.)

We come finally to the task of describing the effects on the plume radius a, of ambient turbulence, specifically the scales of motion smaller than the grid size Δ in Deardorff's model. We will assume that effects on the plume segment length L are negligible compared to the stretching effects. For 'point' source plumes, the initial plume radius is smaller than Δ ($\simeq$ = 90 m) - a value that falls in the inertial subrange of convective turbulence. Batchelor (1950) showed that two particles separated initially by a distance ℓ_o in the inertial subrange move apart, on the average, such that after a travel time t

$$\overline{\ell^2} = \frac{10}{3} C_1 \ell_o^{2/3} \varepsilon^{2/3} t^2, \quad t < \ell_o^{2/3} \varepsilon^{-1/3} \tag{5.75}$$

where C_1 is a constant of order one and ε is the energy dissipation rate of the turbulence. If we associate ℓ with the plume radius a, then (5.75) suggest

$$\frac{d}{dt}(\overline{a^2})^{\frac{1}{2}} \sim a_o^{1/3} \varepsilon^{1/3} . \tag{5.76}$$

It is by no means certain that the effects of small scale turbulence on the spread of a buoyant plume can be simulated simply by adding (5.76) to the rates of plume expansion caused by the stretching and the internal vorticity distribution processes. Nevertheless, we must adopt this simple model due to the lack of better approximations. We will implement it in the stochastic form

$$(da/dt)_{turb} = (0.7Ga_o/\Delta)^{1/3} \dot{\chi} \tag{5.77}$$

where $G = 1 + 2\{(z/\Delta z + 3/2)^2 - 3.3\}^{-1}$, $z > \Delta z/2$ and $\dot{\chi}(t) = \alpha\dot{\chi}(t - 2\Delta t) + \beta\dot{\chi}(t - \Delta t) + \gamma\rho$. Equation (5.77) is based on the two-particle velocity scheme presented in Lamb (1981a) (see page 344 of the reference) and it is consistent with (5.76); $\dot{\chi}$ is a stochastic variable in which the parameters α, β, γ and ρ have the forms given earlier in Equations (5.34) and (5.35). In the simulations we will use $\alpha = -0.8$ and $\beta = 0.6$ to achieve an oscillatory autocorrelation in $\dot{\chi}$, as is consistent with the character of subgrid turbulence.

In summary, to generate an ensemble of trajectories for buoyant particles emitted by a point source, we will use (5.47) with

$$u''_i = (u'', v'', \overline{w''}) \tag{5.78a}$$

$$u'_i = (5.32) \qquad \text{with } \alpha = -0.8,\ \beta = 0.6,\ \gamma = 0.566 \tag{5.78b}$$

$$\hat{u}_i = \text{from Deardorff's data} \tag{5.78c}$$

In (5.78a) $u'' = \dot{\chi}_x$, $v'' = \dot{\chi}_y$ where $\dot{\chi}_x$ and $\dot{\chi}_y$ are Gaussian random numbers with standard devation $\sigma(\dot{\chi}_x) = \sigma(\dot{\chi}_y) = \dot{a}^2 (1 + 2a/\dot{a}\Delta t)$. This formulation results in $\overline{x^{2''}} = \overline{y^{2''}} \simeq (0.5a)^2$ for all travel times. The vertical component of u''_i is derived from

$$\overline{w''} = \frac{K}{4\pi a}, \qquad \frac{dK}{dt} = \gamma_1 \frac{g}{T_s} a \overline{T''},$$

$$\overline{T''} = \frac{1}{V}\{\overline{T''_0}\,V_0 - \gamma_2 \int_0^t (\overline{w''}\,V\,[\frac{\partial T_0}{\partial z} + \frac{g}{c_p}])dt'\}$$

$$V = \gamma_3 \pi a^2 L, \qquad \frac{dL}{dt} = \hat{\Delta u}\ , \tag{5.79}$$

$$\dot{a} \equiv \frac{da}{dt} = \eta\overline{w''} - \frac{a}{2L}\hat{\Delta u} + \left(\frac{0.7Ga_0}{\Delta}\right)^{1/3}\dot{\chi}\ .$$

Note that in the circulation equation we have replaced T_0 in the denominator with T_s, the temperature of the emissions, to make the model consistent with the definition (7.35) of the buoyancy flux.

We can get help in choosing values for the constants, γ_1 and γ_2 and at the same time examine the consistency of our model with other formulations by applying (5.79) to the case in which the ambient fluid is neutrally stratified $(\partial T_0/\partial z = - g/c_p)$, free of both shear and turbulence ($\dot{\chi} = \hat{\Delta u} = 0$), and moving with a uniform horizontal speed U. These are the conditions under which the '2/3 law' (7.34) applies. In this situation our model Equations (5.79) reduce to

$$\overline{w''} = \frac{K}{4\pi a}, \qquad \frac{da}{dt} = \eta\,\overline{w''}, \qquad \frac{dK}{dt} = \frac{g}{T_s}\overline{T''_o}\left(\frac{\gamma_1}{\gamma_3}\right)\frac{V_0}{\pi a L_0}\ . \tag{5.80}$$

The initial volume V_0 of emissions in a plume segment of length L_0 is simply: $V_0 = \pi w_s r_s^2 L_0/U$, where w_s and r_s are the exhaust velocity and radius of the source and $\overline{T''_0} = T_s - T - T_0 = T_s - T_a$, where T_a is the ambient temperature at the source (see (5.52)). Using these expressions we can write (5.80) in the form

$$\frac{da}{dt} = \frac{\lambda_1 K}{a}\ , \qquad \frac{dK}{dt} = \frac{\lambda_2}{a} \tag{5.81}$$

where $\lambda_1 = \eta/4\pi$, $\lambda_2 = (\gamma_1/\gamma_3)F/U$ and F is the buoyancy flux (7.35).

Solving the second member of (5.81) for 1/a and substituting the result in the first member we get

$$\frac{d}{dt}\left(a - \frac{\lambda_1}{2\lambda_2}K^2\right) = 0 \qquad \text{or} \qquad a - \frac{\lambda_1}{2\lambda_2}K^2 = a_0 - \frac{\lambda_1}{2\lambda_2}K_0^2 \tag{5.82}$$

where the subscript o denotes the initial value. Using (5.82) in the second member of (5.81) we find that K is a root of the cubic equation

$$\frac{\lambda_1}{2\lambda_2} K^3 + 3\kappa_o K - \left(\frac{\lambda_1}{2\lambda_2} K_o{}^3 + 3\kappa_o K_o + 3\lambda_2 t\right) = 0 \qquad (5.83)$$

where κ_o denotes the right-hand side of (5.82). Equations (5.82) and (5.83) are the general solutions of (5.81).

Let's apply these solutions to the case where $a_o = K_o = 0$, as implied by the '2/3 law'. We then have $\kappa_o = 0$ and, hence, from (5.83) and (5.82)

$$K = \left(\frac{6\lambda_2^2}{\lambda_1}\right)^{1/3} t^{1/3} \quad \text{and} \quad a = \frac{1}{2}\,(36\lambda_1\lambda_2)^{2/3} t^{2/3}. \qquad (5.84)$$

Combining these expressions with the $\overline{w''}$ equation results in

$$\overline{w''} = \left(0.295\,\frac{\gamma_1}{\gamma_3}\right)^{1/3} \left(\frac{F}{U}\right)^{1/3} t^{-1/3}. \qquad (5.85)$$

where we have assumed that $\eta_o = 0.6$, as recommended by Briggs (1975). Equation (5.85) is identical to the '2/3 law' (7.35) provided that $\gamma_1/\gamma_3 = 4.11$. Lacking a value for either γ_1 or γ_3 we will assume simply $\gamma_1 = 1.0$ and $\gamma_3 = 0.24$.

Thus, we have shown that in neutral flows free of both turbulence and shear our model (5.79) is consistent with the '2/3 law', which is based inherently on a two-dimensional model of the buoyant plume. However, in applications of (5.79) to turbulent flows, this model will acquire three-dimensional characteristics, since then $dL/dt \neq 0$; and Moore (1974) claims that the predictions of a three-dimensional formula are in 'marginally' better agreement with observations than those of the '2/3 law'.

If we apply our model to the problem of a jet, i.e. $F = 0$, in a neutral, laminar flow, we immediately get from the second member of (5.81) that $K =$ constant; and from the first member of (5.81) that $a \propto t^{1/2}$. The latter result is not in agreement with the observation $a \propto t^{1/3}$. It appears that the discrepancy is due to our neglect during the derivation of (5.79) of the viscous term in (5.66). In buoyant plumes the production term in this equation overwhelms the viscous destruction term in the early stages of rise; but in jets, where buoyant production is absent, the effects of viscosity are manifested early. According to our model, if $a \propto t^{1/3}$, then the circulation K, or more precisely the product $\eta_o K$, must decay at the rate $K \propto t^{-1/3}$. By contrast in the buoyant plume $K \propto t^{1/3}$ (see 5.84).

The constant γ_2, which recall is roughly the ratio of the volume of fluid in which $w'' \neq 0$ to that of the fluid in which $T'' \neq 0$, should have a value larger than unity. This constant appears to be analogous to the ratio M_{eff}/M in Briggs' (1975) model, which he sets equal to 2.3. When we used this value in our model (5.79) in simulations of stably stratified flows, we found through experimentation that the predicted plume rise Δz to the level of equilibrium satisfied

$$\Delta z = C_2 \left(\frac{F}{Us}\right)^{1/3} \tag{5.86}$$

where s is the stability parameter: $s = (g/T)[\partial T_0/\partial z + g/c_p]$ and $C_2 = 3.8$. Briggs (1975) found that expression (5.86) fits a variety of observational data but with C_2 in the range 2.3 to 3.1. By increasing the value of γ_2 to 4.0 the predictions of the model satisfy (5.86) with $C_2 = 3.1$, and therefore this value of γ_2 was adopted. Note that it implies that $\overline{w''}$ is non-zero in a cylindrical volume of fluid twice the radius of that in which $T'' \neq 0$. As we pointed out earlier, the value of γ_2 is immaterial when T_0 has an adiabatic profile (see 5.79) and, therefore, in our simulations it will affect only the penetration of the emissions into the inversion layer aloft.

In our model we have not taken into account the effects of momentum issuing from the source. These effects are usually significant only near the source and are manifest in a rapid initial rise of the plume and in a more rapid initial rate of entrainment than would occur were the momentum flux zero. We will approximate the momentum effects in our model simply by replacing the $\overline{w''}$-equation in (5.79) with

$$\overline{w''} = \frac{K}{4\pi a} + 2.03 \left(\frac{F_m}{U}\right)^{1/3} t^{-2/3} \tag{5.87}$$

where F_m is the source momentum flux defined by

$$F_m = \frac{T_0}{T_s} w_s^2 \; r_s^2 \tag{5.88}$$

and T_s, w_s, and r_s have the same meanings as in (7.35). The last term on the right-hand side of (5.87) is derived from the '1/3 law' of bent-over jets as given by Briggs (1975).

As one final point we want to emphasize that the basic

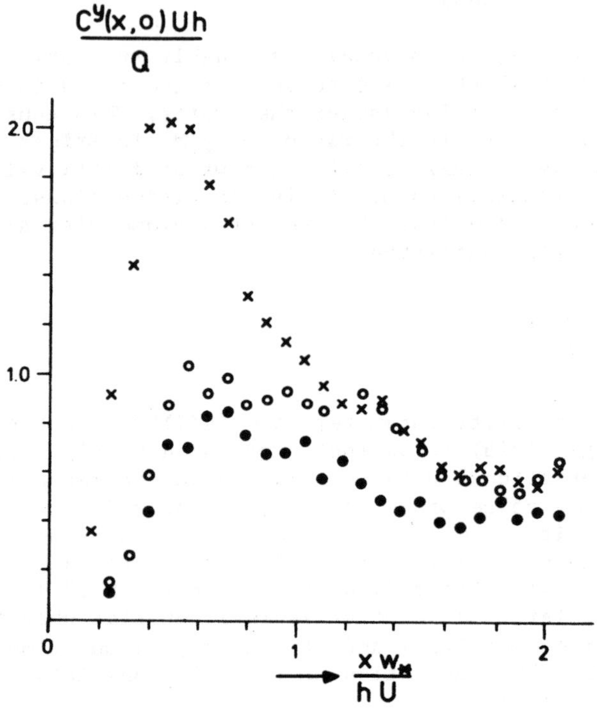

Figure 5.24. Comparison of simulated cross-wind integrated concentration at ground-level in plumes of buoyant ($F = 11.8\ m^4\ s^{-3}$, open and closed circles) and non-buoyant material (crosses) emitted from a source at elevation $z^s = 0.26h$. Closed circles were derived from the buoyancy parameterization (5.79) and (5.87) which neglects vorticity dissipation. Open circles are from the same model with dissipation effects approximated by (5.89).

Equation (5.62) ignores the effects of the vertical motion associated with adjacent plume segments. Equation (5.56) indicates that these effects are small when the plume is nearly horizontal. However, when the ambient fluid speed U is so low that the plume is nearly vertical, a given plume segment is pulled upward by plume material above and pushed upward by the material below. The result is a larger vertical speed $\overline{w''}$ than that given by (5.62) for an isolated thermal and, according to observations, a decrease in the value of the entrainment parameter η_o. We will ignore these effects in our present '1½ order' model rather than burden it with additional complexities that are not certain to improve its performance.

Let us look now at preliminary results of simulations that we have performed with this model. Figure 5.24 compares the ground-level normalized, cross-wind integrated concentration $C^y(x,o)\ Uh/Q$ of non-buoyant material from a source of elevation $z^s = 0.25h$ (x symbols) with the corresponding values (closed circles) from a source with buoyancy flux $F = 11.8\ m^4\ s^{-3}$ and $F_m = 2.8\ m^4\ s^{-2}$. We see immediately that with this '1½ order' model the effects of buoyancy on ground-level concentration are just as pronounced as they were with the "first order" model discussed earlier (Figure 5.22). (Maybe these models are trying to tell us something.) With a buoyancy flux of only $11.8\ m^4\ s^{-3}$, which is less than one-tenth the value of the emissions from moderate sized power plants, the peak ground-level concentration is less than one-half the value of passive material; and the location of the maximum concentration is farther downstream. Before searching for physical explanations for these results, it is prudent to investigate first whether significant features of the simulated concentration profiles are not merely artefacts of the model itself.

One aspect of the buoyant plume results shown in Figure 5.24 (closed circles) that is questionable is the prolonged duration of the buoyancy effect. If we consider that the peak ground-level concentration of passive material is due mainly to material that was released into the cores of downdrafts, then at points downstream of the peak, say $x > hU/w_*$ in Figure 5.24, the material observed was emitted either into the transition zones between updrafts and downdrafts or into updrafts and was quickly detrained. Since in both these regions the intensity of small scale turbulence is high, one would expect that the rise of weakly buoyant plume in such an environment would be quickly attenuated, rather than persisting as the results, shown in Figure 5.24, imply. Thus, this aspect of the simulations is probably an artefact of our neglect in the buoyancy model (5.79) of viscous dissipation of vorticity.

To explore this possibility we modified the circulation equation of (5.79) in two ways. First, to account for enhanced mixing (and subsequent annihilation) of vorticity, we assume

$$\frac{dK}{dt} = \frac{g\ \overline{T''}\ \gamma_1 a}{T_s} - \begin{cases} \Delta\hat{u}\ K/2L, & \text{if } \Delta\hat{u} > 0 \\ 0, & \text{otherwise} \end{cases} \tag{5.89a}$$

(See (5.73)). Secondly, to account for viscous dissipation generally, we assume

$$K = 0 \quad \text{if} \quad \|\overline{w''}\| < 0.1\ [\frac{1}{3}\ \overline{u_i'^2}]^{\frac{1}{2}} \tag{5.89b}$$

and u_i' is the subgrid scale velocity field.

With these modifications incorporated into the model, different results for the buoyant plume simulation are obtained, as indicated by the open circles in Figure 5.24. As expected the concentration profiles of the buoyant and passive plumes converge more rapidly with downstream distance than they did with the non-dissipative model; but at the point of maximum concentration and upstream of this point, the effects of the dissipation are much more subdued - the maximum concentration in the buoyant plume at ground-level is still only one-half that of a passive plume. Clearly these results may be artificial to some extent because since (5.89) are speculative approximations, we could 'force' the buoyant and passive plume results to agree as closely as we liked simply by manipulating the coefficients in (5.89). The question, then, is whether the degree of dissipation imposed by (5.89) in its current form is correct.

The only method immediately available to me to obtain even a rough answer to this question is to compare the simulated cross-wind integrated concentration patterns with the time averaged outlines of power plant plumes obtained from photographic studies by Crawford and Coleman (1979). Figure 5.25a shows such a comparison for the Gallatin power plant $F = 750\ m^4\ s^{-3}$, $z^s = 152$ m) on a day (4 May 1973) with unstable stratification and relatively light, shear-free winds ($U \simeq 5.0\ ms^{-1}$) in the lowest 1000 m. The figure shows the reported mean bottom and top of the observed plume superposed on the isopleths of the predicted C^y distribution. Neither h nor w_* was measured in the field study so we have indicated by error bars the probable range of the plume edge in the non-dimensionalized space depicted in Figure 5.25a. The available evidence indicates that the most likely position of the upper plume boundary is along the top of the error bars. It should also be noted that the bottom edge of the observed plume is virtually parallel to the ground out to the farthest observation point, x = 1300 m. Judging from the fact that the model reproduces this behavior quite well, we conclude that the magnitude of dissipation imposed by (5.89) is in the right range; because if it were excessive, the bottom edge of the simulated plume would slope downward near the source, and conversely. Another factor supporting the soundness of this model is that it does not predict the excessive penetration of buoyant plumes into the capping inversion that was a problem with the 'first order' model of Figure 5.22.

Having found that the predictions of our model with its parameterized plume buoyancy are not unreasonable, we turn finally to an examination of the physical process embodied in this model that is responsible for the acute effects of buoyancy on the ground-level concentration. The process is revealed by a

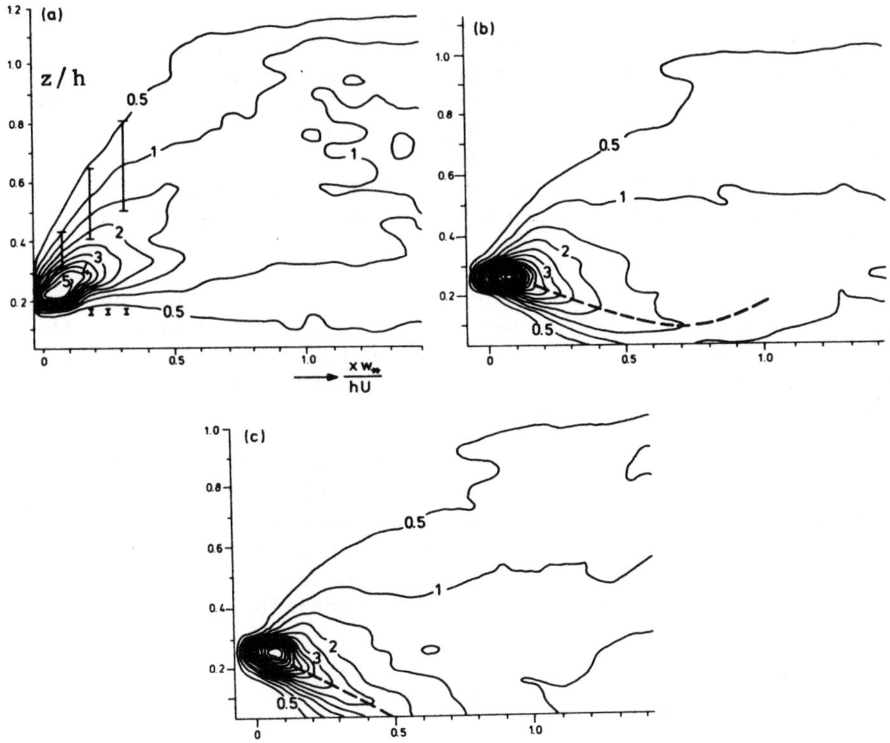

Figure 5.25.(a) Simulation of cross-wind integrated concentration in the Gallatin power plant plume. Vertical bars denote the location of the observed boundaries of the time averaged plume (see text for details). (b) Simulation of a hypothetical source of elevation $z^s = 0.26h$ with $F = 11.8\ m^4s^{-3}$ and $F_m = 2.8\ m^4s^{-2}$. (c) Same as (b) except non-buoyant material(as in Figure 5.5).

comparison of Figures 5.25(b),(c). The former shows the simulated distribution of cross-wind integrated concentration $C^y(x,z)hU/S$ for a source of height $z^s = 0.26\ h$ with buoyancy flux $F = 11.8\ m^4s^{-3}$ and $F_m = 2.8\ m^4s^{-2}$. Figure 5.25(c) shows the corresponding results for a passive plume, $F = F_m = 0$ (as presented earlier in Figure 5.5).

Look at the centerline (dashed lines in Figures 5.25(b) and (c)) of the two plumes. Following our earlier analyses we can express the elevation $z_{c\ell}$ of the plume centerline approximately by

$$\frac{dz_{c\ell}}{dt} = U\frac{dz_{c\ell}}{dx} = -\frac{1}{2}w_* + w_B \qquad (5.90)$$

where $\frac{1}{2}w_*$ represents the mode of the distribution p_w of vertical velocity fluctuations and w_B represents the plume's self-induced vertical speed, i.e., $\overline{w''}$. For a passive plume, $w_B = 0$ and (5.90) gives $dz_{c\ell}/dx = -\frac{1}{2}w_*/U$, which is confirmed by Figure 5.25(c). For the buoyant plume depicted in Figure 5.25(b), the slope of the initially straight portion of its centerline is $dz_{c\ell}/dx = -w_*/(3U)$. According to (5.90) this more gradual slope is due to an effective buoyancy speed $w_B \simeq 0.17\ w_*$. In effect, buoyancy has rotated the plume centerline upward as we saw earlier.

However, the centerline of the buoyant plume does not remain straight beyond $x \simeq \frac{1}{2}hU/w_*$ but rather it turns parallel to the ground and then shortly afterward it turns upward. As a result, ground-level is not subjected to the highest concentrations in the plume as it is when the plume is non-buoyant.

The bending of the buoyant plume's centerline is apparently due to the variation with height of the ambient fluid's vertical speed. Specifically, near the ground w must be a function of z such that at $z = 0$, $w = 0$. Assuming a constant divergence of the horizontal flow at levels $z < h/4$ within downdrafts, we can write (5.90) in the more proper (yet approximate) form

$$\frac{dz_{c\ell}}{dt} = -\frac{2w_* z}{h} + w_B , \qquad z < h/4. \tag{5.91}$$

If w_B can be represented by a constant, then (5.91) indicates that

$$z_{c\ell,min} \sim \frac{w_B h}{2w_*} . \tag{5.92}$$

Using the value $w_B \simeq 0.17w_*$ derived above for the plume shown in Figure 5.25(b), we get $z_{c\ell,min} \simeq 0.1h$, which is exactly what this figure shows.

The overall phenomenon can be described as follows. As buoyant material emitted into a downdraft approaches the ground, the vertical speed of the downdraft fluid begins to decrease. Since the lifetimes of downdrafts are generally long, the buoyant material will eventually arrive at an altitude, given by (5.92), at which its speed w_B relative to the environment has the same magnitude as that of the downdraft, but with opposite sign. The plume material moves laterally at this level until it is swept into an updraft and carried aloft. We expect that if $z_{c\ell,min}$ is well above the internal, turbulent shear layer at the base of the downdrafts, say $z_{c\ell,min} > \|L\|$, where ambient turbulence levels are kept low by the stabilizing effects of subsidence, then a large part of the buoyant plume will not reach ground-level until it has undergone at least one traverse of the mixed layer and its buoyancy velocity w_B has been obliterated.

Whether this phenomenon actually occurs in the atmosphere is not known at the present time but the mere possibility of its existence raises the need for caution both in modeling buoyant plumes in convective conditions and in applying numerical and empirical data for passive material dispersion to estimate buoyant plume concentrations. On the modeling side we can see that entry of buoyant material into the surface layer can depend critically on whether the dissipation of the plume's vorticity has been modeled accurately. Our model and other existing models treat this process in an ad hoc way. Regarding applications of non-buoyant plume data, we saw earlier (see Section 5.3 on comparison of model results with observations) that the maximum ground-level concentration of passive material varies as $1/(z^s h)$. However, with buoyant plumes, we found that with buoyancy fluxes small compared to those of conventional power plant plumes, plume material may not reach ground-level until it has coursed at least once through the depth of the mixed layer. In this case dependence of ground-level concentration on $1/(z^s h)$ is rather quickly transformed into dependence on $1/(h^2)$. Further study is needed to quantify these relationships.

ACKNOWLEDGEMENT

I am indebted to John Irwin for making available the data presented in Figure 5.1, and to William B. Norris and his colleagues in the Tennessee Valley Authority for lending me a number of time-lapse movies of power plant plumes which provided valuable insights into the phenomenon of buoyant plume dispersion.

6. DIFFUSION IN THE STABLE BOUNDARY LAYER

J.C.R. Hunt

University of Cambridge, Cambridge CB3 9EW, U.K.

6.1. INTRODUCTION

The diffusion of a substance released into a turbulent flow cannot be described by any one model or theory. In this chapter I shall aim to introduce and extend some of the main theoretical ideas and models that have been found to describe various aspects of turbulent diffusion in the atmospheric boundary layer. Most of the models tend to have a limited range of application (which can be more or less specified), but between them they do reveal most of the general features and trends. In some cases they also lead to useful quantitative predictions. Special emphasis is placed on statistical models because of their clear physical and theoretical basis and because they indicate some of the limitations of practical models. Statistical models also have the advantage of enabling the diffusion of contaminants to be calculated in terms of local atmospheric measurements, which is not possible in many practical models.

This chapter is mainly concerned with diffusion from small sources in the atmospheric boundary layer over flat terrains in steady conditions. However, the theoretical ideas introduced here can be applied to a number of complex flows in changing topographic or atmospheric conditions. (Turbulent diffusion specifically around buildings and hills has been reviewed by Hunt et al., 1979.)

6.2. BASIC IDEAS ABOUT MOLECULAR AND FLUID ELEMENT MOTION AND PROBABILITY DISTRIBUTIONS

The main object of the study of the diffusion of a substance (S) through a turbulent flow is to relate the development of concentration distributions and fluxes to the turbulent velocity field and to the molecular transport processes. Following Saffman (1960), it is important to distinguish between the velocity and displacements of fluid elements and the velocity and displacement of molecules. Fluid elements are control volumes moving at the local fluid velocity. Their size is large compared with the molecular scale ($\leqslant 10^{-6}$ m) and small compared with the scale of the smallest fluid motion, the Kolmogorov microscale $(\nu^3/\varepsilon)^{1/4}$, which is $\sim 10^{-3}$ m in the atmosphere. As fluid elements (control volumes) move, molecules (or heat in the form of thermally-agitated molecules) pass into and out of the control volume. Consequently a fluid element containing some substance (S) in a concentration C_0 at time $t = 0$, does not have the same concentration at time t, unless the diffusivity of the 'marked molecules' or particles can be neglected (Figure 6.1).

As we shall see, in stratified flows the molecular transport of density as well as momentum affects the fluid element position, $X_i(t)$. The distinction between $X_i(t)$ and the displacement of molecules, $X_i^{(m)}(t)$ is important when data on $X_i^{(m)}$ is used to infer the statistics of X_i (e.g., heat transfer to infer the spread of smoke particles).

As an ideal it is useful to begin a discussion of turbulent diffusion, by neglecting molecular diffusion. Then if $p(x_i,t;x_i',t_0)\,dx_i'$ is the probability of a fluid element moving from near x_i' at t_0 to x_i at t in an ensemble of similar turbulent flows, and if $C_0(x_i)$ is the concentration at time t_0, the mean and mean square concentration at x_i,t are

$$C(x_i,t) = \int p(x_i,t;x_i',t_0)\, C_0(x_i')\, dx_i' \qquad (6.1a)$$

$$\overline{\underset{\sim}{c}^2}\,(x_i,t) = \int p(x_i,t;x_i',t_0)\, C_0(x_i')^2\, dx_i' \qquad (6.1b)$$

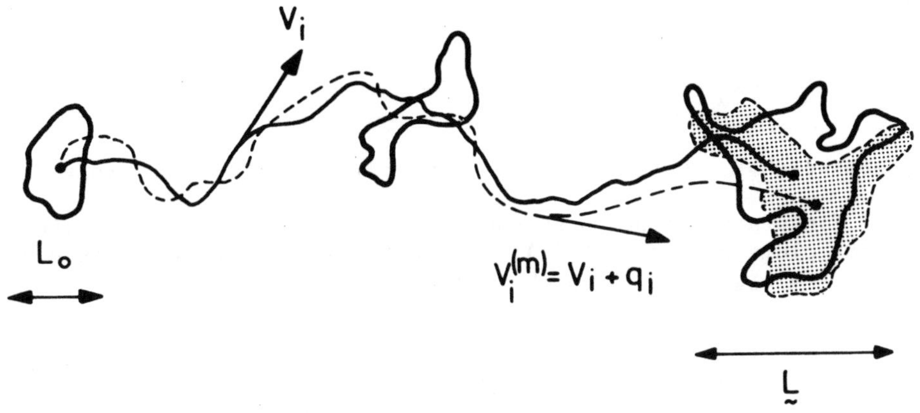

Figure 6.1. Movements of a small patch of pollutant released at a source; dashed lines are paths of marked molecules, denoted by $X_i^{(m)}(t)$; solid lines are fluid element paths, denoted by $X_i(t)$. (Parallel to the local velocity vector.) Note how these paths may eventually diverge. Instantaneous cloud dimensions are indicated by L_o and $\underset{\sim}{L}$ at t_o and t, respectively. The molecular velocity is denoted by q_i.

where $\underset{\sim}{c}$ denotes the instantaneous concentration, which is related to C by $\underset{\sim}{c} = C + c$ (c is the fluctuation). Since the probability of finding an element somewhere is one: $\int p(x_i,t;x_i',t_o)\, dx_i = 1$. Thence

$$\int C(x_i,t)\ dx_i = \int C_o(x_i')\ dx_i' \tag{6.2a}$$

$$\int \overline{\underset{\sim}{c}^2}(x_i,t)\ dx_i = \int C_o^2(x_i')\ dx_i'\ . \tag{6.2b}$$

For the important special case of an instantaneous release of mass M_o at $t = t_o$ from a point source at $x_i{}^s$, $C_o(x_i) = M_o\delta(x_i - x_i{}^s)$, so that according to (6.1), $C(x_i,t) = M_o\, p(x_i,t;x_i{}^s,t_o)$. Now the mean moments, $\overline{X_i{}^n}$ of the fluid element displacement are related to the mean concentration distribution and probability distribution by

$$\int x_\alpha^n\, C(x_i,t)\ dx_i = M_o \int x_\alpha^n\, p(x_i,t;x_i^s,t_o)\ dx_i \equiv M_o\, \overline{X_\alpha^n}. \tag{6.3a}$$

In the following we shall denote the rms particle displacement with respect to the source by $\sigma_\alpha = (\sigma_x, \sigma_y, \sigma_z)$, which is defined as

$$\sigma_\alpha^2 = \overline{(X_\alpha - x_\alpha^s)^2}. \tag{6.3b}$$

Remember that a Greek index means that no summation is involved.

For practical purposes one is interested in the mean position of the plume or cloud, $\overline{X}_i(t)$, (i = 1,2,3 or X,Y,Z), its spread or width $\Sigma_i = (\Sigma_x, \Sigma_y, \Sigma_z)$ which is defined as

$$\Sigma_\alpha^2 = \overline{(X_\alpha - \overline{X}_\alpha)^2} = \overline{X_\alpha^2} - \overline{X}_\alpha^2 = \sigma_\alpha^2 - (\overline{X}_\alpha - x_\alpha^s)^2. \tag{6.3c}$$

An average width is given by $L = (\Sigma_1\ \Sigma_2\ \Sigma_3)^{1/3}$. In cases where the tails of the concentration distribution need to be known (e.g., plumes originating from ground level releases, or explosive regions in inflammable clouds) higher moments, e.g., $\overline{X_\alpha^3}$, are also important.

If a cloud of marked fluid elements or particles is released, at some initial time t_o, then one wants to know the position of the instantaneous cloud centre $\tilde{X}_i$, and the instantaneous dimensions of the cloud $\tilde{\Sigma}_i$, or $\tilde{L} = (\tilde{\Sigma}_1\ \tilde{\Sigma}_2\ \tilde{\Sigma}_3)^{1/3}$. $\tilde{X}_i$ and $\tilde{\Sigma}_i$ are random functions defined by (see also Figure 6.2)

$$\tilde{X}_\alpha = \int x_\alpha\, \tilde{c}(x_i, t)\, dx_i / \int \tilde{c}\, dx_i,$$
$$\tilde{\Sigma}_\alpha^2 = \int (x_\alpha - \tilde{X}_\alpha)^2\, \tilde{c}(x_i, t)\, dx_i / \int \tilde{c}\, dx_i\ . \tag{6.4}$$

Batchelor (1952) showed that

$$\overline{\tilde{X}_\alpha} = \overline{X}_\alpha, \qquad \overline{\tilde{\Sigma}_\alpha^2} + \overline{(\tilde{X}_\alpha - \overline{X}_\alpha)^2} = \Sigma_\alpha^2,$$
$$\overline{\tilde{\Sigma}_\alpha^2} = \tfrac{1}{2}\overline{(X_\alpha^{(a)} - X_\alpha^{(b)})^2}$$

where a and b refer to any pair of particles in the cloud. These expressions imply that (i) the mean cloud centre coincides with the mean position of individually-released particles at some initial time t_o, (ii) the spreading of the cloud relative to its own centre plus the meandering of the cloud's centre equals the spread of individually released particles, and (iii) the spreading of the cloud relative to its own centre is equal to one-half the mean square dispersion of a pair of particles. Eventually the

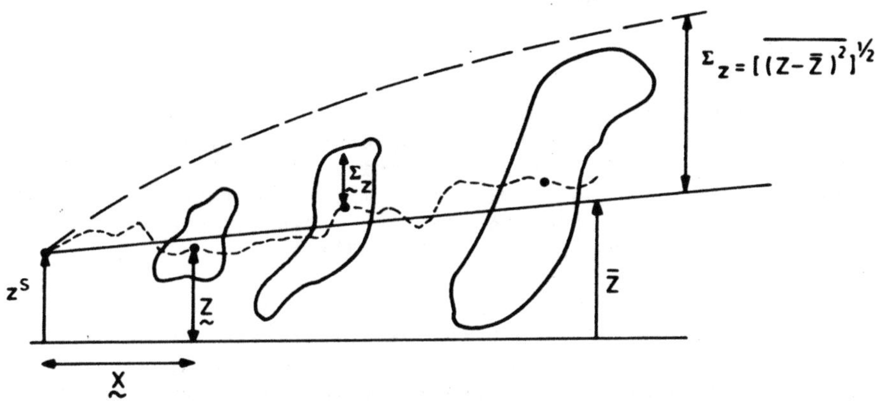

Figure 6.2. Notation for turbulent cloud diffusion. $\underset{\sim}{Z}$ = height of the instantaneous cloud center, $\overline{Z}$ = mean fluid element height, $\underset{\sim}{\Sigma}_z$ = the vertical dispersion of the instanteneous cloud relative to $\underset{\sim}{Z}$, Σ_z = the mean vertical fluid element spread relative to $\overline{Z}$. The source height is denoted by z^s. The dashed curve is the path of a single cloud. The solid curve represents the average path of the ensemble of clouds.

cloud becomes much larger than the scale of turbulent motions and particle displacements become decorrelated, so that

$$\overline{(\underset{\sim}{X}_\alpha - \bar{X}_\alpha)^2} \to 0, \qquad \Sigma^2_\alpha = \overline{\underset{\sim}{\Sigma}^2_\alpha},$$

$$\overline{\left(X^{(a)}_\alpha - \overline{X}^{(a)}_\alpha\right)\left(X^{(b)}_\alpha - \overline{X}^{(b)}_\alpha\right)} \to 0.$$

Using the conservation condition (6.2) the second moments provide an estimate of the mean concentration C_σ in the mean cloud (cloud centre $\overline{X}_i$) with representative volume $L^3 = \Sigma_x\, \Sigma_y\, \Sigma_z$, $C_\sigma \sim M_0/L^3$. On the other hand, the mean concentration as averaged over the centre of each cloud (cloud centre $\underset{\sim}{X}_i$) is $\overline{\underset{\sim}{C}_\sigma} \sim M_0/\underset{\sim}{L}^3$, where $\underset{\sim}{L}^3 = (\underset{\sim}{\Sigma}_x\, \underset{\sim}{\Sigma}_y\, \underset{\sim}{\Sigma}_z)$.

For a continuous point source (source strength Q) in a uniform windfield (U) the mean concentration can be estimated by $\overline{C}_\sigma \sim Q/(U\, \Sigma_y\, \Sigma_z)$. Similarly (following Chatwin and Sullivan (1979)), the mean square of the fluctuation in concentration and its ratio to the square of the mean concentration can be estimated. Since according to (6.2)

$$\overline{\underset{\sim}{c}^2}\, L^3 \sim (M_0/L^3_o)^2\, L^3_o \quad \text{and} \quad (C\, L^3)^2 \sim M_0^2 \tag{6.5a}$$

it follows that the relative variance

$$\overline{\underset{\sim}{c}^2}/C^2 \sim \overline{c^2}/C^2 \sim (L/L_o)^3 \qquad \text{(for } L/L_o > 1). \qquad (6.5b)$$

Here, L_o^3 denotes the initial cloud volume. Thus in the absence of molecular diffusion, the turbulence spreads out any initial distribution so that the variance relative to the square of the mean concentration increases without limit. (The rate of increase is at its largest when dL/dt is greatest - usually the initial stages of a plume's growth (Pasquill, 1974).)

When the substance S is also diffused by molecular motion, the moments of $\underset{\sim}{c}$, M_o and X_i are not exactly related by equations (6.1) and (6.3). But except near rigid surfaces and density interfaces, and in stable conditions, it is a good approximation to assume that (6.1a) and (6.3) are valid.

However, mean square concentration fluctuations <u>are</u> sensitive to molecular effects even in homogeneous flows. By considering the continuity equation for concentration $\underset{\sim}{c}$ (= C + c)

$$\frac{d\underset{\sim}{c}}{dt} = \frac{\partial \underset{\sim}{c}}{\partial t} + \underset{\sim}{u}_i \frac{\partial \underset{\sim}{c}}{\partial x_i} = D \frac{\partial^2 \underset{\sim}{c}}{\partial x_i \partial x_i} \qquad (6.6)$$

it follows (Green's theorem) that in an infinite medium or one bounded by impervious walls

$$\frac{1}{2} \int d\overline{\underset{\sim}{c}^2}/dt \; dx_i = - D \int \overline{(\partial \underset{\sim}{c}/\partial x_i)^2} \; dx_i < 0. \qquad (6.7)$$

Several recent experimental and theoretical investigations have shown that in clouds or plumes initially $\overline{c^2}/C^2$ increases as indicated by (6.5b). But after a time of the order of T_L, molecular diffusion smears out the increasing concentration gradients between adjacent fluid elements and leads to $\overline{c^2}$ remaining of the same order as C^2 in contrast to the result (6.5b) for no molecular diffusion.

6.3. TURBULENT DIFFUSION IN IDEALISED FLOWS

Marked Fluid Elements in Unstratified Turbulence Away from Boundaries

Consider the displacements X_i = (S,Y,N) of an ensemble of fluid elements passing through a 'source' at x_i^s at time t = 0. The orthogonal components S,Y,N of the displacement vector X_i are

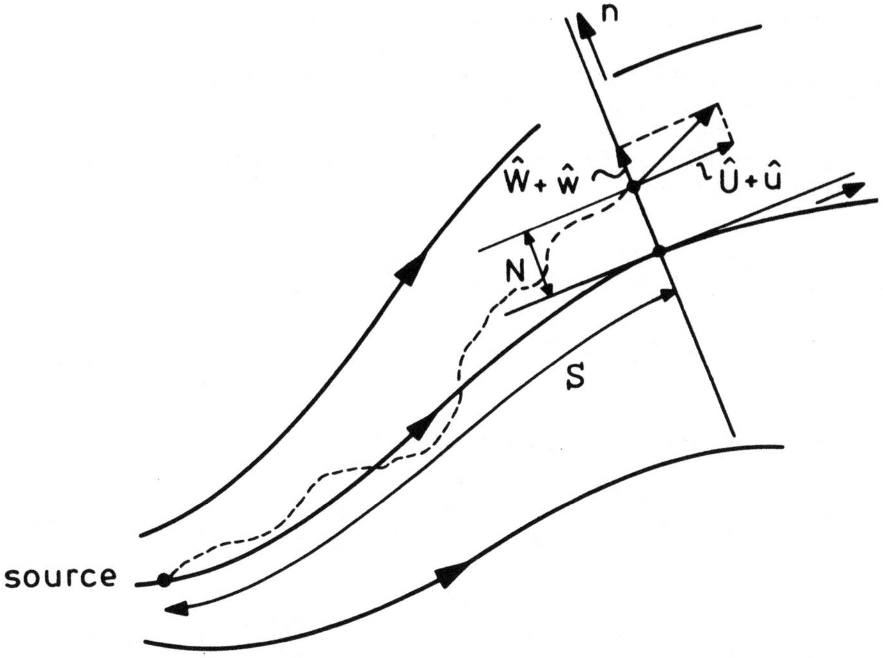

Figure 6.3. Frame of reference and notation for the analysis of fluid element motion in a general straining or shear flow. The dashed line represents a fluid element path.

defined with respect to a streamline frame of reference (s,y,n), where the s- and n-axes are respectively parallel and perpendicular to the mean stream line through the source (Figure 6.3).

The two-dimensional mean velocity flow field and the turbulence flow have components (U,0,W), (u,v,w) and $(\hat{U},0,\hat{W})$, $(\hat{u}, \hat{v}, \hat{w})$ in the Cartesian and streamline frames of reference respectively.

Then

$$\frac{dS}{dt} = \hat{U} + \hat{u}, \qquad \frac{dY}{dt} = (\hat{U} + \hat{u})\frac{dY}{ds} = \hat{v},$$

$$\frac{dN}{dt} = (\hat{U} + \hat{u})\frac{dN}{ds} = \hat{W} + \hat{w} \; . \tag{6.8}$$

The effects of different mean and turbulent flow fields on turbulent diffusion can readily be seen by expanding $\hat{U}_i$ and $\hat{u}_i$ in Taylor series about the source on the same streamline.

Unidirectional shear flow. In this case the velocity is defined as $U_i = (U(z),0,0)$. Let $(S,Y,N) = (X,Y,Z)$ and $(s,y,n) = (x,y,z)$. Thus $\hat{U}_i = (U(z),0,0)$ and $\hat{u}_i = (u,v,w)$. Then (6.8) becomes, with $U^s = U(z^s)$ and $dU^s/dz = (dU/dz)_{z=z^s}$, for a Taylor expansion around z^s

$$\frac{dX}{dt} = \{U^s + Z\frac{dU^s}{dz} + u + X\frac{\partial u}{\partial x} + Y\frac{\partial u}{\partial y} + Z\frac{\partial u}{\partial z} + \dots\}, \tag{6.9a}$$

$$\frac{dZ}{dt} = \{w + X\frac{\partial w}{\partial x} + Y\frac{\partial w}{\partial y} + Z\frac{\partial w}{\partial z} + \dots\}. \tag{6.9b}$$

In the case of weak turbulence ($\sigma_u/U^s \ll 1$),

$$\frac{dZ}{dx} = \frac{dZ/dt}{U+u} = \frac{w + Z\frac{\partial w}{\partial z} + X\frac{\partial w}{\partial x}}{U^s}\left(1 - \frac{Z}{U^s}\frac{dU^s}{dz} - \frac{u}{U^s} + ..\right), \tag{6.9c}$$

$$\frac{dY}{dx} = \frac{dY/dt}{U+u} = \frac{v + Z\frac{\partial v}{\partial z} + X\frac{\partial v}{\partial x}}{U^s}\left(1 - \frac{Z}{U^s}\frac{dU^s}{dz} - \frac{u}{U^s} + ..\right). \tag{6.9d}$$

Thence very close to the source $t \ll L_x/\sigma_u \sim T_L$ the particle displacement becomes

$$X = x = U^s t, \qquad Z = wt = xw/U^s \tag{6.10}$$

and the probability that the particle lies between z and z + dz is

$$\{\int p(x,y,z,t;x_i^s,0)\,dxdy\}\,dz = \{\frac{1}{t}\,p^s(w)\}\,dz \tag{6.11}$$

where $p^s(w)$ is the probability distribution of the vertical velocity at x_i^s. So for an instantaneous point source of strength M_0 (using (6.1a))

$$\int C\,dxdy = M_0\,p^s(Z/t)/t \tag{6.12a}$$

and for a continuous point source of strength Q

$$C^y(x,z) = \int C\,dy = \frac{Q}{U^s}\,\frac{p^s(U^s Z/x)}{(x/U^s)} \tag{6.12b}$$

Thus, by measuring the concentration distribution close to the source, the turbulence can be inferred (see Section 6.3 on

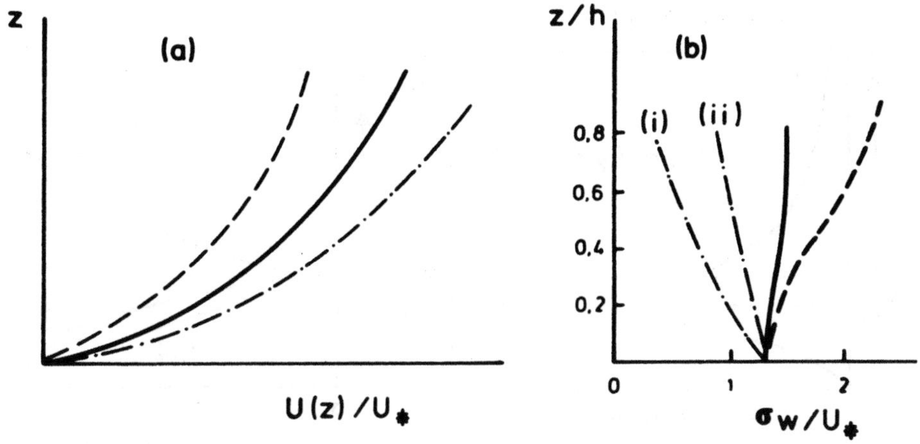

Figure 6.4. Velocity distribution and particle motion in the atmospheric boundary layer. (a) Typical mean velocity profile in the surface layer; dashed line unstable, solid line neutral, dashed-dotted line stable conditions. (b) Typical profiles for the rms of the vertical velocity component (σ_w), The curves (i)-(ii) indicate the range in stable conditions (see Section 6.4).

diffusion in a stably stratified fluid) and by examining $p^s(w)$, some paradoxical results of plume behaviour in convective conditions can be explained (Sections 6.4, 5.4 and Figure 6.4(c)).

From the first-order expression of Z in (6.10), the higher-order term involving Z in (6.9) can be evaluated and thence the rates of growth of the first three moments of X, Y, Z near the source (Figure 6.4). (See Monin and Yaglom, 1971, pp. 557 - 558.)

The longitudinal growth of a cloud relative to its <u>mean centroid</u> (which does not move at a velocity U^s) is (using continuity)

$$\frac{1}{2}\frac{d}{dt}\Sigma_x^2 = \overline{u^2}t + \frac{3}{2}t^2(\overline{uw}\frac{dU^s}{dz} + \frac{1}{2}\overline{w\frac{\partial u^2}{\partial z}}) + \frac{1}{2}t^3\overline{w^2}(\frac{dU^s}{dz})^2. \qquad (6.13)$$

The mean vertical displacement of the cloud or plume (Figure 6.4(d)) is

$$\frac{d\overline{Z}}{dt} = t\,\frac{d\overline{w^2}}{dz} \qquad (6.14)$$

$$\frac{d\overline{Z}}{dx} = \frac{-\overline{uw}}{(U^s)^2} - \frac{x}{(U^s)^3}\left(\overline{w^2}\frac{dU^s}{dz} - U^s\frac{d\overline{w^2}}{dz}\right). \qquad (6.15)$$

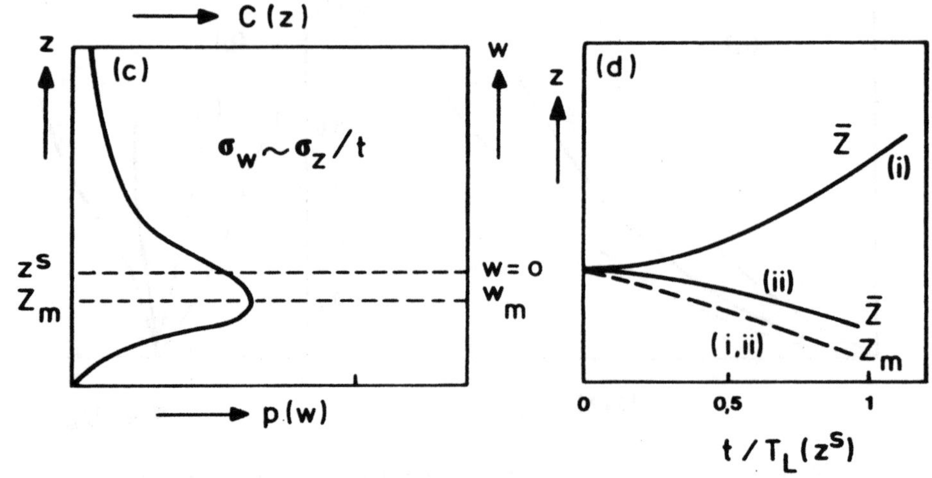

Figure 6.4. (c) Probability distribution of the vertical velocity close to the source (the same as the C(z) near the source). Typically skewed p(w) are found in the unstable boundary layer and in the wake of hills (in both $\overline{w^3} > 0$). (d) Mean plume or cloud height $\overline{Z}$ (solid line) and height of the maximum mean concentration Z_m in: (i) convective conditions $d\sigma_w^2/dz > 0$, $\overline{w^3} > 0(z^s/h \sim 0.1)$; (ii) wake of a hill $d\sigma_w^2/dz < 0$, $\overline{w^3} > 0$. (Elevated source just above the wake.)

The mean vertical growth of a plume (or total diffusion of a cloud, i.e., centroid movement plus cloud growth):

$$\frac{1}{2}\frac{d}{dx}\Sigma_z^2 = \frac{\overline{w^2}\, x}{(U^s)^2}\left[1 - \frac{2}{U^s}\frac{\overline{uw^2}}{\overline{w^2}} - 0(x/U^s T_L)\right] + \\ - \frac{x^2}{(U^s)^4}\left[\frac{3}{2}\overline{w^3}\frac{dU^s}{dz} - U^s\frac{d\overline{w^3}}{dz} + \ldots\right]. \qquad (6.16)$$

The lateral growth is given by

$$\frac{d\overline{Y}}{dx} = 0, \qquad (6.17)$$

$$\frac{1}{2}\frac{d}{dx}\Sigma_y^2 = \frac{\overline{v^2}\, x}{(U^s)^2}\left[1 + 0\left(\frac{x}{U^s T_L}\right) + 0\left(\left\{\frac{\overline{v^2}}{(U^s)^2}\right\}^2\right)\right]. \qquad (6.18)$$

The Σ_x, Σ_y and Σ_z in (6.13), (6.16) and (6.18) are defined by

(6.3c). Since v remains correlated as the particle moves vertically through the shear layer, and v and w are uncorrelated (for small spacing), the local horizontal transverse width of a plume, defined as

$$\Sigma^2_{y,\ell} = \int y^2 \, C(x,y,z) \, dy / \int C \, dy,$$

is then (from (6.9d)) given as a function of z by

$$\Sigma^2_{y,\ell} = \frac{\overline{v^2} \, x^2}{(U^s)^2} \left[1 - \frac{z}{U^s} \frac{dU^s}{dz}\right]^2 (1 + O(x/U^s T_L)). \tag{6.19}$$

An important general conclusion from these expansions is that as a plume or cloud travels from its source, the non-homogeneity (e.g., $d\overline{w^2}/dz \neq 0$) or non-Gaussianity (e.g., $\overline{w^3} \neq 0$) of the turbulence and the non-uniformity of the profile affects the diffusion as much as the decreasing correlation of the turbulent velocity from its value at the source. Secondly, note how marked fluid elements can have a mean vertical velocity, even when the mean vertical velocity measured by a stationary observer, is zero. (Think of marking the edges of an agitated region of fluid; the marked elements move preferentially into the agitated region, even if there is no mean flow in that direction.)

Over longer periods of time, of the order of T_L, the diffusion is more appropriately analysed in terms of the time rather than distance from the source. The interpretation of the diffusion as a function of distance can then only be approximate. Thus

$$\overline{Z} = \int_0^t \overline{V}_z \, dt,$$

$$\Sigma_z^2 = 2 \int_0^t \left\{ \int_0^{t'} \sigma_w^L(t') \, \sigma_w^L(t'') \, \rho_L^w(t',t'') \, dt'' \right\} dt' \tag{6.20}$$

where $V_z'(t) = V_z - \overline{V}_z$ and $\sigma_w{}^L = (\overline{V_z'^2})^{\frac{1}{2}}$ are the fluctuating velocity of marked fluid elements and its rms value, respectively. The relations between the Lagrangian and Eulerian parameters depend on the structure and homogeneity of the turbulence and the mean flow. In all turbulent flows velocity fluctuations of a fluid element eventually become decorrelated,

$$\lim_{t' - t'' \to \infty} \rho_L(t',t'') \to 0 .$$

Even in many inhomogeneous flows, one can assume that $\rho_L(t',t'') = \rho_L(\tau)$, where $\tau = t' - t''$.

The Lagrangian time scale is then defined as

$$T_L^w = \tfrac{1}{2} \int_{-\infty}^{+\infty} \rho_L^w(t',t'')\, dt' \; . \tag{6.21}$$

If $\sigma_w{}^L(t') \simeq \sigma_w{}^L(t'')$ when $\| t' - t'' \| \lesssim T_L$ and if $\rho_L = \rho_L(\tau)$, then for estimating diffusion the Lagrangian velocity V_z can be regarded as effectively a stationary random function. So the vertical diffusion of fluid elements is given by (see also Section 2.7 and 7.2).

$$\Sigma_z^2 = 2(\sigma_w^L)^2 \int_0^t (t - \tau)\, \rho_L^w(\tau)\, d\tau \; . \tag{6.22a}$$

A long time after the elements have left the source, $t >> T_L$, the vertical diffusion has the asymptotic form (Taylor, 1921):

$$\Sigma_z^2 \simeq 2(\sigma_w^L)^2 \left[t\, T_L - \int_0^\infty \tau\, \rho_L^w(\tau)\, d\tau \right] \tag{6.22b}$$

where T_L may vary slowly with time.

Only when $(\sigma_w{}^L)^2 T_L$ is constant can we conclude that $\Sigma_z \sim \sqrt{t}$. In many conditions of the atmospheric boundary layer, this is not the appropriate conclusion.

In a shear flow, as the vertical diffusion carries the marked particles into faster and slower-moving fluids above and below the source, the along wind diffusion is increased. Any mean upward drift $\overline{V}_z$ causes an increased mean along wind drift (Figure 6.4). From (6.9a) follows

$$\frac{d\overline{X}}{dt} \equiv \overline{V}_x = U^s + t\overline{V}_z \frac{dU^s}{dz} + \ldots \; . \tag{6.23a}$$

Further

$$\frac{d(X - \overline{X})}{dt} = u + (Z - \overline{Z}) \frac{dU^s}{dz} \quad \text{and} \quad Z - \overline{Z} = \int_0^t V_z'(t'')\, dt'' .$$

Integration by parts then leads to

$$\Sigma_x^2 = \overline{\left\{ \int_0^t \left[(t - t'') V_z'(t'')\, dU^s/dz + u(t'') \right] dt'' \right\}^2} \tag{6.23b}$$

The special case where $d^2U/dz^2 = 0$ and $\sigma_w{}^2$ is constant shows how shear increases the growth of a cloud in the flow direction, for when $t \gg T_L$ (6.23b) leads to

$$\Sigma_x^2 \sim \frac{2}{3} (\sigma_w^L)^2 \; T_L \; t^3 (\frac{dU^s}{dz})^2 \tag{6.24}$$

(Tennekes and Lumley, 1972, p. 232). So this dispersion of a cloud is primarily produced by the combination of vertical diffusion and the mean velocity gradient.

The expressions (6.22) and (6.23) shows how most problems of turbulent diffusion eventually reduce to finding the Lagrangian correlation functions (e.g., $\rho_L{}^w(\tau)$) and the Lagrangian integral time scale T_L for the various velocity components. Physical and dimensional arguments (e.g. Corrsin, 1963) suggest that, if the Eulerian turbulence properties are varying slowly a time scale $T_L{}^w$ may be defined by

$$T_L^w = \beta \; T_E^w = \beta \; L_x^w / \sigma_w \tag{6.25a}$$

where β is a constant which is found by experiment to be about 1.0 in grid turbulence (Snyder and Lumley, 1971), and about 0.5 near the surface ($z < 50$ m) of the atmospheric boundary layer in neutral conditions (Pasquill, 1974).

The second empirical assumption that is often made is to assume that the functional form of $R_L{}^w(\tau)$ is approximately the same as that of the autocorrelation of w taken at a fixed point. In other words, if $T_E{}^w$ is the Eulerian fixed-point time scale,

$$\rho_E^w(\tau/T_E^w) \simeq \rho_L^w(\tau/T_L^w) \tag{6.25b}$$

In strongly convective or stable conditions the result (6.25b) should, as we shall show, be used with great caution. The results (6.25) enable T_L and $R_L(\tau)$ to be estimated from measurements of the turbulence taken at one point. At present this assumption is necessary because of the difficulty of Lagrangian turbulence measurements. In the future with remote sensing equipment, such assumptions may become redundant.

Straining flows. In many complex atmospheric boundary layer flows the stream-lines converge and diverge. As in the case of the shear flow, it is useful to expand the mean velocity in terms of its gradients on the mean streamline through the source. Then for a two-dimensional flow, since $\hat{W} = 0$ on $n = 0$, and by continuity

$$\hat{W}(x_i, t) = N \left(\frac{\partial \hat{W}}{\partial n}\right)_{n=0} + \ldots \simeq -N \left(\frac{\partial \hat{U}}{\partial s}\right)_{n=0} .$$

With this result (6.8) reduces for the case of weak turbulence to

$$\hat{U}\frac{dN}{ds} + N\frac{d\hat{U}}{ds} = \hat{w} \quad \text{and} \quad \hat{U}\frac{dY}{ds} = \hat{v}\ , \tag{6.26}$$

leading to

$$\begin{aligned} \sigma_n^2 &\equiv \overline{N^2}(s) = \frac{1}{\hat{U}(s)^2}\int_0^s\int_0^s \overline{\hat{w}(s')\hat{w}(s'')}\, ds'ds''\ , \\ \sigma_y^2 &\equiv \overline{Y^2}(s) = \int_0^s\int_0^s \frac{1}{\hat{U}(s')\hat{U}(s'')}\overline{\hat{v}(s')\hat{v}(s'')}\, ds'ds''\ . \end{aligned} \tag{6.27}$$

Note how the effects of convergence and divergence of streamlines on σ_n reduce mathematically to the transposing of the mean velocity from inside to outside the integrals. It is the local value of $\hat{U}(s)$ that controls σ_n whereas it is the integrated effect of $\hat{U}(s)$ that controls $\sigma_y(s)$ (Figure 6.5). (Hunt and Mulhearn, 1973). Note how in straining flows the plume width is less sensitive to the modelling of the turbulence than in homogeneous or shear flows, so that solutions to the diffusion equation are often comparable with results of statistical theory.

The main effects of such complex flows on a plume are to distort the path of the plume to follow the source streamline (perhaps bringing the plume close to the surface) and to broaden or narrow its width. Since $\sigma_n \sim \hat{U}^{-1}(s)$, the center line concentration $\sim[\sigma_y\sigma_n\hat{U}(s)]^{-1}$ does not change much, if we neglect the change in σ_y. (See also Section 6.5 on complex atmospheric conditions.)

Flux gradient relations - when are they likely to go wrong? Only in rather simple turbulent flows is it practical to estimate the diffusion of contaminants by calculating the statistics of the displacement of marked particles in terms of the turbulent velocity field. In some quite complicated diffusion problems, it may be appropriate to use a diffusion equation, i.e., to model the turbulent diffusion flux F_i by $F_i \equiv -\overline{u_i c} = K_{ij}\partial C/\partial x_j$, where C is described by the diffusion equation

$$\frac{\partial C}{\partial t} + U_i\frac{\partial C}{\partial x_i} = \frac{\partial}{\partial x_i}\left(K_{ij}\frac{\partial C}{\partial x_j}\right). \tag{6.28}$$

Such a model is most useful when K_{ij} can be wholely defined in terms of the turbulent velocity field $\underset{\sim}{u_i}$. However, in general, K_{ij} is also a function of the concentration distribution (see for example Berkowicz and Prahm, 1980). For instance, in homogeneous turbulence, with a Gaussian probability distribution

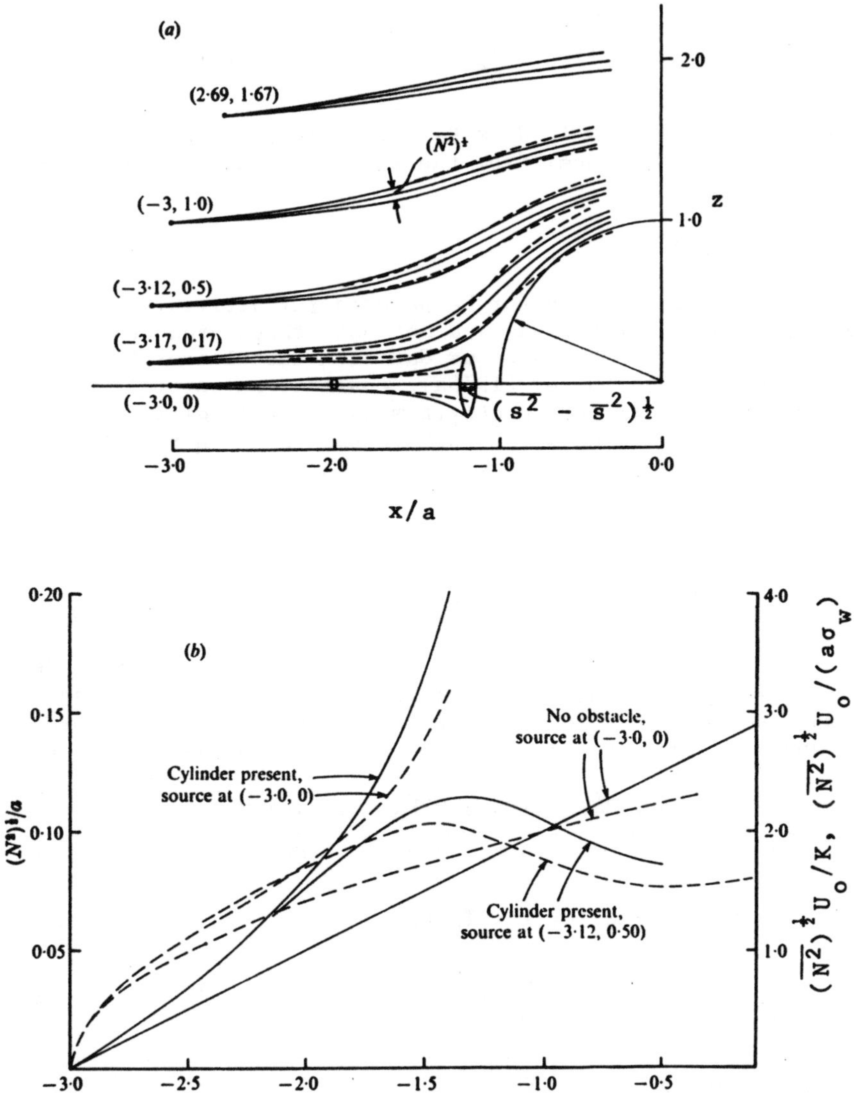

Figure 6.5. Turbulent diffusion in straining flow (Hunt and Mulhearn, 1973): calculation of widths $(\overline{N^2})^{\frac{1}{2}}$, $(\overline{Y^2})^{\frac{1}{2}}$ of plumes from sources placed in weak homogeneous (large scale) turbulent flow around a circular cylinder. The source coordinates are given by the numbers in parentheses. (For experimental confirmation see Hunt et al., 1979.) (a) Solid lines ——— = $(\overline{N^2})^{\frac{1}{2}}$; →0← = $(\overline{S^2} - \overline{S}^2)^{\frac{1}{2}}$; dashed lines - - - - - - - = $(\overline{Y^2})^{\frac{1}{2}}$. (b) $(\overline{N^2})^{\frac{1}{2}}$ divided by the cylinder radius, a, calculated from statistical theory ($t \ll T_L$) (solid lines), difusion equation (dashed lines) (left-hand axis shows values for $\sigma_w/U_0 = 0.05$, right-hand axis shows normalized values in terms of σ_w and of the diffusivity K).

$$F_z = K_z \frac{\partial C}{\partial z} \quad \text{with} \quad K_z = K_{33} = (\sigma_w^L)^2 \int_0^t \rho_L^w(\tau)\, d\tau \tag{6.29}$$

so that only when $t > T_L$ does K_z become independent of the traveltime (or distance) from the source and just a function of the turbulence. Then

$$K_z = (\sigma_w^L)^2\, T_L. \tag{6.30}$$

Can this expression for K_z be used for arbitrary concentration distributions and not just those produced by isolated sources in homogeneous Gaussian turbulence? Following Corrsin (1952) and Lumley (1975a), consider an initial concentration distribution $\underset{\sim}{c}(x_i, t = 0) = C_0(z)$ diffusing in a homogeneous but non-Gaussian turbulence. Then, in the absence of molecular diffusion, the flux in the z-direction

$$F_z(z,t) = \overline{w(\underset{\sim}{c} - C)} = - \overline{V_z\left(Z \frac{dC_o}{dz} + \frac{1}{2} Z^2 \frac{d^2C_o}{dz^2} + \frac{1}{6} Z^3 \frac{d^3C_o}{dz^3} \ldots\right)} \tag{6.31}$$

where $Z = \int_0^t V_z(\tau)\, d\tau$ is the particle displacement. F_z can be reexpressed entirely in terms of Z or in terms of higher-order Lagrangian autocorrelations. From (6.31), assuming $\overline{V_z} = 0$, we find

$$F_z = - \frac{1}{2} \frac{d\overline{Z^2}}{dt} \frac{dC_o}{dz} - \frac{1}{6} \frac{d\overline{Z^3}}{dt} \frac{d^2C_o}{dz^2} - \frac{1}{24} \frac{d\overline{Z^4}}{dt} \frac{d^3C_o}{dz^3} - \ldots$$

which may be reduced for $t < T_L$ to

$$F_z = -\overline{V_z^2}\, t \frac{dC_o}{dz} - \frac{1}{2} \overline{V_z^3}\, t^2 \frac{d^2C_o}{dz^2} - \frac{1}{6} \overline{V_z^4}\, t^3 \frac{d^3C_o}{dz^3} + \ldots \tag{6.32}$$

or for $t > T_L$ to

$$F_z = -(\sigma_w^L)^2 T_L \frac{dC_o}{dz} - \lambda_3 (\sigma_w^L)^3 T_L^2 \frac{d^2C_o}{dz^2} - \lambda_4 (\sigma_w^L)^4 T_L^3 \frac{d^3C_o}{dz^3} \ldots . \tag{6.33}$$

The constant λ_3 is given by

$$\lambda_3 = \int_0^t \int_0^t \frac{\overline{V_z(t)\, V_z(t')\, V_z(t'')}}{2\, (\sigma_w^L)^3\, T_L^2}\, dt'dt'' .$$

It depends on the skewness and is zero for Gaussian turbulence. The constant λ_4 is O(1) even in Gaussian turbulence.

Since most turbulent flows are non-Gaussian (a property usually associated with non-isotropic and inhomogeneous turbulence, e.g., Wyngaard, 1979), F_z depends on the higher derivatives of the concentration field almost as much as on the local concentration gradient. Only if the scale ℓ_c of the local concentration distribution is large enough so that

$$(\lambda_3, \lambda_4)\, \sigma_w^L\, T_L/\ell_c < 1 \quad \text{or} \quad (\lambda_3, \lambda_4)\, L_x/\ell_c < 1 \qquad (6.34)$$

and if the turbulence is fairly homogeneous and Gaussian can (6.29) and (6.30) be assumed. (Recent laboratory experiments by Sreenivasan et al. (1981) have shown how, with homogeneous Gaussian turbulence, the flux gradient relation (6.29) is a good approximation, even with very sharp concentration gradients, whereas with non-Gaussian and inhomogeneous turbulence (6.29) can be quite wrong.)

An important practical consequence of being able to assume the validity of the eddy diffusivity concept is that K_{ij} can be inferred from the flux in one situation (e.g., heat in the atmospheric boundary layer), and used to estimate the flux of a different substance with quite a different concentration distribution (e.g., a point source of pollution).

Diffusion in Stably-Stratified Turbulence

In this section we concentrate on the vertical growth of a plume or cloud in a stably stratified flow. Although the horizontal components of turbulence are also changed by stable stratification, the general features of a diffusing cloud or plume are unaffected (unless there is a marked change in wind direction with height, which can have a large effect).

In turbulent flow with neutral stability, fluid elements from a source move vertically over an ever increasing distance as time proceeds, as shown by (6.22). However, in stably-stratified flows increasing vertical displacement of fluid elements requires a steady input of energy (assuming the density of the fluid element remains unchanged). Given (i) a finite total amount or a very weak continuous supply of energy in the turbulence, (ii) an average kinetic energy of fluid elements of $\frac{1}{2}\rho(\sigma_w^L)^2$ and (iii) a roughly equal partition of energy between kinetic and potential energy (demonstrable by measurement and calculation), it follows that fluid elements can only diffuse through a vertical distance of order σ_w^L/ω_B, where ω_B is the Brunt-Vaisala frequency. However, as an element rises into warmer air, molecular diffusion results in a rise of temperature and density decreases. Thus by changing their temperature, fluid elements can diffuse above and below the limiting depth of σ_W^L/ω_B. But such molecular exchange is also likely to dilute the concentration of contaminant marking the fluid.

These physical ideas have been developed into an heuristic mathematical model for fluid-element vertical displacements in a homogeneous shear-free turbulent flow under the action of (a) the buoyancy forces ($\sim g\Delta\theta/\Theta$), $\Delta\theta$ being the difference between the fluid elements potential temperature and local mean potential temperature, (b) a damping force ($\sim -\mu\omega_B\, dZ/dt$) produced by waves radiated by fluid elements moving in the density gradient (typically $\mu \simeq \frac{1}{2} \pm \frac{1}{4}$), and (c) a random pressure gradient $H(t)$ ($= 1/\rho\ \partial p/\partial z$) produced largely by the acceleration of other particles. The fluid element height is described by the differential equations

$$\frac{d^2Z}{dt^2} + 2\mu\,\omega_B\frac{dZ}{dt} - \frac{g}{\Theta}\Delta\theta = H(t)$$

and (6.35a)

$$\frac{d\Delta\theta}{dt} = -\gamma\,\omega_B\,\Delta\theta - \frac{dZ}{dt}\frac{d\Theta}{dz}$$

(Pearson et al. (1983), Csanady (1964)), where $\gamma\omega_B$ is 'diffusivity' of the fluid element. (Typically $0.1 \lesssim \gamma \lesssim 0.4$; see Section 6.4.) Evidently the magnitude of $\overline{Z^2(t)}$ depends on the spectrum $E_H(\omega)$ of $H(t)$. It is chosen so that $\overline{V_Z^2} = \overline{w^2}$, or $\sigma_W{}^L = \sigma_W$, the Eulerian mean square velocity. On the basis of the observed insensitivity of velocity spectra to stratification, we assume that $E_H(\omega)$ is also a weak function of the stratification characterized by the parameter $\omega_B L_x/\sigma_W$. Thence, after normalizing E_H and ω, we find

$$\overline{Z^2}(t) = 2(\sigma_W/\omega_B)^2 \int_{-\infty}^{+\infty} [1 + (\gamma/\omega')^2]\, E_H'(\omega')\, p^{-1}(\omega') \times [1 - \cos(\omega'\omega_B t)]\, d\omega' , \qquad (6.35b)$$

$$1 = \int_{-\infty}^{+\infty} (\gamma^2 + \omega'^2)\, E_H(\omega')\, p^{-1}(\omega')\, d\omega',$$

where,

$$p(\omega') = (1 + 2\mu\gamma - \omega'^2)^2 + (\gamma + 2\mu)^2\,\omega'^2 ,$$

and

$$E_H = \sigma_W{}^2\,\omega_B\,E_H'(\omega'), \qquad \omega' = \omega/\omega_B.$$

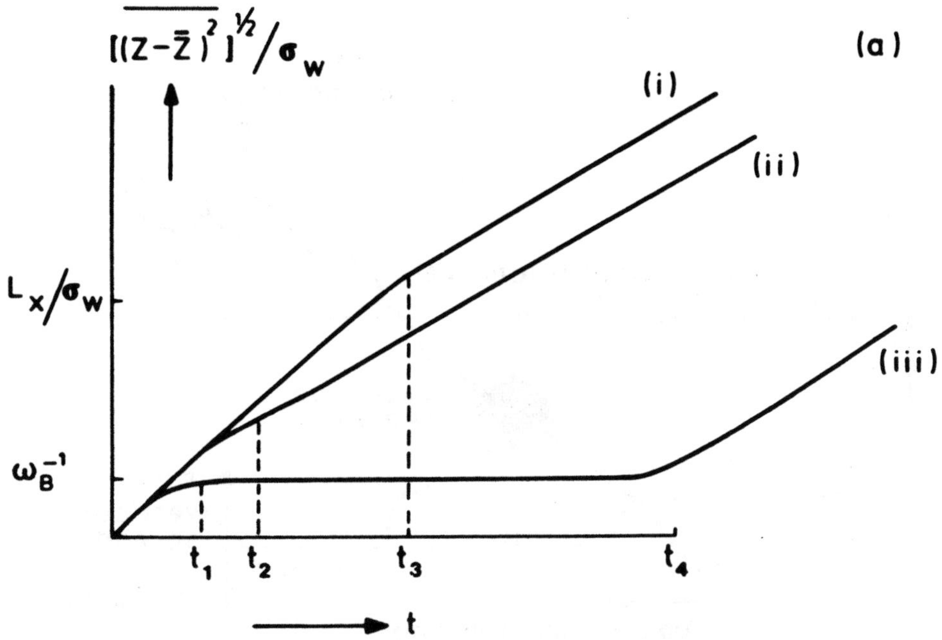

Figure 6.6. Vertical diffusion (log scale) in stably stratified turbulent flows: (a) rms displacement of fluid elements as a function of time in (log scale) effectively homogeneous turbulence; (i) neutral stability; (ii) moderate stable statification, with mixing of density by fluid elements ($T_L\omega_B \sim \gamma \sim 1$); (iii) strong stable stratification ($\omega_B L_x/\sigma_w > 1$, $\gamma \ll 1$). $t_1 \sim \omega_B^{-1}$, $t_2 \sim \omega_B^{-1} \sim T_L$, $t_3 \sim T_L$ and $t_4 \sim \gamma^{-2}\omega_B^{-1}$.

Also analytical expressions for $\rho_L{}^w(\tau)$ and T_L can be obtained from (6.35a,b)

This model helps explain some of the known effects of stratification on diffusion.

(i) Near a source when $\omega_B t \ll 1$ (6.35b) reduces to the well-known result (see Section 2.7)

$$\overline{Z^2} = \sigma_w^2 t^2. \tag{6.36}$$

(ii) Far downwind, when $\omega_B t > 1$

$$\overline{Z^2} = \left(\frac{\sigma_w}{\omega_B}\right)^2 [\zeta_z^2 + 2\, t\, T_L\, \omega_B^2] , \tag{6.37a}$$

where

$$\zeta_z^2(\omega_B L_x/\sigma_w) = 2\int_{-\infty}^{+\infty} \frac{E_H'(\omega')}{p(\omega')}\, d\omega'$$

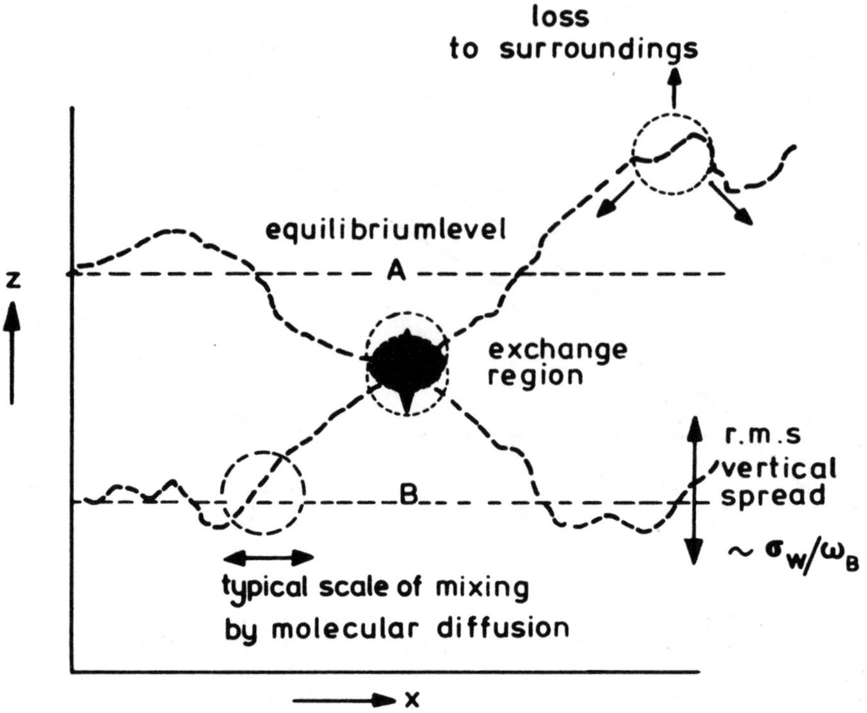

Figure 6.6.(b) Schematic diagram showing how a vertical heat or density flux is produced by a transfer of heat (or density) between fluid elements when they have restricted vertical motions in strong stable stratification. ($\omega_B L_x/\sigma_w \gtrsim 1$, $\gamma < 0.1$.)

is a constant of order one for the most likely forms of E_H. Figure 6.6(a) shows that when $\omega_B L_x/\sigma_w > 1$, the plume thickness $(\overline{Z^2})^{\frac{1}{2}} \sim \sigma_w/\omega_B$. Also (See Pearson et al.,1983)

$$T_L\ \omega_B = \frac{\pi\ \gamma^2\ E_H'(0)}{p(0)} = 2\ \mu\gamma^2 \sim \gamma^2 \tag{6.37b}$$

and therefore the diffusivity for fluid elements becomes

$$K_z = T_L\ \sigma_w^2 \sim \gamma^2\sigma_w^2/\omega_B. \tag{6.37c}$$

Thus when the effects of molecular diffusion are negligible, ($\gamma = 0$), as $t\omega_B \to \infty$, $d\overline{Z^2}/dt \to 0$. Consequently, in this limit $T_L \to 0$. For the integral of $\rho_L{}^w(\tau)$ to be zero, $\rho_L{}^w(\tau)$ must be negative over some range of τ. It also follows from (6.22b) that

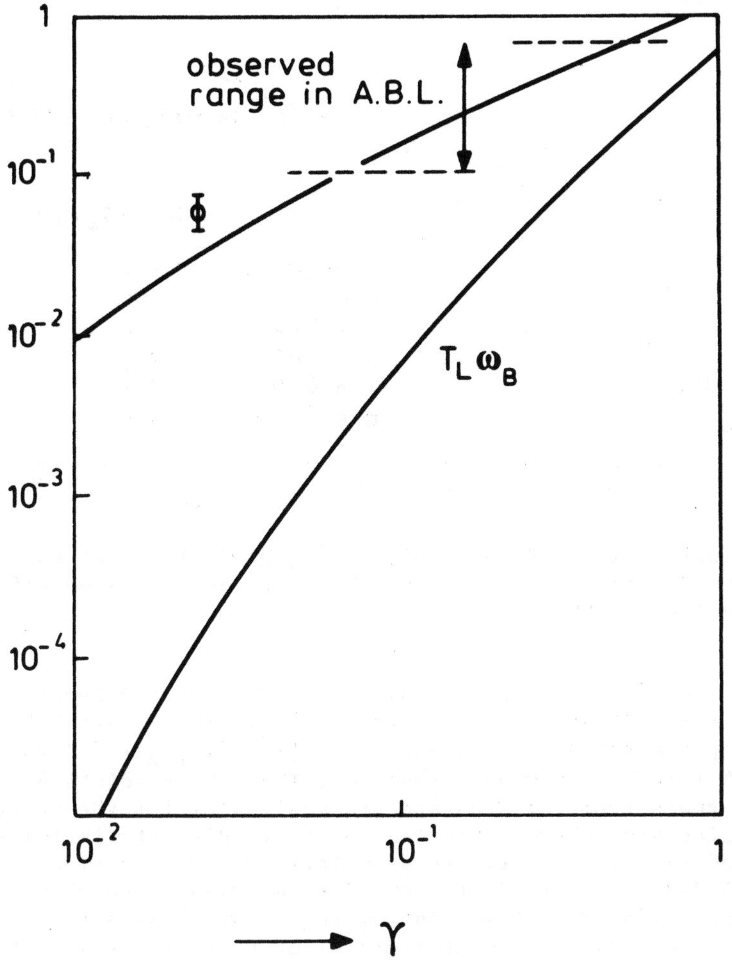

Figure 6.6.(c) Estimated relations between dimensionless vertical density flux Φ, Lagrangian timescale $T_L\ \omega_B$ and the parameter γ characterizing the mixing between fluid elements (Pearson et al., 1981).

if $T_L \to 0$,

$$\overline{Z^2}(t \to \infty) = -2\ \sigma_w^2 \int_0^\infty \tau\ \rho_L^w(\tau)\ d\tau.$$

and since $\overline{Z^2} > 0$, this result gives another reason why $\rho_L{}^w(\tau)$ must be negative over some range of τ.

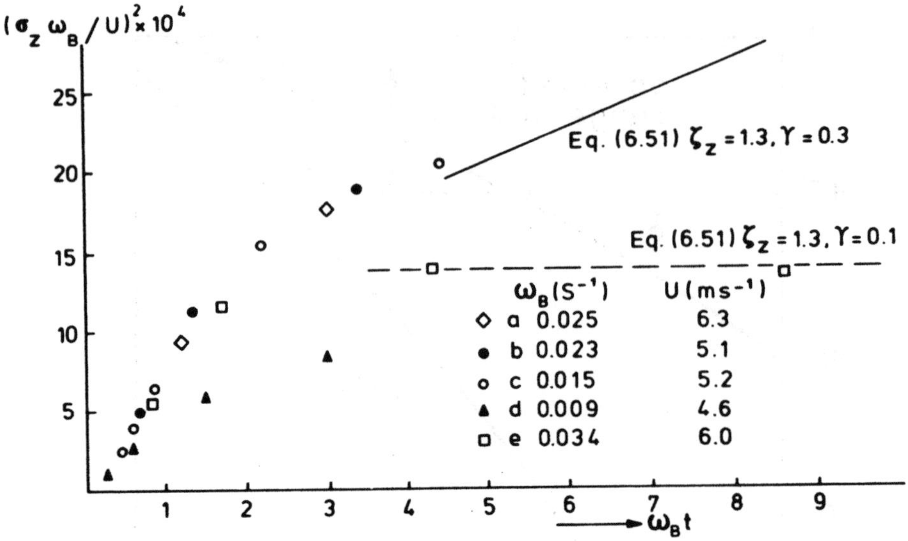

Figure 6.6.(d) Field measurements of $(\sigma_z \omega_B/U)^2$ by Hilst and Simpson (1958) plotted as a function of $\omega_B t$ and compared with theoretical estimates.

(iii) An important practical consequence of the sensitivity of T_L to stratification is that the ratio of the Eulerian to Lagrangian timescale $\beta = (L_x/\sigma_w)/T_L \sim O[\gamma^2 \sigma_w/(\omega_B L_x)] \ll 1$ when $\omega_B L_x/\sigma_w \gg 1$ in homogeneous turbulence. Thus the prediction of a plume's growth from Eulerian spectra must allow for the change in the relationship between Eulerian and Lagrangian statistics in stratified flows. However, the asymptotic depth of a plume $\overline{Z^2}(t \to \infty)$ can be inferred from another Eulerian measurement, namely the variance of the temperature σ_θ^2 which is generated by the random vertical motion of fluid elements in the temperature gradient. The model suggests that

$$\sigma_\theta^2 = \tfrac{1}{2}\,\overline{Z^2}(t \to \infty)\,(d\Theta/dz)^2 \tag{6.38a}$$

whence

$$\sigma_\theta^2 = \tfrac{1}{2}\,(\sigma_w^2/\omega_B)^2\,\zeta_z^2\,(d\Theta/dz)^2 \;. \tag{6.38b}$$

These relations can be used to estimate $\overline{Z^2}$ from measurements of σ_θ^2, or to estimate σ_θ^2 from σ_w^2 and ω_B^2. (6.38) is a demonstration of the approximate equipartition of kinetic and potential energy. (6.38b) is also consistent with Brost and Wyngaard's (1978) model. In their case $\zeta_z \simeq 1.0$.

(iv) In a stably stratified flow, the temperature (or

density) flux $F_\theta = -\overline{w\theta}$ is caused by the random motion of fluid elements with different temperatures (as described in Section 6.3 on flux gradient relationships) and by the transfer of temperature between fluid elements due to molecular diffusion. Thus, in a homogeneous uniformly-stratified flow,

$$F_\theta = -\frac{1}{2}\frac{d\overline{Z^2}}{dt}\frac{d\Theta}{dz} + \overline{\Delta\theta(t)\, V_z(t)}$$

where $\Delta\theta(t)$ is the change in temperature of a fluid element. An estimate of these terms with our heuristic model leads to the following dimensionless expression for the heat flux

$$\Phi \equiv \frac{g\, F_\theta}{\Theta\, \sigma_w^2\, \omega_B} = T_L\omega_B + \tfrac{1}{2}\gamma\zeta_z^2 \simeq \tfrac{1}{2}\gamma\zeta_z^2 \sim \gamma \tag{6.39}$$

for $\gamma < 1$, since $\omega_B T_L = O(\gamma^2)$.

Physically this means that in strong stratification, the transfer between fluid elements dominates over the transport by fluid elements (Figure 6.6(b)). In terms of a diffusivity for heat,

$$K_h = \frac{F_\theta}{d\Theta/dz} = \Phi\frac{\sigma_w^2}{\omega_B} \simeq \tfrac{1}{2}\gamma\zeta_z^2\frac{\sigma_w^2}{\omega_B} \simeq \gamma\frac{\sigma_w^2}{\omega_B}\,. \tag{6.40}$$

In this case, it is not correct to use this value of K_h to estimate $\overline{Z^2}$ for fluid elements. However, from measurements of temperature flux we can estimate γ and thence the growth of $\overline{Z^2}$ from (6.37).

6.4. TURBULENCE DIFFUSION IN THE STABLY-STRATIFIED ATMOSPHERIC BOUNDARY LAYER

Some Properties of the Stably-Stratified Atmospheric Boundary Layer

The turbulence in the stably stratified atmospheric boundary layer has been measured and analysed most completely in the lowest 50 or 100 m, where most variables scale on z and local fluxes. For the purpose of discussing the general features of turbulent diffusion in the stable atmospheric boundary layer, simple formulae for the turbulence are used wherever possible. We also refer to Section 4.3 and the papers by Kaimal (1973), Caughey et al. (1979), Busch (1973) and recent measurements at the Boulder Atmospheric Observatory (BAO) (Hunt et al.,1983).

Mean velocity profile

$$\frac{dU}{dz} = \frac{u_*}{kz} (1 + \beta_1 \frac{z}{L}),$$

whence (6.41)

$$U = \frac{u_*}{k} (\ln (\frac{z}{z_o}) + \beta_1 \frac{z}{L})$$

where L is the Obukhov length ($L > 0$ in stable conditions), u_* the friction velocity, k the Von Karman constant and the constant $\beta_1 \simeq 5 \pm 0.5$.

Mean temperature profile

$$\frac{d\Theta}{dz} = \frac{\theta_*}{\alpha kz} (1 + \beta_2 \frac{z}{L}),$$

whence (6.42)

$$K_h = \frac{\alpha k u_* z}{(1 + \beta_2 z/L)}$$

where θ_* is the temperature scale and $\alpha = 1.35 - 1.0$ and $\beta_2 \simeq 6 \pm 0.5$.

Vertical turbulence and heat flux

$$\sigma_w/(-\overline{uw})^{\frac{1}{2}} = 1.3 \pm 0.3 \qquad (6.43a)$$

Caughey et al. (1979) suggest that σ_w decreases almost linearly with height through the stable layer (with typical values of $L <$ 50 m). There is a good deal of other data where σ_w remains quite constant with height, e.g., that of BAO observation in April 1980). Also,

$$-\overline{w\theta}/\sigma_\theta\sigma_w \simeq 0.35 \pm 0.1 \qquad (6.43b)$$

(Wyngaard et al., 1971; Okamoto and Webb, 1970; Caughey et al., 1979 and BOA). Combining (6.41), (6.42) and (6.43a) shows that when $\beta_1 z \gg L$, L, σ_w, $d\Theta/dz$, and dU/dz become independent of z with

$$dU/dz \sim 2\omega_B \quad \text{and} \quad L \simeq 5\, \sigma_w/\omega_B. \qquad (6.44)$$

Note that in this case the temperature flux can be normalized as in Equation (6.39). Then, in the surface layer, when $\beta_1 z/L > 1$

$$\Phi = \frac{-g\ \overline{\theta w}}{\Theta\ \sigma_w^2\ \omega_B} \sim 0.3 \ . \tag{6.45}$$

The horizontal integral scale of w is defined by

$$L_x^w = \frac{1}{\sigma_w^2} \int_0^\infty \overline{w(x)w(x + x')}\ dx' . \tag{6.46}$$

In the near neutral case

$$L_x^w \propto z$$

where the constant of proportionality is 0.5 ± 0.2.

From the equation of turbulent kinetic energy (1.57), (assuming stationary conditions and that the diffusion of energy is negligible) and from (6.41) we find

$$\varepsilon = \frac{u_*^3}{kz}[(1 + (\beta_1 - 1)\ \frac{z}{L}] \ .$$

Thus in stably stratified turbulence kinetic energy is converted more into increased dissipation than into buoyancy flux. The ratio between both terms is $\beta_1 - 1$. (The increased dissipation is caused by the increased velocity gradient.) Thence, with the assumption $L_x{}^w \sim \sigma_w{}^3/\varepsilon$, (see also (2.40))

$$L_x^w = \frac{L_x^w(z/L=0)}{1 + (\beta_1 - 1)\ z/L} \ . \tag{6.47a}$$

In general, (6.47a) may be as good an approximation as Kaimal's (1973) suggestion

$$L_x^w = 0.015\ \frac{z}{Ri} \tag{6.47b}$$

In the case of strong stratification, (6.47a) reduces to

$$L_x^w \simeq \frac{0.6\ \sigma_w/\omega_B}{\left(1 + \frac{5}{4}\ (\sigma_w/\omega_B z)\right)} \ . \tag{6.47c}$$

Horizontal components of turbulence. Following Busch (1973) and recent BAO observations, it appears that as the stability increases, σ_u/u_*, σ_v/u_* can increase from their values of about

2.5 or 2.0 in neutral conditions to as much as 3.5 or 3.0, when $z/L \sim 0.5$. (Though, in Caughey et al. (1979) σ_u/u_* is quoted as about 2.3 at ground level in very stable conditions.) However, measurements of σ_v or the spectrum of v are usually quite ill-defined in strongly-stable conditions because the wind is usually slowly changing direction all the time. This low-frequency 'swinging' of the wind, which can be also confined to the lowest 30 m (Yanskey et al., 1966), often leads to low-level shear of plumes.

Kaimal (1973) suggested that near the surface the integral scales of u and v in stable conditions have the form

$$L_x^{u,v} = \frac{z}{\beta_{u,v}\,Ri} = \frac{(\sigma_w/\omega_B)^2(1+\beta_1\, z/L)}{0.27\,\beta_{u,v}\, z} \tag{6.48}$$

where $\beta_u \simeq 0.09$ and $\beta_v \simeq 0.15$. This suggests that in strongly stable conditions $L_x^{u,v} \simeq (4/\beta_{u,v})\,\sigma_w/\omega_B$, independently of z. However, Caughey et al. (1979) found that when $h > z \geq L/10$, L_x^u and L_x^v steadily increase to a value of about h/3 and h/6 respectively at $z \simeq h$.

Turbulence at heights of 50 to 300 m. In this height range, where many sources of airborne pollution are located, stable stratification often occurs, but seldom in steady atmospheric conditions. Consequently, the structure of turbulence for given mean conditions of velocity and temperature gradients (say defined every 20 min) cannot be defined very precisely. Another reason for this uncertainty is that the occurrence of velocity fluctuations at these heights is often intermittent in stable conditions and associated with internal wave motions.

Kofoed-Hansen (1972) recorded both the unsteadiness and intermittency of the turbulence along with turbulent diffusion measurements; more of such data would be valuable. Finnegan and Einaudi (1981) have shown how large internal waves distort the turbulence in phase with the waves - another aspect of the variability of turbulence structure in these conditions. Recent measurements at BAO and Cabauw (Nieuwstadt,1984) suggest that given say σ_w and ω_B there are not unique values but limited ranges of values of all the turbulence parameters needed to determine turbulent diffusion.

Diffusivities and temperature fluctuations. In the surface layer the ratio K_h to the product $\sigma_w L_x^w$ is almost invariant with stable stratification, since from (6.42), (6.43a) and (6.47a)

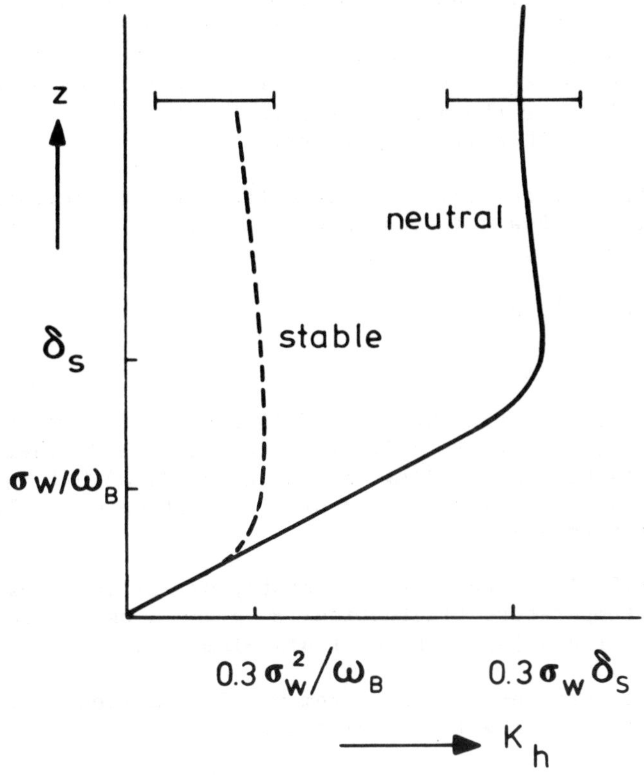

Figure 6.7. Profiles of vertical diffusivity of heat, K_h, in the surface layer in neutral and strongly stable conditions. (Note that K_h may differ from K_z, the diffusivity for fluid elements). The thickness of the surface layer is denoted by δ_s.

$$\frac{K_h}{\sigma_w L_x^w} = 0.6 \frac{[1 + (\beta_1 - 1) z/L]}{1 + \beta_2 z/L} \tag{6.49}$$

the ratio varies between about 0.6 and 0.5. Note that as the stratification increases, $L_x{}^w$ changes from 0.5z to about $0.6\sigma_w/\omega_B$, so that K_h tends to a constant ($\simeq 0.3\ \sigma_w{}^2/\omega_B$) when $z > \sigma_w/\omega_B$ (Figure 6.7). For reasons given in Section 6.3 on diffusion in stably-stratified turbulence, the invariance of the ratio in (6.49) does not necessarily imply that the same mechanism of transport is involved in the heat transfer, as the stratification increases. Our model of Section 6.3 suggests that the heat transfer or mixing between fluid elements can be expressed by the parameter Φ which in strong statification is given by (6.39). From (6.45) the measurements show that in the surface layer. Thence, from the computed graphs in Figure 6.6(c), $\gamma \simeq 0.7$, $T_L \omega_B = 0.15$ and the vertical diffusivity for fluid elements $K_z \simeq 0.15\ \sigma_w^2/\omega_B$. Thus, in the surface layer our estimated value of K_z (using a homogeneous turbulence model without allowing for the mixing induced by shear), is only about $\frac{1}{2}K_h$, the observed thermal diffusivity.

Above the surface layer, where data is rather sparse, BAO and Cabauw data, on a few occurrences when $L \lesssim 100$m $(\sigma_w/\omega_B \lesssim 20)$ suggest that $0.1 \lesssim \Phi \lesssim 0.3$, where

$$\Phi = \frac{K_h}{\sigma_w L_x} \simeq \frac{K_h\,\omega_B}{\sigma_w^2} \tag{6.50}$$

and $0.1 \lesssim \gamma \lesssim 0.4$. From (6.37b) and (6.37c) it follows that in these stable conditions $T_L \gtrsim 0.01/\omega_B$ and that the diffusivity for fluid elements $K_z \gtrsim 0.01\ \sigma_w^2/\omega_B$ may be as small as $0.1\ K_h$. (It is important to know how γ varies in order to estimate the long-range diffusion of plumes.)

From (6.38a) we see that in a stably stratified flow the variance of temperature fluctuations σ_θ gives a useful indication of the vertical displacements of fluid elements.

From the surface-layer measurements (summarised in (6.42), and (6.43), for $\beta_1 z/L > 1$, it follows that the ratio

$$\zeta_\theta = \frac{\sigma_\theta}{d\Theta/dz} / \left(\frac{\sigma_w}{\omega_B}\right) \simeq 0.5\ .$$

From BAO observations on 15 April 1981 at 10 m and 100 m ($\sigma_w/\omega_B \lesssim 20$ m), in rather variable density gradients (ω_B typically halving from 10 to 100 m),the values of ζ_θ were found to range from 0.5 to 0.7. This range of values is consistent with the theoretical model of fluid elements being largely restricted in their vertical movement to a depth of about σ_w/ω_B.

Elevated Sources Above the Surface Layer.

In our present state of knowledge of the atmospheric boundary layer and of diffusion from elevated sources, there is little more we can do than to show how the wide range of known diffusion phenomena (especially of plumes) are consistent with the known variability in turbulence structure. For sources below about 30 m, the more consistent behavior of the atmospheric boundary layer and also of diffusion experiments, make theoretical estimations of plumes more possible and useful.

There are many observations of smoke plumes from chimneys travelling many kilometers with negligible vertical diffusion, yet with some horizontal diffusion. I know of no such case where all the relevant turbulence and diffusion data are available, but two cases are quite instructive.

(i) Kofoed-Hansen (1972) described how a chimney plume emitted at 150 m from Kynby travelled without any appreciable vertical spread 20 km past Risø, Denmark. Time average measurements were made at 123 m. At the time of the observed plume behavior $U \simeq 8 \text{ m s}^{-1}$, $Ri \simeq 1$, $\omega_B \simeq 3\ 10^{-2}\ \text{s}^{-1}$ and the plume thickness $\sigma_z \simeq 7$ m. The wind speed and temperature gradient did not change appreciably at the 56 m level where the vertical turbulence was measured. $\sigma_w/U \simeq 0.02$) If we assume that σ_w was also constant with height over this range, then $\sigma_w/\omega_B \simeq 5$ m at the plume's elevation.

The very slow vertical growth of the plume is consistent with the predictions (6.37) if it is assumed that the parameter γ (which characterizes the rate of exchange of temperature between fluid elements) has the low value of 0.1, a value observed at BAO. Then the prediction of (6.37) shows that

$$\sigma_z \simeq 5\ [\zeta_z^2 + 6\ 10^{-4} t]^{\frac{1}{2}}. \tag{6.51}$$

Typically with $\zeta_z \simeq 1 \pm 0.5$ and $t = 2500$ s for a distance of 20 km, Equation (6.51) leads to the prediction that $\sigma_z \simeq 6.6 \pm 1.6$ m, quite consistent with the observation of 7 m.

(ii) Hilst and Simpson (1958) measured plume growth on five occasions from a source at 56 m. Their results can be plotted (following Pearson et al. (1983)) in the form $\sigma_z{}^2\omega_B{}^2/U^2$ as a function of $\omega_B t$ (Figure 6.6d). These measurements of σ_z and $U(z)$ were taken over a range of stability from $\omega_B = 0.009$ to the strongest at $\omega_B = 0.034\ \text{s}^{-1}$. Using the fact that close to a source ($t < \omega_B^{-1}$) $\sigma_z^2 \propto \sigma_w^2 t^2$ (see (6.36)), σ_w^2/U^2 can be extracted from the first data points: the values are given in Figure 6.6(d). In one case (e) the plume growth ceased when $\sigma_z \simeq 1.3\ \sigma_w/\omega_B$. In three other cases (a), (b), (c), the growth of σ_z^2 was markedly lower (but linear) for $t \gtrsim 2\omega_B^{-1}$ than for $t \lesssim 2\omega_B^{-1}$. In the last case (d) the data are inconclusive. These measurements, which have often been cited as good examples of how T_L is reduced in stable stratification (e.g., by Pasquill 1974, Csanady, 1964), are consistent with the variability of γ or Φ. If we take for case (e), $\zeta_z = 1 \pm 0.5$, and $\gamma \simeq 0.1$, the model predicts $\sigma_z \simeq (1 \pm 0.5)\sigma_w/\omega_B$. In case (a), (b) and (c) we assume that γ has the higher value of 0.3 associated with typical surface-laver conditions and $\zeta_z \simeq 1.0$. It then follows that

$$\sigma_z^2 \simeq (\sigma_w/\omega_B)^2(1.0 + 0.18\ t\omega_B) \tag{6.52}$$

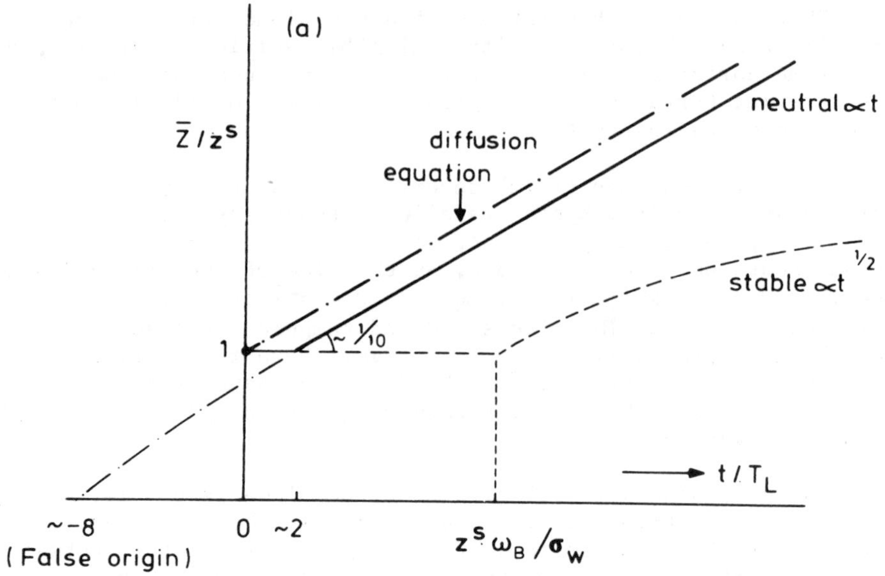

Figure 6.8. Main features of the mean vertical position $\bar{Z}$ and the vertical width, σ_z, of plumes and clouds released in the surface layer in neutral and strongly stable conditions.
(a) Mean vertical height of a cloud released from an elevated source (at z^s). The plume behavior is similar in neutral but not in stable conditions, c.f. Figure 6.9. (The scales are linear.)

when $t\omega_B > 1$. Note that in these experiments we assume that the small particles released follow the motion of fluid elements.

Sources in the Surface Layer

An essential feature of diffusion in the surface layer is that the properties of the turbulence within the plume or cloud change appreciably over its traveltime (Figure 6.8).

Elevated sources in the surface layer ($t < T_L$). Consider the initial diffusion from a source at a height z^s above level ground.

In an unstable atmospheric boundary layer the mean height $\bar{Z}$ of a puff or plume begins to rise at a rate $d\bar{Z}/dt = t\ d\sigma_w^2/dz > 0$ (6.14), while the height of the maximum concentration decreases at a rate $dZ_m/dt = w_m < 0$ (See Figure 6.4c), where w_m is the mode of the probability distribution of w. This slightly paradoxical result is a consequence of the skewness of w (which means

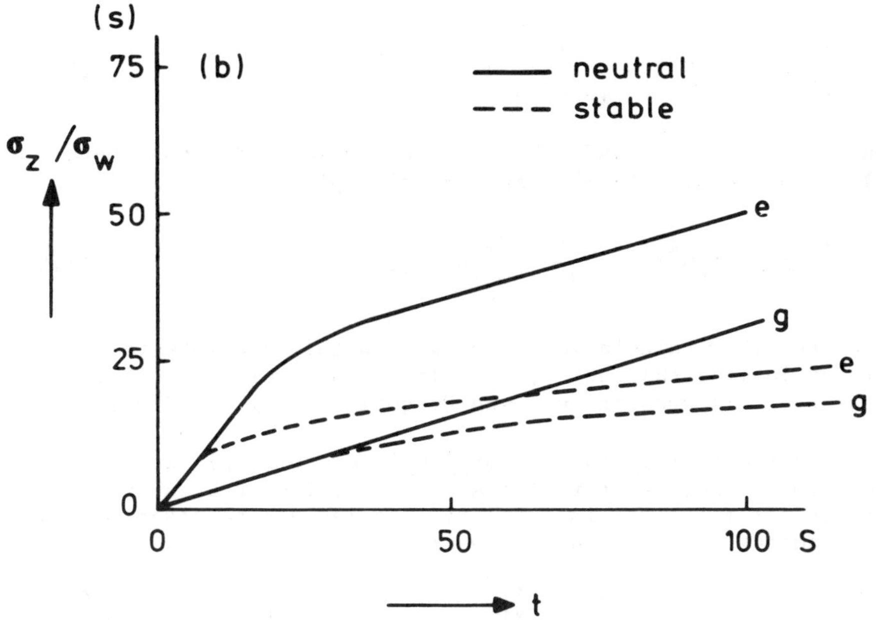

Figure 6.8.(b) Ratio of vertical width, σ_z, to vertical turbulence, σ_w, as a function of t for clouds released from elevated (e) and ground level (g) sources. (Numerical values: $\omega_B \simeq 0.035\ s^{-1}, \sigma_w \simeq 0.3\ m\ s^{-1}$ and $z^s \simeq 30$ m.)

$d\overline{Z}^3/dt > 0$). See Figure 6.4(d) where $\overline{Z}$ and Z_m are also shown for a wake in which $d\overline{Z}/dt < 0$ and $dZ_m/dt < 0$. (See also Chapter 5.)

In neutral and stable surface layers, when $t < T_L$, $d\overline{Z}/dt \simeq 0$ because the turbulence is more nearly Gaussian and usually $d\sigma_w{}^2/dz \simeq 0$. However, Caughey et al (1979) found that in some stable conditions σ_w^2 decreased to nearly zero in a distance of order 30 to 100 m. Then elevated plumes would initially descend.

However, for a steady source, $d\overline{Z}/dx$ is of more interest. Using (6.15) to estimate the change $\Delta\overline{Z}$ in $\overline{Z}$, near the source, say where $t \sim T_L$, we find

$$\frac{\Delta\overline{Z}}{z^s}(x \sim T_L U^s) \simeq \int_0^{T_L U^s} \frac{d\overline{Z}}{dx}\, dx \simeq \gamma_1 \left[\frac{u_*^2}{U^s \sigma_w} - \gamma_1 \frac{z^s}{2U^s} \frac{dU^s}{dz} \right] \quad (6.53)$$

where $\gamma_1 = \beta L_x^w/z^s$. For neutral conditions this leads to

$$\frac{\Delta \overline{Z}}{z^s} (x \sim T_L U^s) \sim \frac{\gamma_1}{\ln(z^s/z_0)} \ll 1 \qquad (6.54a)$$

with $\gamma_1 \simeq \frac{1}{4}$ if $\beta = \frac{1}{2}$ and $L_x^w = \frac{1}{2}z$. In stable conditions where $z^s > \sigma_w/\omega_B$, (6.41), (6.47c) and (6.53) imply

$$\frac{\Delta \overline{Z}}{z^s} (x \sim T_L U^s) \sim \frac{0.1}{(z^s)^2} \left(\frac{\sigma_w}{\omega_B}\right)^2 \ll 1. \qquad (6.54b)$$

These and other calculations show why, within one integral time scale of leaving the source, the diffusion develops as if the turbulence were homogeneous (see Pasquill (1974), p. 203).

<u>Elevated sources in the surface layer</u> ($t > T_L$). In a homogeneous, Gaussian turbulence the diffusion of a substance from a source after several integral time scales can be described by a flux gradient relation (6.29), when

$$t > T_L \quad \text{or} \quad \sigma_z > L_x^w, \qquad (6.55)$$

These criteria can also be applied in inhomogeneous (but approximately Gaussian) turbulence if L_x^w is also the length scale of the inhomogeneity, as is the case in the surface layer. If the u, v components of turbulence have much larger scales than w, then (6.29) can only be applied if F_z, C are integrated over x and y.

For an elevated source in neutral conditions, $\sigma_z \gtrsim L_x^w$ when $t/T_L(z = z^s) \gtrsim 2$ and $x/z^s \gtrsim \ln(z^s/z_0)$. In strongly stratified conditions $\sigma_z \gtrsim L_x^w$ when $x/z^s \gtrsim 0.6$. This transition time or distance is about equal to that given by random flight computations of turbulent diffusion obtained by Durbin and Hunt (1980) and Reid (1979). It is slightly greater than that used by Smith and Blackall (1979) in their modified diffusion equation models.

When these conditions are satisfied, the mean concentration C for a line source (or the concentration averaged over y for a point source) can be calculated from the diffusion equation

$$\frac{\partial C}{\partial t} + U(z) \frac{\partial C}{\partial x} = \frac{\partial}{\partial z} \left(K_z(z) \frac{\partial C}{\partial z}\right) + K_x \frac{\partial^2 C}{\partial x^2} . \qquad (6.56)$$

Nieuwstadt and Van Ulden (1978) have shown that the diffusion equation describes ground releases very well.

The salient features of a cloud's development at traveltime t or plume's development at distance x can most usefully be expressed in terms of the first and second moments of the concentration distribution.

For a cloud the definitions of these moments are (see also (6.3))

$$\overline{Z^c} = \int z\ C\ dxdydz\ /M_o$$
$$(\Sigma_z^c)^2 = \overline{(Z - \overline{Z^c})^2} = \int (z - \overline{Z^c})^2\ C\ dxdydz/M_o \qquad (6.57)$$
$$(\sigma_z^c)^2 = \overline{(Z - z^s)^2} = \int (z - z^s)^2\ C\ dxdydz/M_o$$

where $M_0 = \int C\ dxdydz$. The integration extends over the whole cloud volume.

For a plume these definitions become

$$\overline{Z^p} = \int z\ C\ dydz/S$$
$$(\Sigma_z^p)^2 = \int (z - \overline{Z^p})^2\ C\ dydz/S \qquad (6.58)$$
$$(\sigma_z^p)^2 = \int (z - z^s)^2\ C\ dydz/S$$

where $S = \int C\ dydz$. The integration extends over the cross-section of the plume. The mean height of a plume based on the concentration profile may differ from the mean height of the fluid elements.

(i) Effects of the diffusivity gradient: The mean height $\overline{Z^c}$ of a puff is increased by the gradient of K_z

$$\frac{d\overline{Z^c}}{dt} = \left(\frac{dK_z}{dz}\right)_{z=z^s} + (\overline{Z^c} - z^s)\left(\frac{d^2K_z}{dz^2}\right)_{z=z^s} + \ldots. \qquad (6.59)$$

In the absence of molecular processes $d\overline{Z^c}/dt = \overline{V}_z$, the mean Lagrangian velocity of fluid elements. Thus in the <u>neutral</u> boundary layer $d\overline{Z^c}/dt \simeq (0.35 \pm .05)u_*$ while in a <u>stable boundary</u> layer, $z^s > \sigma_w/\omega_B$, $d\overline{Z^c}/dt = 0.6\ \sigma_w^3/(\omega_B z^s)^2$.

Combination of the criteria for $\sigma_z > L_x^w$, discussed above, and of (6.57) show why, in neutral conditions, diffusion from an elevated source, though shifted in time by $\sim 8\ T_L(z^s)$, has the the same form as that from a ground-level source (Figure 6.8(c)). In stable conditions, the small variation in diffusivity (above a height σ_w/ω_B) leads to a negligible rise in $\overline{Z}$ until the material diffuses down to the ground. When $\sigma_z^c \gg z^s$ or $t \gg T_L z^s \omega_B/\sigma_w$ the material rises as if it was reflected at the surface. In that case $\overline{Z} \simeq (\sqrt{2/\pi})\ \sigma_z{}^c$ (see (6.64)).

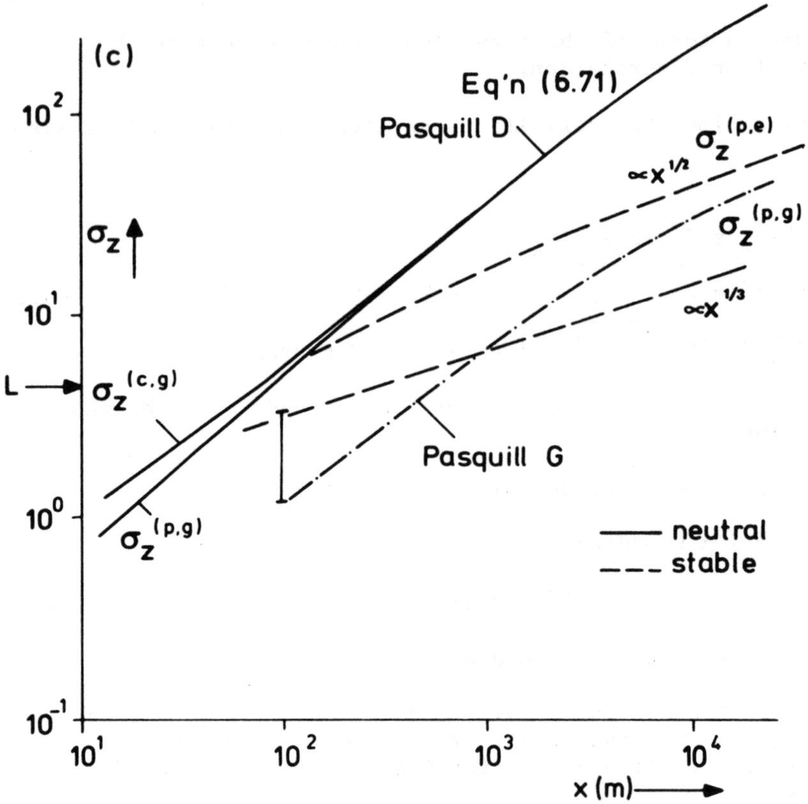

Figure 6.8.(c) Vertical width (in m) as a function of distance x for clouds (c) and plumes (p) released from elevated (e) and ground level (g) sources in the surface layer. Comparisons are made with Pasquill's curves (1974) and curves computed from (6.71) (taking $\sigma_z{}^c = \sqrt{2}\ \overline{Z}$) for neutral stability, and (6.73), (6.74) for stable conditions (taking $\sigma_w/\omega_B \simeq 1$ m and L = 5m). The vertical bar indicates experimental values from Thompson's Porton measurements of diffusion from ground level sources analyzed by Nieuwstadt and Van Ulden (1978) (solid lines denote neutral conditions; dashed lines denote stable conditions).

The second moment $\sigma_z{}^c$ (in the vertical direction) of the concentration distribution of a puff is affected by the change in $\overline{Z}^c$ and by the spread relative to the mean height. In the neutral surface layer when $\overline{Z}^c \gg z^s$ (i.e., when a puff behaves as if it had come from the ground) the solution to (6.56) and (6.57) shows, following Chatwin (1968), that

$$\int_{-\infty}^{+\infty} C\ dx \propto \frac{1}{t} \exp(-\frac{z}{ku_* t}), \qquad (6.60)$$

whence

$$\overline{Z}^c = ku_* t \simeq 0.3\ \sigma_w t \text{ and } \sigma_z^c = \sqrt{2}\ \overline{Z}^c \simeq 0.43\ \sigma_w\ t \tag{6.61a}$$

and

$$\Sigma_z^c = 0.3\ \sigma_w\ t. \tag{6.61b}$$

In the first part of Section 6.3 it was argued that, if T_L varies slowly, Taylor's (1921) statistical theory can be applied to inhomogeneous turbulence. For a puff in the neutral surface layer this is the case because the rate of change of T_L with time at the centroid of the puff is

$$\frac{dT_L}{dt}(t) = \frac{dT_L}{dz}\ \frac{d\overline{Z}^c}{dt} \simeq \frac{0.2}{u_*} \times 0.4\ u_* \simeq 0.08\ .$$

Therefore (6.22b) can be used to compare with (6.61b). After making suitable estimates for $(\sigma_w L)^2 \simeq 0.9\ \sigma_w^2$ and $T_L (\simeq \frac{1}{4}\ \overline{Z}^c/\sigma_w)$, the statistical theory gives (Hunt and Weber, 1979)

$$\Sigma_z^c = 0.37\ \sigma_w t\ . \tag{6.62}$$

Note that during this vertical diffusion, the puff reaches the surface at $z = 0$ but does not diffuse across it.

In stable conditions (if $z^s > \sigma_w/\omega_B \sim L/5$) and if $t > 2T_L$, then the diffusivity K_z is approximately constant and the vertical growth of a puff is given simply by

$$\sigma_z^c = (2K_z t)^{\frac{1}{2}} \tag{6.63}$$

until the puff diffuses down to within about σ_w/ω_B of the surface. For a gaseous substance diffusing like heat in the surface layer, $K_z \sim K_h \simeq 0.3\ \sigma_w^2/\omega_B$, but if the molecular or Brownian difusion of a substance is very slow (e.g. smoke) it only follows the motion of fluid elements, and then K_z may be slightly lower (as previously suggested in this section).

When $\sigma_z(t) \gtrsim z^s - \sigma_w/\omega_B$, the downward diffusion is halted by the decreasing diffusivity near the surface; the concentration increases and effectively the cloud or plume is reflected. In fact, for a puff in strongly stable conditions the solution for the horizontal average

$$C^* = \iint C\ dxdy/M_o$$

is exactly that of a simply reflected puff.

From (6.56) and (6.63) it follows that

$$C^* = \frac{1}{\sqrt{(2\pi)}\,\sigma_z^c}\left[\exp\left(-\tfrac{1}{2}\left(\frac{z-z^s}{\sigma_z^c}\right)^2\right) + \exp\left(-\tfrac{1}{2}\left(\frac{z+z^s}{\sigma_z^c}\right)^2\right)\right]. \qquad (6.64)$$

(ii) Velocity gradient effects: The first effect of the shear in the atmospheric boundary layer is to lower the mean height of a cloud or plume. This effect can be estimated by the statistical analysis for $t < T_L$, when it is a small effect, and by the diffusion equation when $t > T_L$. In neutral conditions, the plume height is more determined by the surface or dK_z/dz than by dU/dz. But in stable conditions, it is useful to consider a steady line source in a weak, uniform shear before the plume reaches the ground. An approximate solution to (6.56) near the centre of the plume is (according to Lauwerier's (1954) solution of the diffusion equation (6.56) with K_z constant)

$$C = \frac{Q}{\sqrt{(2\pi)}\,U^s}\left[1 - \frac{z-z^s}{U^s}\frac{dU^s}{dz}\left(\frac{1}{2} + \frac{1}{3}\zeta^2 + \ldots\right)\right]\exp(-\tfrac{1}{2}\zeta^2)$$

where $\zeta = (z - z^s)/\sigma_z^P$. So the value of z at the position of maximum concentration is about

$$z_m \simeq z^s - \frac{(\sigma_z^P)^2}{2\,U^s}\frac{dU^s}{dz}. \qquad (6.65a)$$

This result is applicable to a stably-stratified boundary layer where the shear may have a strong effect after $t > T_L$ and before the plume reaches the surface. Substitution of (6.41), (6.42) and (6.63) into (6.65a) shows that when $z^s > \sigma_w/\omega_B$

$$z_m \simeq z^s - \frac{1}{6}\, x\, [\sigma_w/(\omega_B z^s)]^2\ . \qquad (6.65b)$$

Thus, taking values found in the field measurements of Doran et al. (1978), where $\sigma_w/\omega_B \simeq 10$ m (or $L \simeq 50$ m), and $z^s = 26$ m, the estimate of (6.65) is that z_m decreases by about 5 and 30 m when x = 200 and 800 m, respectively. The observed values are about 3 and 20 m (the reflected Gaussian profiles show a reduction in z_m of less than 2 m) (Figure 6.9).

The second effect of the wind shear in the atmospheric boundary layer is to increase the length σ_x of a cloud in the direction of the wind, by the mechanism described in section 6.3 on unidirectional shear flows, so that

$$\sigma_x^c \sim \sigma_z^c\, t\, dU/dz\ . \qquad (6.66)$$

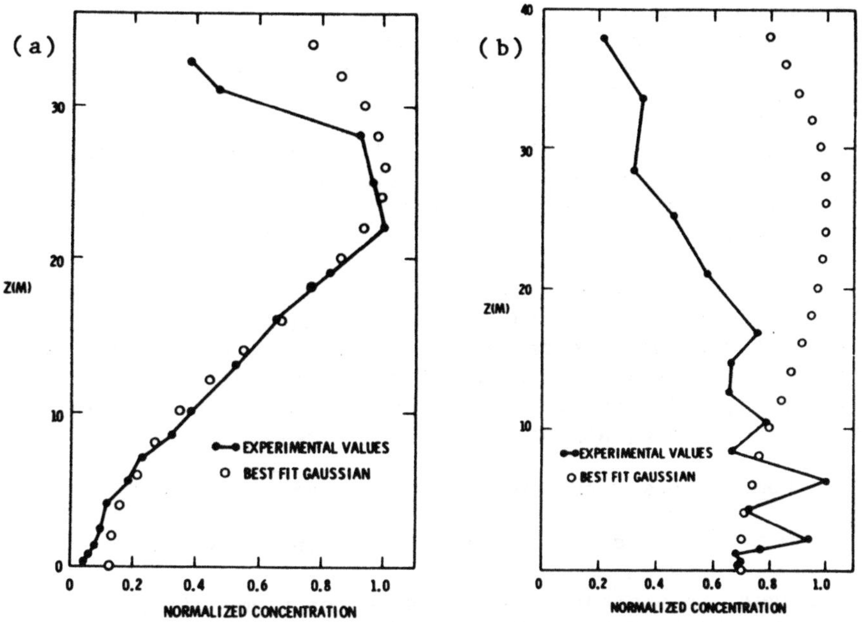

Figure 6.9. Field measurements of vertical mean concentration profiles at 200 m (a) and 800 m (b) downwind of an elevated continuous source at 26 m in a stable surface layer (L ≃ 50 m). Note the rapid decrease of the height of the maximum concentration z_m (Doran et al., 1978).

The solution to (6.56) in the neutral surface layer gives (Chatwin, 1968)

$$\sigma_x^c \simeq 1.2\ \sigma_w\ t\ . \tag{6.67}$$

Since even the vertical turbulent velocities are correlated in the cloud, (because $L_x{}^w \sim \sigma_z{}^c \sim z$), the cloud's centre moves randomly and the cloud's average length relative to its center $(\overline{\underset{\sim}{\Sigma}_x{}^2})^{\frac{1}{2}}$ is less than the total particle spread Σ_x (cf. (6.4)) (Hunt and Weber, 1979).

In the stable surface layer (6.65) and the solution of (6.56) show that until the cloud touches the ground

$$\sigma_x^c \sim (K_z/6)^{\frac{1}{2}} \left(\frac{dU^s}{dz}\right) t^{3/2} \sim 0.4\ \sigma_w \omega_B^{\frac{1}{2}}\ t^{3/2}. \tag{6.68}$$

Thus, in neutral conditions, by comparing (6.67) and (6.61a), we see that far downwind of the release point $\sigma_x/\sigma_z \simeq 3$, so the cloud's shape remains approximately the same. But in stable conditions it follows from (6.63) and (6.68) that $\sigma_x/\sigma_z \propto (dU/dz)t$ and therefore, the length to depth ratio steadily increases (Wilson, 1981).

The third effect of shear dU/dz is on the relations between the dimensions and concentration of clouds and plumes, especially their relative variation with traveltime or with distance from the source, respectively.

In particular, one needs to know (a) how far (i.e., the value of $\overline{X}$) a cloud travels in time t, or how long it takes for the center of the cloud $\overline{X}$ to reach a given distance x, and (b) how the scales $\sigma_y{}^c$, $\sigma_z{}^c$ of a cloud at $x = \overline{X}$ relate to the scales of a plume $\sigma_y{}^p$, $\sigma_z{}^p$ released from a steady source at the same point (x = 0) (see also Equations (6.57) and (6.58)).

In homogeneous turbulence with no shear

$$\overline{X}^c = t\,U^s \quad \text{and} \quad \sigma_z^c(t) = \sigma_z^p(x), \text{ when } x = \overline{X}^c(t) \tag{6.69}$$

However, in a shear flow, the cloud or plume diffuses vertically into faster- or slower-moving fluid which affects its average speed $d\overline{X}/dt$.

First we consider the $\overline{X}^c(t)$ for a ground-level source (or far downwind of an elevated source)

$$d\overline{X}^c/dt = U(z \sim \overline{Z}^c). \tag{6.70}$$

For the neutral surface layer it can be shown that

$$d\overline{X}^c/dt = U(0.56\,\overline{Z}^c) \quad \text{and} \quad \overline{Z}^c \simeq 0.4\,u_* t. \tag{6.71}$$

Thence $\overline{X}^c \simeq \overline{Z}^c[\ln(0.56\,\overline{Z}^c/z_0) - 1]/0.16$ or approximately $\overline{Z}^c \propto t \propto (\overline{X}^c)^{0.9}$, (Chatwin, 1968).

For a very stable surface layer ($\overline{Z}^c > \sigma_w/\omega_B$) follows with $\overline{Z}^c = \sqrt{2/\pi}\,\sigma_z$ and (6.63)

$$\frac{d\overline{X}^c}{dt} \simeq U(\overline{Z}^c) \simeq \frac{2}{\sqrt{(\pi)}}\frac{dU}{dz}(K_z t)^{\frac{1}{2}} \tag{6.72a}$$

so that

$$\overline{X}^c \simeq \sigma_w \omega_B^{\frac{1}{2}} t^{3/2} \text{ and } \overline{Z}^c \propto t^{\frac{1}{2}}. \tag{6.72b}$$

Next we turn to the relation between $\sigma_z{}^c$ and $\sigma_z{}^p$. Even if $\overline{X}^c(t)$ is calculated by allowing for the vertical diffusion and the shear, $\sigma_z{}^p(x)$ is not exactly equal to σ_z^c ($\overline{X}^c = x$). Detailed calculations show that for an effectively ground-level source in neutral flows,

$$\sigma_z^p/\sigma_z^c \simeq \left[1 - \frac{3}{4\ \ln(\overline{Z}^c/z_o)}\right], \tag{6.73}$$

where $\sigma_z{}^c = \sqrt{2}\ \overline{Z}^c$ and $\overline{Z}^c$ is given by (6.71). The ratio $\sigma_z{}^p/\sigma_z{}^c$ varies slightly with distance, so that at about 300 m from a ground-level source over grassland, $\sigma_z{}^p$ is only about 13% less than $\sigma_z{}^c$ for a cloud whose centre has reached this point. In very stable flows, the solution for a ground-level source to (6.56), when K_z is constant and $U = z dU/dz$, leads to

$$\sigma_z^p \simeq 1.63\left(\frac{x\ K_z}{dU/dz}\right)^{1/3} . \tag{6.74}$$

Using $K_z \simeq 0.3\sigma_w{}^2/\omega_B$ and $dU/dz \simeq 2\omega_B$ we find $\sigma_z{}^p \sim 0.87(x\sigma_w{}^2/\omega_B{}^2)^{1/3}$. The ratio $(\sigma_z{}^p/\sigma_z{}^c)$ at $\overline{X}^c = x$ is constant with distance and approximately equal to about 0.8. Note how in stable conditions $\sigma_z{}^p \propto x^{1/3}$ which is a much slower growth rate with x than in neutral conditions.

A similar procedure can be used for calculating $\sigma_y{}^p(x)$ in terms of $\sigma_y{}^c(x = \overline{X}^c)$. In neutral and stable conditions

$$\sigma_y^c(t) \simeq \sigma_v\ t \tag{6.75}$$

for distances of up to 1–10 km. Therefore, $\sigma_y{}^c(x = \overline{X}^c)$ can be estimated from (6.71) and (6.72) in neutral and stable conditions. (This suggests that $\sigma_y \propto x^{2/3}$ in stable conditions, which is not consistent with the recommended tables for σ_y, e.g., Pasquill (1974, pp. 368). Why not?)

Comparison of vertical diffusion from ground-level and elevated sources. It is commonly assumed that measurements of turbulent diffusion obtained from sources at one height can be used to infer the rate of diffusion at different heights. Such differences which are greatest near the source can be analysed by comparing the vertical diffusion from a ground-level source, $\sigma_z{}^g$, and the vertical diffusion from an elevated source, $\sigma_z{}^e$ during the initial stage of diffusion ($t < T_L$). From (6.61) and from Taylor's statistical diffusion theory, it follows that for neutral conditions

$$\sigma_z^{c,g} \simeq 0.43\ \sigma_w t\ , \tag{6.76a}$$

$$\sigma_z^{c,e} \simeq \sigma_w t \qquad (6.76b)$$

so at given time from release there is a large difference. But the ground-level cloud travels downwind more slowly than the elevated cloud, so at a given distance the difference between $\sigma_z^{c,g}$ and $\sigma_z^{c,e}$ might be expected to be smaller. For neutral conditions we express t in terms of $\overline{X}^c$ by (6.71)

$$\frac{\sigma_z^{c,g}\ (\overline{X}^c = x)}{\sigma_z^{c,e}\ (\overline{X}^c = x)} \simeq \frac{0.43\ U(z^s)}{U(0.56\ \overline{Z}^c) - U(e\ z_o)} .$$

Calculating this ratio with z^s = 30 m and z_0 = 0.30 m for x = 30 m and x = 300 m, we find 1.12 and 0.64, respectively, which for most practical purposes is an insignificant difference.

In stable condition $\sigma_z^{c,g}$ is given by (6.60) when $\sigma_z < \sigma_w/\omega_B$ and by (6.63) when $\sigma_z > \sigma_w/\omega_B$. The latter equation can be written with (6.72b) as

$$\sigma_z^{c,g} \simeq (2\ K_z)^{\frac{1}{2}} \left(\frac{x}{\sigma_w\ \omega_B^{\frac{1}{2}}}\right)^{1/3} , \qquad (6.77a)$$

when $\sigma_z > \sigma_w/\omega_B$. For the elevated source we find

$$\sigma_z^{c,e} \simeq (2\ K_z t)^{\frac{1}{2}} \simeq \left[\frac{K_z\ x}{2\ z^s \omega_B}\right]^{\frac{1}{2}} . \qquad (6.77b)$$

before the plume reaches the ground, i.e. $x \lesssim 2(z^s)^3(\omega_B/\sigma_w)^2$. So

$$\sigma_z^{c,g}/\sigma_z^{c,e} \simeq \frac{(2\ z^s)^{\frac{1}{2}}}{x^{1/6}\ (\sigma_w/\omega_B)^{1/3}} \gtrsim 1 . \qquad (6.78)$$

In fact, it is rare for ω_B to be quite constant with height; usually it decreases so that $K_z(z = z^s) > K_z\ (z = 0)$ leading to σ_z^e being greater than σ_z^g (Doran et al., 1978).

The vertical concentration profiles of clouds and plumes vary considerably with stratification. If the profile is approximated by $C(z) = C_{z=0}\exp(-Az^s)$, then for puffs released from the ground s changes from 1.0 to 2.0 as the flow becomes more stable (see Section 7.5). For steady sources s changes from about $\{1 + 1/\ln(\overline{Z}/z_o)\}$ to 3.0 as the stability changes from neutral to strongly stable. (Measurements at 100 m from a source indicate s changing from 1.5 to 2.2 as L changes from ∞ to 1.6 m, (Nieuwstadt and Van Ulden, 1978).)

6.5. CONCLUDING REMARKS

Concentration Distributions

We have noted that the distribution of mean concentration C from a point or line source can be defined close to a source in terms of probability distributions of the velocity field (see Section 6.3), and that far downwind (or after a long enough time $t \gg T_L$, if the turbulence is not too inhomogeneous), $C(x_i,t)$ is described by the diffusion equation. Alternatively, given U(z) and estimates for Σ_x, Σ_y, Σ_z from the motion of fluid elements C_σ can be estimated from the conservation of the diffused substance. It follows

$$C_\sigma \sim \chi_c \frac{M_o}{\Sigma_x \Sigma_y \Sigma_z} \text{ (for clouds)}$$
$$C_\sigma \sim \chi_p \frac{Q}{U \Sigma_y \Sigma_z} \text{ (for plumes)} \qquad (6.79)$$

where the constants χ_c, χ_p depend on the concentration distributions.

For many practical purposes, C has to be calculated when $t \sim T_L$. An example is the maximum ground-level concentration, which for an elevated plume in neutral conditions occurs when $t \sim 3\ T_L$.

To calculate C in this region, various heuristic adjustments to the diffusion equation have been proposed to match the two zones of diffusion (e.g., Smith and Blackall, 1979). Other approximate methods have been proposed, such as the random flight models (Durbin and Hunt, 1980), or such as those based on truncated equations of concentration fluctuations and velocity moments (El Tahry et al., 1981).

The maximum value of C on a surface is probably uncertain to within a factor of 2, partly because it depends on the 'tail' of the concentration profile C(z). The maximum value in the plume is less uncertain.

The location of the surface maximum is worse defined than the value of the maximum (Scriven, 1969).

In strongly stable conditions, by the time the plume has reached the surface, $t \gg T_L$. Since the turbulence is quite homogeneous (for $z \gtrsim \sigma_w/\omega_B$), the vertical diffusion of a plume or cloud is then adequately described by the diffusion equation. (The lateral spreading is, however, highly unpredictable).

For ground-level sources (since $T_L(z^s)=0$), the diffusion equation should be a reasonably good approximation. Extensive laboratory measurements have now confirmed this to be the case (e.g., Robins and Fackrell, 1979).

Complex Atmospheric Conditions

As often as not, atmospheric motions are more irregular than those considered here, especially in N.W. Europe, where fronts and other changeable conditions are a marked feature of the weather. In practice, these changeable conditions are usually ignored as a first approximation. In a second approach very simple overall estimates of the effects are made or moderately simple calculations of the diffusion equation based on eddy diffusivities are used. Examples are the estimates, made in some prediction schemes, for the concentration during the changes in the height of a mixed layer or the passage of a plume through a front or an internal boundary layer. The changes in σ_z, σ_y and U are being guessed by plausible physical arguments and then the concentrations are calculated by continuity. The principles described in Section 6.3 may be able to help in assessing these parameters and also the mean height of the plume or cloud, an important parameter for estimating the ground-level concentration. For the third approach we use eddy diffusivity models. The arguments in Section 6.3 on flux-gradient relationships can help in assessing the accuracy of these models in complex conditions, where one may find large scales and large gradients of turbulence with consequential non-Gaussianity. For example, when a stable or neutral boundary layer flow over the sea meets convective conditions at the shore, the use of an eddy diffusivity model (as suggested by Section 6.4 on sources above the surface layer) may give erroneous predictions.

Topography

As an example of the effect of changes in the surface topography, consider the motion and turbulent diffusion of plumes or puffs released from an elevated source upwind of a three-dimensional hill on a level plain (Figure 6.10). The scaling arguments for this simple case can then be applied to more complex topography. (For a fuller description, see Hunt (1981), Hunt et al. (1979), Hunt (1980), Smith (1980)).

In neutral conditions, the mean streamline through the source at height z^s, which we take to be greater than about $L_1/10$, passes over the hill (height H), but its distance above the hill surface is affected by the the perturbation to the approach flow. This perturbation can usually be approximated by a potential flow of a uniform stream of strength $(H/L_1)U(z = L_1)$ except in a thin internal boundary layer over the hill. This is similar to that

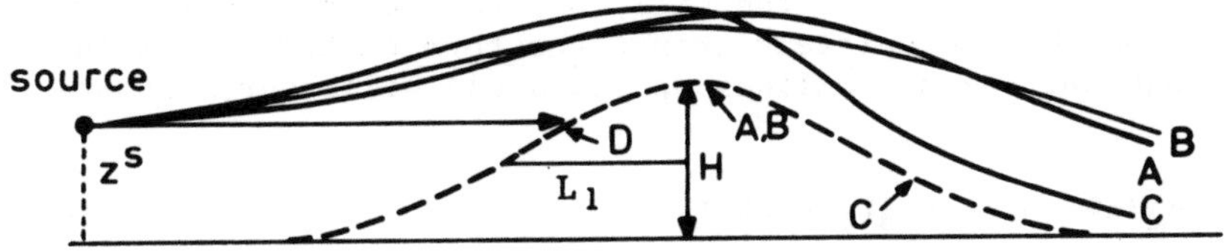

Figure 6.10. Plume behavior over a isolated round hill as the approach flow becomes more stable. The arrow indicates where the plume trajectory is closest to the surface and where the surface concentration is likely to be greatest. (A) Neutral flow $(L_1\omega_B)/\sigma_w \ll 1$; (B) Slightly stable flow $(L_1\omega_B)/\sigma_w > 1$, $(L_1\omega_B)/U \ll 1$; (C) Moderately stable flow $(L_1\omega_B)/U > 1$, $(H\omega_B)/U < 1$; (D) Strongly stable flow $(H\omega_B)/U > 1$. (Source height is such that $H > z^s > H - U/\omega_B$.)

over a roughness change and where the shear stress and turbulence are significantly changed. Typically, the ratio of the minimum distance n_m from the hill to the source streamline, divided by z^s is $n_m/z^s = 1 - 2H/L_1$. On the lee side a wake region of increased turbulence develops, $\partial\sigma_w^2/\partial z < 0$, so that the mean height of the plume or cloud decreases, $d\overline{Z}/dt < 0$, below the mean streamline (the process of 'entrainment'). The hill only affects the maximum surface concentration (located at x_{max}) if the hill is upwind of the location of the maximum ground-level concentration in the absence of the hill. In that case the location of the maximum surface concentration is on the hill top ($x_{max} = x_H$).

In slightly stable conditions defined by $\sigma_w/\omega_B \lesssim L_1$ and $U(H)/\omega_B \gg L_1$ the upwind velocity gradient is appreciably increased (see Section 6.4 on turbulence properties of the stable boundary layer). This increases the magnitude but not the form of the perturbation of the streamlines, and increases the thickness of the inner boundary layer and wake regions. The latter effect, caused by the increase in surface turbulence, is relatively greater in stratified flow where the incident turbulence is weaker. The entrainment into the wake is stronger for the same reason. A rough estimate of the increased deflection of the streamline is given by

$$\frac{n_m}{z^s} \simeq 1 - \frac{2H}{L_1}\left(1 + \frac{L_1\ \omega_B}{u_*\ \ln(L_1/z_o)}\right) . \tag{6.80}$$

In this case, too, the condition on the location of the maximum surface concentration, discussed above, applies.

In moderately stable conditions, defined by $U(H)/\omega_B \lesssim L_1$ and $U(H)/\omega_B > H$, the flow still has enough kinetic energy for most streamlines to pass over the hill, but the stratification is strong enough to completely change the form of the streamline perturbation. The streamlines approach the hills surface more closely

$$n_m/z^s \simeq 1 - (\omega_B L_1/U)(H/L_1) \qquad (6.81)$$

and this occurs on the leeward side, so $x_{max} \sim x_H + L_1$. The near wake is suppressed by this accelerating stable flow (Snyder et al., 1979). The effect of the streamline almost touching the surface (a greater effect over a three-dimensional hill because the streamlines spread out in the y-direction), is that the maximum ground-level concentration rises to within a factor of about 2 of the mean center line concentration of a plume in the absence of the hill.

In strongly-stable conditions given by $U(H)/\omega_B \lesssim H$, the streamlines from sources below the height of the hill may either directly impact on the hill if $z^s < (H - U(H)/\omega_B)$ or can pass over the hill and approach the surface on the lee side if $(H - U/\omega_B) \lesssim z^s \lesssim H$. In the former case, if the source is directly upwind of the hill, the plume impinges directly on the hill. Diffusion calculations and laboratory measurements (Snyder et al., 1979) show that the maximum ground-level concentration equals the centre line concentration. So it is very important first to be able to estimate σ_z, σ_y and $U(z)$ over level ground.

In the most stable conditions where $U/\omega_B \lesssim L_1/(\ln(L_1/z_0))^{\frac{1}{2}}$ the temperature difference between the surface of the hill and the air drives motions (katabatic winds) which are comparable with the airflow over the hills driven by the approach wind.

We have seen how the airflow pattern and the location and magnitude of surface concentrations are sensitive to the relative strength of stable stratification in the oncoming flow. One consequence of this is that when $U/\omega_B \sim H$ small variations in stability and wind speed have a large effect on the diffusion pattern from an upwind source. Then averaging U and ω_B over an hour or so can give very misleading results, if these values are used to estimate average concentrations.

ACKNOWLEDGEMENT

This lecture draws on unpublished work and many helpful conversations at Cambridge with Drs Puttock, Pearson and Britter and at Boulder with Drs Gaynor, Kaimal and Wyngaard and on correspondence with dr Arya. I am grateful for support from NOAA while I was working at the University of Colorado, Boulder, and, while preparing this chapter, for support by a contract from the State of Maryland Department of Natural Resources to Flow Analysis Associates of Ithaca, N.Y.

7. APPLICATIONS IN AIR POLLUTION MODELING

S.R. Hanna

E.R.T., Concord, Mass. 01742, U.S.A.

7.1. INTRODUCTION

In the first four chapters, up-to-date information was given on the meteorological structure of the planetary boundary layer (PBL). The purpose of the last three chapters was to discuss how this new knowledge of the PBL could be used to develop new and improved diffusion theories and models. Chapters 5 and 6 dealt specifically with diffusion in the convective boundary layer and the stable boundary layer, respectively. Clearly they have pre-empted much of what I might say, and I will only briefly cover these areas for the sake of completeness. The emphasis of this chapter will be on more general applications of PBL results to diffusion prediction techniques.

Atmospheric diffusion is a direct result of atmospheric turbulence. There are several theories for atmospheric diffusion, but they all depend on the same set of PBL turbulence parameters. Vertical profiles through the PBL of the following parameters are helpful in the development of diffusion theories: U, V, W, σ_u, σ_v, σ_w, ε, $T_E{}^u$, $T_E{}^v$, $T_E{}^w$. In addition, the following external parameters are often useful: G, z_o and h.

As we have seen in Chapters 1, 2 and 4 scaling velocities and temperatures can be constructed from surface flux observations, such as u_* (2.1), θ_* (2.33), L (1.61) and w_* (2.54).

Each variable and each parameter is associated with a precise sampling time. Because the atmosphere contains eddies at a wide range of scales, it is important to choose the proper sampling for a particular problem.

All the variables and parameters defined above are Eulerian; i.e., they are measured from a platform that is stationary or moves through the air with a constant speed (e.g., on an airplane). In contrast, diffusion is a Lagrangian process in which the atmosphere is perceived by an inertialess particle moving exactly with the wind. A fundamental problem, then, is the relation between the above Eulerian variables and the corresponding Lagrangian variables. It is standard practice to assume that all variables except the time scales are the same in the Eulerian and Lagrangian systems. For example, turbulent energies are the same in both systems. Our meager knowledge of the relation between Eulerian and Lagrangian time scales will be reviewed later.

The final step in relating diffusion and turbulence depends on the sampling time, travel time, and release time of the pollutant material. Sampling time is the length of time that the variable is sampled or observed, travel time is the time elapsed since the pollutant material left its source, and release time is the total time period over which pollutants were released from the source. These times define 'windows' that are used to filter the turbulence in order to produce the observed diffusion. If travel time is much greater than release time or if sampling time is much less than travel time, then puff diffusion is appropriate. If sampling time and release time are much greater than travel time, then continuous plume diffusion is important. Methods of filtering the turbulent energy spectrum to account for diffusion in the proper way are outlined in Section 7.2.

The following discussions are arranged according to the traditional outline of diffusion theories: statistical models, Gaussian models, K-diffusion models, and similarity models. A few new special research areas will be discussed, such as the probability distribution function for turbulent vertical velocities, the natural variability of observed pollutant concentrations, and the representativeness of wind speed observations.

7.2. STATISTISTICAL MODELS OF DIFFUSION

The best way to point out the importance of time scales and turbulent energy spectra on diffusion is through the derivation of statistical models of diffusion. These models have a long history and are still relevant. The concept of release time, travel time, and sampling time are very well illustrated by statistical models.

Taylor's Statistical Theory

In 1921, Taylor derived the following equation for the standard deviation σ_y of the crosswind spread of material released continuously from a point source in a homogeneous and stationary atmosphere:

$$\sigma_y^2 = 2\sigma_v^2 \int_o^t \int_o^t \rho_L(\tau)\, d\tau\, dt' \tag{7.1}$$

The derivation of this equation is reviewed in Section 2.7, where we have written $\sigma_y{}^2 = \overline{y^2}$. In (7.1) σ_v is the standard deviation of lateral turbulent velocity fluctuations, t is the travel time of the plume from the source to the receptor of interest, and ρ_L is the Lagrangian correlation function for lateral turbulence. It is assumed here that sampling time and release time are much larger than the travel time.

To solve Equation (7.1) it is necessary to know the turbulence parameters $\sigma_v{}^2$ and $\rho_L(t)$. Since it is assumed that Lagrangian and Eulerian turbulent energies are equal, then $\sigma_v{}^2$ can be obtained directly from a tower measurement or from PBL theory. The problem of specifying $\rho_L(t)$ cannot be solved so easily. Gifford (1955) and Hay and Pasquill (1959) assumed that the Lagrangian and Eulerian correlation functions $\rho_L(t)$and $\rho_E(t)$ and energy spectra $E_L(n)$, $E_E(n)$ were similar in shape but were displaced by a scale factor β:

$$nE_L(n) = \beta n E_E(\beta n) , \tag{7.2}$$

$$\rho_L(\beta t) = \rho_E(t) . \tag{7.3}$$

The factor β is defined as the ratio of the Lagrangian and Eulerian time scales

$$\beta = T_L/T_E . \tag{7.4}$$

The assumption regarding similar shapes has been tested using the results of a few experiments in which turbulent fluctuations are measured concurrently on a tower (Eulerian) and by a floating neutral balloon (Lagrangian). Spectra observed in one such

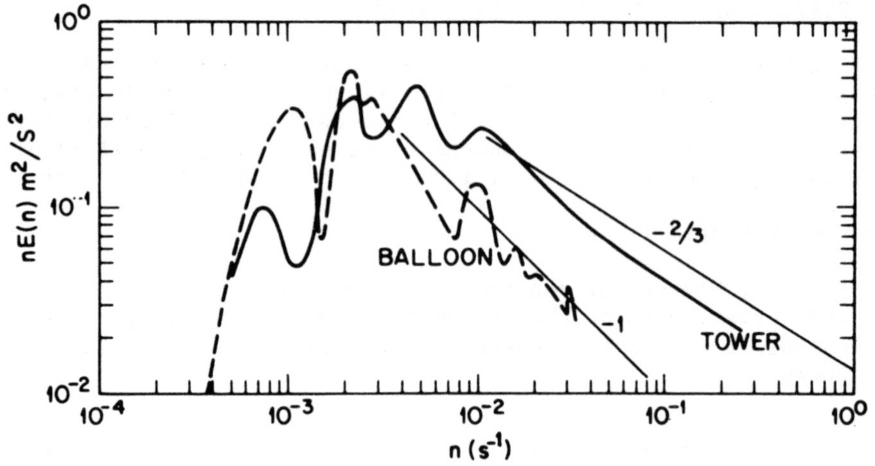

Figure 7.1. Lagrangian (balloon) and concurrent Eulerian (tower) energy spectra for the w component at a height of 300m. (Hanna, 1981a)

experiment reported by Hanna (1981a) are plotted in Figure 7.1. The slight difference in slopes at high frequencies is predicted by similarity theory. In this figure, E_L is displaced to the left of E_E by a factor of about four.

The ratio β is inversely proportional to turbulence intensity according to a simplified theory and set of observations reported by Pasquill (1974)

$$\beta = a/(\sigma_v/U) \; . \tag{7.5}$$

The 'constant' a is found to equal about 0.6, giving $\beta \simeq 2$ in the daytime PBL, $\beta \simeq 4$ in neutral conditions, and $\beta \simeq 10$ in stable conditions. Knowing β, ρ_E and E_E, it is possible to estimate the required Lagrangian ρ_L and E_L through Equations (7.2) and (7.3).

Pasquill (1974) shows that Equation (7.1) can be transformed, using the definition of an energy spectrum, to a form, which was also briefly discussed in Section 2.7, Equation (2.77).

$$[\sigma_y^2]_{T_s,t} = [\sigma_v^2]_{\tau,0}\, t^2 \int_0^\infty E_L(n)\left(1 - \frac{\sin^2(\pi n T_s)}{(\pi n T_s)^2}\right)\left(\frac{\sin^2(\pi n t)}{(\pi n t)^2}\right) dn \tag{7.6}$$

where T_s is sampling time, t is travel time, and τ is a time long enough to include most of the turbulent energy spectrum. If an

Eulerian spectrum must be used, replace $E_L(n)$ by $E_E(n)$ and divide the time variables by β in both sine terms. The sine terms are filters that effectively block out turbulent fluctuations with time scales less than the travel time t and greater than the sampling time T_s. The spectral window is therefore between periods of t and T_s. Initially, the continuous plume is acted on by very small eddies, but as it grows the small eddies have less and less influence. They merely oscillate parcels deep within the plume and do not contribute to total diffusion.

The spectrum filtering function is less clear for instantaneous releases of puffs. Smith (see Pasquill, 1974, p. 143) has derived a filter function of the form

$$\frac{1-e^{-\sigma^2 n^2/U^2}}{\sigma n/U} \tag{7.7}$$

for puffs, where σ is the dimension of the cloud. This function has the characteristic that long period eddies as well as short period eddies are filtered out for puffs, since large eddies can transport the puff but do not diffuse it. Diffusion of puffs is accomplished by eddies with sizes roughly as large as the puff itself.

Doran et al. (1978) applied Equation (7.6) for continuous plumes assuming β = 4 and an Eulerian spectral shape that matches several recent observations of spectra. In choosing a spectral shape, they were interested in spectrum behavior at low frequencies, since that is the region of the spectrum that eventually dominates diffusion for continuous plumes. They chose the functions

$$\text{stable:} \quad nE_E(n) = \frac{6.25\times10^{-6}}{f} + \frac{f}{1+15.2f^{5/3}}, \tag{7.8}$$

$$\text{unstable:} \quad nE_E(n) = \frac{3.6\times10^{-9}}{f} + \frac{f}{1+10^4 f^{5/3}}, \tag{7.9}$$

where f=nz/U is a dimensionless frequency. The ultimate goal of their research was determination of the function S defined by

$$S = \frac{\sigma_y}{x\sigma_\theta} = \frac{[\sigma_v]_{T_s/\beta,\, t/\beta}}{[\sigma_v]_{T_s/\beta,\, 0}} \tag{7.10}$$

where σ_θ is the standard deviation of wind direction fluctuations ($\sigma_\theta \simeq \sigma_v/U$). Doran et al. integrate Equation (7.6) numerically after substituting Equation (7.8) or Equation (7.9), and obtain

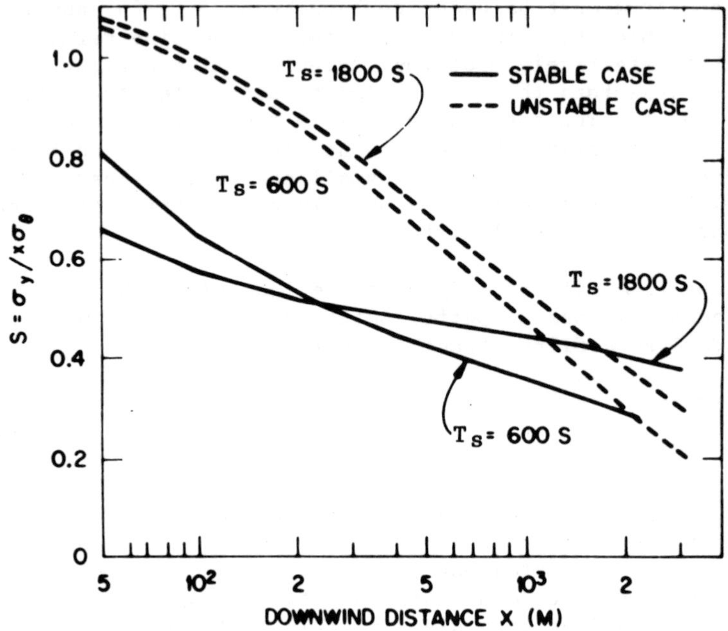

Figure 7.2. Variation of $S=\sigma_y/x\sigma_\theta$ with distance. Sampling times of 600s and 1800s are used (Doran et al., 1978).

the results in Figure 7.2 for sampling times T_s of 600s and 1800s. These sampling times are typical for PBL experimental runs and are the basis for U.S. Environmental Protection Agency (USEPA) σ_y and σ_z curves (Turner, 1969). In this manner, simple methods of estimating σ_y can be derived from integration of the most recent PBL spectra.

Monte Carlo Diffusion Models

Another statistical procedure for calculating diffusion involves writing an equation for the random fluctuations of turbulent speed of a particle and then constructing trajectories of thousands of individual particles (Reid, 1979; Lamb, 1979; Hanna, 1978). At certain times or distances downstream, the resulting concentration distribution can be studied. This technique was impractical before the advent of high-speed computers, and even now is marginally effective. But as computer speed and capacity increase, the Monte Carlo technique will probably be used more often.

Model description. The model separates the velocity field into a mean and a turbulent component, with a dividing line between the components based on the averaging time for the mean wind velocity field observations or predictions. The equation for total particle speed in direction i is

$$V_i = \overline{V}_i + V'_i \ . \tag{7.11}$$

Usually winds are averaged over a one-hour period. If a network of wind monitoring stations is used, then the observed mean winds can be altered so that they are mass-consistent (Sherman, 1978). If the predictions of a dynamic model are used (e.g., Pielke, 1974), then the mean wind field can be assumed to be already mass consistent. At any rate, all diffusion at time scales larger than the averaging time for the mean wind field is accomplished by time and space variations in the mean winds.

Diffusion at time scales less than the averaging time for the mean wind is estimated by assuming that the turbulent fluctuations have two components - a correlated component and a random component:

$$V'_i(t + \Delta t) = V'_i(t)\rho_L^i(\Delta t) + \rho_i \tag{7.12}$$

where the parameter Δt is the time step and $\rho_L(\Delta t)$ is the Lagrangian autocorrelation coefficient for component i and time step Δt. The random or Monte Carlo component ρ_i is picked randomly from a Gaussian distribution with mean zero and standard deviation σ_{ρ_i} given by the formula

$$\sigma_{\rho_i} = \sigma_i \{1 - [\rho_L^i(\Delta t)]^2\}^{1/2} \qquad \text{where } \sigma_\alpha = (\overline{V'^2_\alpha})^{\frac{1}{2}} \ . \tag{7.13}$$

This condition assures that turbulent energy σ_α^2 is conserved from one time step to the next.

From Equations (7.12) and (7.13) it is seen that this model needs values of σ_i and $\rho_L{}^i(\Delta t)$ in order to operate. The sampling time T_s, appropriate for these estimates, is the averaging time for the mean wind field. For convenience, an exponential form for $\rho_L{}^i$ is assumed

$$\rho_L^i(\Delta t) = \exp(-\Delta t/T_L^i) \ . \tag{7.14}$$

It is assumed that a particle responds to the local values of σ_i and $\rho_L{}^i(\Delta t)$ at the current position of the particle, and hence this model can accommodate gradients of these parameters in the domain. On the other hand, Taylor's (1921) model, described in the previous section, assumes a homogeneous field of turbulence.

At this point, up-to-date estimates of turbulent energy profiles and time scales from recent PBL observations and theories are crucial to the success of the statistical model. Best current estimates of these variables are given in the following sections.

Turbulent energy and Lagrangian time scales in the unstable PBL. Analyses of the 1973 Minnesota PBL experiment show that the depth of the mixed layer, h, is an important scaling parameter in unstable conditions. Along with the friction velocity u_*, and the Obukhov length, L, it determines the convective scaling velocity, w_* (see Equation (2.56); here we shall take 0.4 for the Von Karman constant). Sampling time for the Minnesota experiment was about 1 hr. The horizontal components of turbulent energy in the PBL σ_u and σ_v are found to be functions only of h/L, which are given by Equation (4.15). Irwin (1979a) suggests power law formulas for the vertical component σ_w. Since his formulas let σ_w approach zero as z/h approaches zero, the surface layer formula by Panofsky et al. (1977) is suggested near the ground (see also Section 4.2 on variances).

$$\frac{\sigma_w}{w_*} = 0.96\left(3\frac{z}{h} - \frac{L}{h}\right)^{1/3} \qquad \frac{z}{h} < 0.03\ ,$$

$$\frac{\sigma_w}{w_*} = \min\left[0.96\left(3\frac{z}{h} - \frac{L}{h}\right)^{1/3},\ 0.763\left(\frac{z}{h}\right)^{0.175}\right] \qquad 0.03 < \frac{z}{h} < 0.4,$$

$$\frac{\sigma_w}{w_*} = 0.722\left(1 - \frac{z}{h}\right)^{0.207} \qquad 0.4 < \frac{z}{h} < 0.96\ , \qquad (7.15)$$

$$\frac{\sigma_w}{w_*} = 0.37 \qquad 0.96 < \frac{z}{h} < 1\ .$$

It is seen from these equations that the turbulent energies at a given height z are completely determined by h, L and w_* in unstable conditions.

Kaimal et al. (1977) give formulas for the peak wavelength λ_m, which was also discussed in Section 4.2 on spectra. It can be shown that the Eulerian time scale is related to λ_m through the formula (Hanna, 1981a).

$$T_E = 0.16\lambda_m/U\ . \qquad (7.16)$$

Using Equations (7.5) and (7.16) and Kaimal et al's (1977) empirical formulas for λ_m, we arrive at the following equations for Lagrangian time scales T_L in unstable conditions

$$T_L^u = T_L^v = 0.15\frac{h}{\sigma_u}$$

and

$$T_L^w = 0.1\frac{z}{\sigma_w}\frac{1}{(0.55 + 0.38(z-z_o)/L)} \qquad \frac{z}{h} < 0.1,\ -\frac{(z-z_o)}{L} < 1,$$

$$T_L^w = 0.59 \frac{z}{\sigma_w} \qquad \frac{z}{h} < 0.1, \quad -\frac{(z-z_o)}{L} > 1 , \tag{7.17}$$

$$T_L^w = 0.15 \frac{h}{\sigma_w} \{1 - \exp(-\frac{5z}{h})\} \qquad z/h > 0.1 .$$

All the parameters needed in the diffusion equations (7.12) and (7.13) can be estimated once h, L and w_* are known.

Turbulent energy and Lagrangian time scales in the stable PBL. During stable conditions there is a layer of mechanical turbulence adjacent to the ground surface. The height at which the turbulence dies off to nothing is denoted by h, which can be as low as a few meters or as high as a few hundred meters. Several researchers, such as Wyngaard (1975) and Rao and Snodgrass (1978) show that the depth h is a function of the ratio of the scaling lengths u_*/f and L, where f is the Coriolis parameter (see also Section 4.3 on the depth of the SBL).

$$h = c(u_* L/f)^{\frac{1}{2}} . \tag{7.18}$$

The parameter c equals about 0.25. If u_* is 0.35 ms^{-1} at sunset and the cooling rate is 1 K hr^{-1}, then the models show that u_* reaches a steady state value of 0.18 m s^{-1} after about 2 hrs. The Obukhov length L would reach a value of about 20 m, and h would be about 30 m at mid latitudes. Model results such as these are plotted by Wyngaard (1975) for cooling rates of 0.2, 0.5, 1.0, and 2.0 K hr^{-1}.

Minnesota PBL observations in stable conditions verify these model predictions. In general, the observed and predicted variations of turbulent energy with height can be approximated by the following linear relations

$$\sigma_w/u_* = \sigma_v/u_* = 1.3(1 - z/h) , \tag{7.19}$$

$$\sigma_u/u_* = 2(1 - z/h) . \tag{7.20}$$

Caughey et al. (1979) give observations of the variation with height of the wavelength λ_m of peak spectral energy for the three turbulence components during stable conditions (see section 4.3 on turbulence spectra in the SBL). The observed points can be fit by the following simple power laws

$$\frac{\lambda_{mu}}{h} = 1.5(\frac{z}{h})^{\frac{1}{2}}, \quad \frac{\lambda_{mv}}{h} = 0.7(\frac{z}{h})^{\frac{1}{2}}, \quad \frac{\lambda_{mw}}{h} = 1.0(\frac{z}{h})^{0.8}. \tag{7.21}$$

The predictions of lateral turbulence energy and timescales may be underestimated since low frequency meanders have been fitered out. Using the relation (7.16), (7.4) and (7.5), we arrive at formulas for T_L

$$T_L^u = 0.15 \frac{h}{\sigma_u} (\frac{z}{h})^{0.5} , \tag{7.22}$$

$$T_L^v = 0.07 \frac{h}{\sigma_v} (\frac{z}{h})^{0.5} , \tag{7.23}$$

$$T_L^w = 0.10 \frac{h}{\sigma_w} (\frac{z}{h})^{0.8} . \tag{7.24}$$

The equations in this section show that turbulent energies and Lagrangian time scales in a stable boundary layer are completely determined by the paramters u_*, f, and L.

Turbulent energy and Lagrangian time scales in the neutral PBL. It is more difficult to prescribe formulas for the neutral PBl because of the lack of observations, especially in the upper part of the PBL. In general, turbulent energy decreases monotonically with height, and the relevant scaling height is u_*/f. Wyngaard et al. (1974) applied a second-order closure model to this problem and obtained results that can be approximated by exponential formulas

$$\sigma_u/u_* = 2.0 \exp(-\frac{3fz}{u_*}) , \tag{7.25}$$

$$\sigma_v/u_* = \sigma_w/u_* = 1.3 \exp(-\frac{2fz}{u_*}) . \tag{7.26}$$

Hanna (1968) used observations from tall towers to show that the peak wavelength in the vertical energy spectrum in a neutral PBL was proportional to height near the surface and asymptotically approached a value proportional to G/f at great height. The parameter G is the geostrophic or free stream wind speed, and is related to u_* through the surface Rossby number G/fz_o (see Section 2.1). To assure consistency with other formulas in this chapter, it is assumed that u_*/G equals 0.03. Thus we obtain the neutral formula

$$T_L = \frac{0.5 \ z/\sigma_w}{1 + 15 \ fz/u_*} \tag{7.27}$$

which is assumed to be valid for all three components of turbulence in neutral conditions.

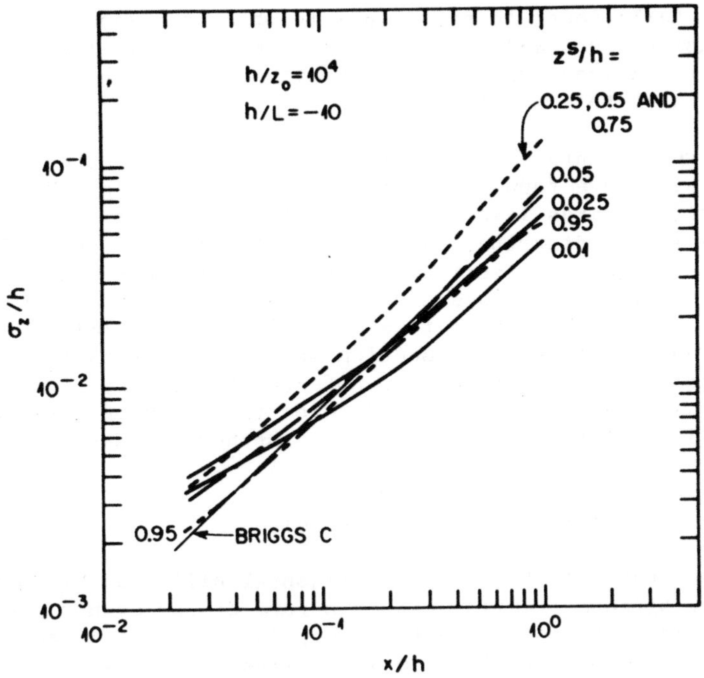

Figure 7.3. Model predictions of σ_z/h as a function of x/h for $h/z_0 = 10^4$, $h/L = -10$, and release heights z^s/h equal to 0.025, 0.01, 0.5, 0.25, 0.5, 0.75 and 0.95. Briggs (1973) empirical formula for C stability is also shown (Hanna, 1980).

Results of application of the Monte Carlo model. In a homogeneous atmosphere with particles released continuously with random initial turbulent speeds, the resulting diffusion, $\sigma_y(t)$, agrees exactly with the analytical solution to Taylor's (1921) statistical model (7.1) with an exponential correlogram, which is given by (2.72).

Reid (1979) applied the Monte Carlo model to vertical diffusion from near surface releases, where there is obviously great wind shear. He found that the resulting σ_z curves were very much in agreement for release heights ranging from 0 to 10 m. Hanna (1980) looked at release heights ranging across the full depth of the PBL during unstable conditions and obtained the curves of σ_z/h plotted in Figure 7.3. At any given downwind distance, σ_z/h is about a factor of two larger for mid-PBL release heights than for release heights near the surface or near the top of the PBL. This result is due to the maximum in σ_w that occurs at mid-PBL heights in unstable conditions.

McNider et al. (1981) and Ohmstede et al. (1980) have applied the Monte Carlo model to situations in which the mean wind field was non-homogeneous and non-stationary. Mountain-valley and sea breeze circulations are examples of situations which are ideal for application of this model.

All researchers who have applied this model to the unstable surface layer have found that the mean particle height has a gentle drift towards levels of low σ_w if there is a gradient of σ_w and if the distribution of vertical speeds is assumed to be Gaussian. The actual distribution of vertical speeds in an unstable PBL is skewed (see also Section 5.4), with a skewness coefficient Sk (7.75) equal to about 0.2 (Chiba, 1978). Proper use of the skewed distribution function would keep the particles at the height to which they belong. Unfortunately, very little information is available on the form or variation of skewness with height and stability. This problem is further discussed in Section 7.6.

The Langevin Horizontal Diffusion Model

A new statistical model of horizontal diffusion has been derived by Gifford (1981) based on a form of Langevin's equation. Given a gross diffusivity coefficient, K_y, and Lagrangian time scale, $T_L{}^y$, appropriate for sampling times of several days and given a measure of the initial turbulent energy, $\sigma_{vo}{}^2$, this model can simulate mean observations of σ_y for travel times from 1 s to 1 yr. The basic of the model is a differential form of Equation (7.12)

$$dV_y'/dt + \beta V_y'(t) = \mu(t) \qquad (7.28)$$

where the parameter β is equivalent to $1/T_L{}^v$, and $\mu(t)$ is a small, random force which arises from the pressure terms.

Using the definition, $V_y' = dY/dt$, and the boundary conditions, $Y = 0$ and $V_y' = V_{yo}'$ at $t = 0$, it is possible to solve Equation (7.28) by the method of variation of parameters. It is helpful to recognize that the large-scale eddy diffusivity, $K_y = \sigma_v{}^2 T_L{}^v$. The solution, in Gifford's (1981) notation, is

$$\Sigma_y^2 = T - (1-e^{-T}) - (\tfrac{c}{2})\,(1-e^{-T})^2 \,,$$

where (7.29)

$$\Sigma_y = \sigma_y/(2K_y T_L^v)^{\frac{1}{2}}, \quad T = t/T_L \quad \text{and} \quad c = (1 - \sigma_{vo}^2/\sigma_v^2).$$

The parameter c is the fraction of the turbulence in the entire flow, $\sigma_v{}^2$ that is <u>not</u> accounted for by the instantaneous

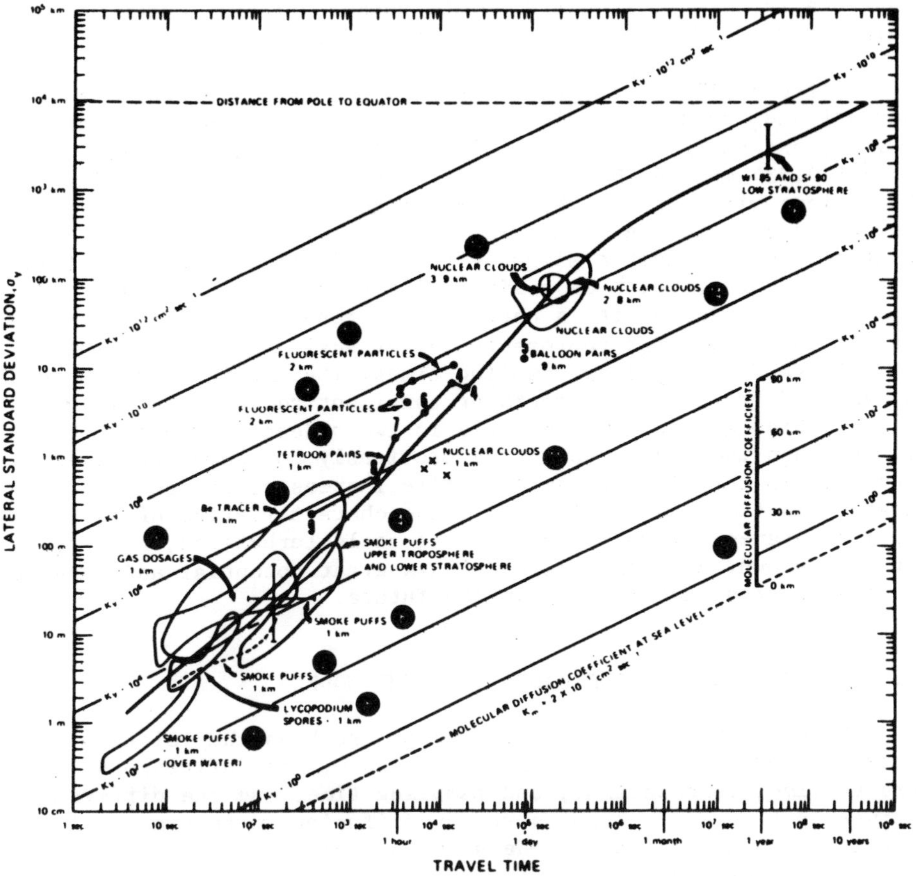

Figure 7.4. Summary of data on horizontal atmospheric diffusion. The solid line is the solution given by Equation (7.29) (Gifford, 1981).

turbulence at the source, $\overline{v_0'^2}$. As c approaches zero (i.e., as σ_{vo}^2 approaches σ_v^2) Equation (7.29) reduces to the solution to Taylor's equation. This solution is appropriate for diffusion in a homogeneous atmosphere from a continuous point source, for large sampling times.

In contrast, as c approaches unity (i.e., as σ_{vo}^2 approaches zero), the solution is appropriate for relative or puff diffusion

$$\Sigma_y^2 = T^3/3 \qquad \text{at small times.} \tag{7.30}$$

For an instantaneous point source, the initial speeds for all particles are clearly equal. The Langevin model has the neat characteristic that it can simulate both continous source and relative diffusion, or a combination of both.

Gifford (1981) next used observed graphs of lateral standard deviation σ_y versus travel time, t, to make best estimates of K_y, T_{Lv} and σ_{vo}^2: $K_y = 1.235\ 10^5\ m^2\ s^{-1}$, $T_L^v = 2.13\ 10^5$ s, $\sigma_{vo}^2 = 0.235\ m^2\ s^{-2}$. His predicted σ_y curve is plotted on Figure 7.4, along with much of the observational data. The predicted curve agrees with the data well as possible with this degree of scatter, and provides a slightly better fit than Heffter's (1975) emperical curve, which is a linear fit to these same data: $\sigma_y = 0.5t(s)$. Again, it is necessary to repeat that the chosen values for K_y and T_L^v represents larger scale averages in the atmosphere for very long sampling times. The natural variations of K_y and T_L^v from month to month or continent to continent is not known. This Lagrangian time scale is two orders of magnitude larger than the ones recommended previously for sampling times of one hour in the PBL. As sampling time increases, so does the Lagrangian time scale. Similarly, the characteristics of the ratio of instantaneous turbulence to large scale turbulence, σ_{vo}^2/σ_v^2, are not known. But this new model has great potential and deserves much more study and testing in the future.

7.3. IMPROVEMENTS TO THE GAUSSIAN PLUME MODEL

Despite the importance of concepts such as the relation between Lagrangian and Eulerian turbulence, the influence of travel time, release time, and sampling time, and the differences between puffs and continuous source diffusion, most practical applications of diffusion models use a single simple model - the Gaussian plume model

$$C = \frac{Q}{2\pi U\sigma_y\sigma_z}\exp\left(-\frac{y^2}{2\sigma_y^2}\right)\left[\exp\left(-\frac{(z-H)^2}{2\sigma_z^2}\right) + \exp\left(-\frac{(z+H)^2}{2\sigma_z^2}\right)\right] \tag{7.31}$$

where C is concentration (kg m^{-3}), Q is continuous point source strength (kg s^{-1}), H is the effective plume height above the ground, y is the lateral distance from the plume centerline and z is the height above the ground. The last term is an effective image source at a distance H below the ground, which accounts for plume reflection at the ground surface.

Wind Speed in the Gaussian Plume Model

PBL theories and observations influence nearly all parameters in Equation (7.31). The job of an air pollution meteorologist is to translate new results into a form that can be easily applied in a model. For example, the wind speed, U, in Equation (7.31) should be measured at the effective plume height, H. Because most industrial facilities have a wind instrument only at a single height near the ground, say 10 m, it is necessary to extrapolate this measurement to the effective plume height. Irwin (1979b) studied the latest wind profile theories in order to develop methods for calculating the power p in the formula

$$U(z) = U(10)\ (z/10)^p\ . \qquad (7.32)$$

He used the wind formulations by Nickerson and Smiley (1975) and relations between Obukhov length L, roughness, and stability class given by Golder (1972). His estimates of p, as a function of stability and roughness are given in Table 7.1.

Table 7.1. Estimates of the Power p in Equation (7.32)

Class	A	B	C	D	E	F
z_0 = 0.01m	0.05	0.06	0.06	0.12	0.34	0.53
z_0 = 0.10m	0.08	0.09	0.11	0.16	0.32	0.54
z_0 = 1.00m	0.17	0.17	0.20	0.27	0.38	0.61
z_0 = 3.00m	0.27	0.28	0.31	0.37	0.47	0.69

The latter stability classes in this table are the familiar Pasquill-Gifford classes. At heights above 100 m, the wind speed should be set equal to its value at 100 m.

Plume Rise Calculations

Current methods for calculating plume rise in stable layers are satisfactory, due to the simplicity of the theory and the relative ease of obtaining observations. In the initial stages of rise of a buoyant plume, internal turbulence of the plume completely dominates ambient turbulence. But in neutral or unstable conditions, the plume eventually reaches a point where ambient turbulence dominates the internal turbulence of a plume, and its vertical progress stops. This situation is more difficult to deal with than the stable situation, but it often occurs in the daytime PBL. Briggs (1979) has developed a 'breakup' model for estimating plume rise in this situation which assumes that plume rise terminates when the plume internal eddy dissipation rate just equals the ambient eddy dissipation rate

$$1.5\ w^3/\Delta z = \varepsilon \qquad (7.33)$$

where w is the vertical speed of the plume centerline and height Δz is the plume rise. For a buoyant, bent-over plume, the height z of the plume centerline is given by the formula $z = z^s + \Delta z$, where

$$\Delta z = 1.6 \frac{F^{1/3}}{U} x^{2/3}, \tag{7.34}$$

where z^s is stack height and F is initial plume buoyancy flux, defined by

$$F = (g/T_s) w_s r_s^2 (T_s - T_a) \tag{7.35}$$

where T_s is the temperature of the emissions, T_a the ambient temperature, w_s the vertical velocity of the emissions and r_s the source radius.

In neutral conditions, Briggs (1975) suggests that the relation $\varepsilon = u_*^3/kz$ is valid through the depth of the boundary layer. Using the definitions $dx = Udt$ and $w = dz/dt$ it is possible to arrive at the following formula for effective plume rise

$$\Delta z_e = 1.3 \frac{F}{Uu_*^2} \left(1 + \frac{z^s}{\Delta z_e}\right)^{2/3}.$$

Briggs (1975) suggest that this equation may be approximated by

$$\Delta z_e = 1.54 \left(\frac{F}{Uu_*^2}\right)^{2/3} (z^s)^{1/3} . \tag{7.36}$$

In convective conditions, Briggs (1975) uses the approximation $\varepsilon = 0.25\ (g/T)\ \overline{w\theta}_0$. The resulting effective plume rise prediction is

$$\Delta z_e = 3\left(\frac{F}{U}\right)^{3/5} \left(\frac{g}{T} \overline{w\theta}_0\right)^{-2/5}. \tag{7.37}$$

This formula is tentative, since plume-rise observations in convective conditions are the least satisfactory due to rapid dilution.

Estimation of σ_y and σ_z using σ_θ and σ_e

In the process of applying the Gaussian plume model it is necessary to provide values for the diffusion parameters σ_y and σ_z as a function of downwind distance. Although an elaborate network of stability class schemes and σ-curves has evolved due to regulatory needs, most boundary layer meteorologists stand behind Pasquill's (1961) recommendation that observations of turbulence be used to estimate σ_y and σ_z if they are available. The following similarity equations express the relation between σ_y and σ_z and turbulence.

$$\sigma_y = \sigma_\theta x\, S_y\left(\frac{x}{UT_L^v}\right) \quad \text{and} \quad \sigma_z = \sigma_e x\, S_z\left(\frac{x}{UT_L^w}\right) \tag{7.38}$$

where σ_θ and σ_e are the standard deviations of horizontal and vertical wind direction fluctuations, respectively, and S_y and S_z are dimensionless functions (see Equation (7.10)). In addition, the discussion around (7.10) suggests that S_y and S_z are functions of β, sampling time, and the shape of the energy spectrum. It is the intention of all scientists who have studied this problem to employ up-to-date estimates of PBL spectra in order to solve (7.6), compare the results with observations, and then attempt to simplify the results so that the final method is easy to apply. Pasquill (1976) gets good results by removing the dependence on the time scale in Equation (7.38) for σ_y. Irwin (1979a) derived the following analytical approximations to Pasquill's table:

$$\begin{aligned} S_y &= (1 + 0.031x^{0.46})^{-1} \qquad x < 10^4 \text{ m}, \\ S_y &= 33x^{-1/2} \qquad x > 10^4 \text{ m}. \end{aligned} \tag{7.39}$$

Draxler (1976) plotted observations of $S_z = \sigma_z/\sigma_w t$ as a function of time after release and suggested the formula

$$S_y = S_z = [1 + 0.40(x/UT_{Ly})^{1/2}]^{-1}. \tag{7.40}$$

This formula is valid for ground level sources, for all stabilities for S_y, and for stable and neutral conditions for S_z. The S_y and S_z curves suggested by Doran et al. (1978) have already been plotted in Figure 7.2. Irwin (1979a) also suggests shapes for the S_y and S_z curves. Actually, there is so much variability in the observed data that none of these curves for S_y and S_z has an advantage over the others, and one might just as well use the simple analytical formulas given in Equations (7.39) and (7.40).

Determination of Stability Class

If turbulence observations are not available, then σ_y and σ_z must be estimated by an empirical technique such as the Pasquill (1961) - Gifford (1961) curves. A separate curve is given for each of seven stability classes, ranging from A (most unstable) to F (most stable). These classes are based on cloudiness, surface wind speed and insolation. The widely used STAR computer routine uses these criteria to convert hourly meteorological observations at a typical weather station into stability classes. Smith (1951) and Cramer (1957) recommend using wind direction fluctuations to estimate stability class. The U.S. Nuclear Regulatory Commission recommends that the vertical temperature gradient measured on a tower be used as an indicator for stability. It was this confusion that led to the workshop on stability classification schemes in 1977 sponsored by the American Meteorological Society (Hanna et al., 1977). The workshop participants recommended the use of a stability parameter based more on relevant PBL parameters, such as the Richardson number, Ri, or the dimensionless paramters h/L or fL/u_*. Briggs and McDonald (1978) reanalyzed vertical spread observed during the 1956 Prairie Grass field experiment using some of these stability parameters. This was a carefully run, research-grade experiment. They found that vertical spread is best ordered by L in all conditions. In stable conditions, it is also fairly well ordered by $u^2/\Delta\theta$, u^2 and $\Delta\theta$. In unstable conditions, $\Delta\theta$ does not help at all, but $u^2/\Delta\theta$ and u^2 work fairly well. The plumes did not reach high enough for h to be useful.

Sedefian and Bennett (1980) compared several turbulence classification schemes and found a fair amount of disagreement. They found that even the PBL parameters such as Ri experience difficulties in practice because of inadequate methods available for observing wind and temperature gradients at most sites where diffusion models must be applied. The PBL stability parameters work well for research-grade towers, but break down at less well-maintained sites.

The EPA STAR routine, on the other hand, is based on easily measured parameters such as wind speed and cloudiness, and so is less susceptible to instrument and calibration problems. Rigorous stability parameters could be better implemented by careful development of empirical relations from existing data sets. For example, Venkatram (1980) analyzed data collected during the Prairie Grass, Kansas, and Minnesota experiments and discovered that a simple relation existed during stable conditions between the Obukhov length, L, and the friction velocity, u_*

$$L = A\,u_*^2, \quad A = 1.1\ 10^3\ s^2\ m^{-1}. \tag{7.41}$$

Thus L can be estimated without making costly heat flux measurements.

Revisions of Pasquill-Gifford (P-G) Sigma Curves

The original Pasquill-Gifford sigma curves were presented graphically, which was fine in the slide rule era. But today the Gaussian plume equation (7.31) can be easily programmed, even for a small pocket calculator, and it is preferable to have analytical formulas for σ_y and σ_z. Briggs (1973) combined the Pasquill-Gifford, TVA, and Brookhaven curves, using theoretical formulas for asymptotic limits, to obtain the set of recommended formulas in Table 7.2. Separate formulas are given for urban and rural conditions. The formulas are limited to downwind distances less than about 10 km, sampling times of about 20 min, and roughnesses of about 3 cm in rural areas and 1 m in urban areas.

Table 7.2. $\sigma_y(x)$ and $\sigma_z(x)$ for $10^2 < x < 10^4$ m

Pasquill type	σ_y(m)	σ_z(m)
Open-Country Conditions		
A	$0.22x(1+0.0001x)^{-\frac{1}{2}}$	$0.20\ x$
B	$0.16x(1+0.0001x)^{-\frac{1}{2}}$	$0.12\ x$
C	$0.11x(1+0.0001x)^{-\frac{1}{2}}$	$0.08\ x(1+0.0002x)^{-\frac{1}{2}}$
D	$0.08x(1+0.0001x)^{-\frac{1}{2}}$	$0.06\ x(1+0.0015x)^{-\frac{1}{2}}$
E	$0.06x(1+0.0001x)^{-\frac{1}{2}}$	$0.03\ x(1+0.0003x)^{-1}$
F	$0.04x(1+0.0001x)^{-\frac{1}{2}}$	$0.016x(1+0.0003x)^{-1}$
Urban Conditions		
A-B	$0.32x(1+0.0004x)^{-\frac{1}{2}}$	$0.24x(1+0.0001x)^{\frac{1}{2}}$
C	$0.22x(1+0.0004x)^{-\frac{1}{2}}$	$0.20x$
D	$0.16x(1+0.0004x)^{-\frac{1}{2}}$	$0.14x(1+0.0003x)^{-\frac{1}{2}}$
E-F	$0.11x(1+0.0004x)^{-\frac{1}{2}}$	$0.08x(1+0.00015x)^{-\frac{1}{2}}$

The sigma curves are the center of continual argument in the USA because of the important role played by modeling in regulatory decisions. The curves are subject to much misapplication to distances, sampling times, and roughness outside of their range of derivation. For example, Briggs' (1973) formulas in Table 7.2 are valid only out to x = 10 km. At greater distances, σ_y becomes proportional to $x^{\frac{1}{2}}$ in this formula, which clearly does not fit observed σ_y at large distances in Figure 7.4. Formulas suggested by Heffter (1975) or Gifford (1981) are better for σ_y at large distances (see Section 7.2 on the Langevin equation). In rough terrain, the sigmas are usually site-dependent and the blind application of Table 7.2 can lead to poor agreement with observations.

Under clear, calm night time conditions (so-called stability category G), recent diffusion experiments suggest that σ_y and σ_θ

are greater than those observed under category F (Sagendorf, 1979). Previous to this work, the Nuclear Regulatory Commission had been assuming that σ_y for G conditions was less than that for F conditions, leading to very high predicted ground level concentrations for sources near the ground. The reason for relatively large σ_y in G conditions is the meandering of the wind direction. The instantaneous concentration under the plume may be high, but the plume wanders about so much during an hour that the average concentration at a fixed point is low.

During convective daytime conditions, sigma values are dependent on the convective velocity scale, w_*, and the mixing depth, h. This subject is covered in Chapter 5. Many users of sigma curves were unhappy with the original Pasquill-Gifford A category σ_z curve because it increased too rapidly with downstream distance. This curve was based on a few observations at limited downwind distances, and should not have been extrapolated beyond 1 km downwind. Briggs' revised A category σ_z curve is more realistic (Table 7.2). Recent studies of diffusion in the convective PBL have led several researchers independently to the simple formula

$$\sigma_y = \sigma_z = 0.6 w_* x/U. \qquad (7.42)$$

In estimating maximum ground-level concentrations, the schizophrenic nature of the convective PBL must be taken into account. It is found that there are strong thermal updrafts covering 40% of the area and weaker downdrafts covering 60% of the area (see Figure 4.14 and Section 5.4). A plume from an elevated source can get caught in a downdraft and be brought to the ground for a short period, causing a high ground-level concentration for a short time. On the other hand, when a plume gets caught in an updraft for a few minutes, it is lifted towards the top of the mixed layer, h. The 20 min or 1 hr average σ_z would reflect both of these processes. Computer simulations of dispersion of neutrally bouyant particles in the convective boundary layer lead to (see Section 5.3)

$$C_{max} = 1.2 \frac{Q}{HhU} \quad \text{at} \quad x_{max} = 2HU/w_*. \qquad (7.43)$$

If we differentiated the Gaussian equation (7.31) to find C_{max}, we would find that $C_{max} \propto H^{-2}$, which is different from the $C_{max} \propto H^{-1}$ relation implied by Equation (7.43). This suggests that tall stacks may not have as much benefit as we thought in convective conditions where downdrafts can quickly bring the plume to the ground. However, one should be cautious to extend these results to buoyant plumes (See Section 5.6).

7.4. K-DIFFUSION MODELS

The turbulent flux of pollution is often assumed to be proportional to the mean gradient

$$-\overline{wc} = K_z \partial C/\partial z \ . \tag{7.44}$$

This assumption, along with the equation of continuity, leads to the gradient transport model, also called the K diffusion model

$$\frac{\partial C}{\partial t} + U_i \frac{\partial C}{\partial x_i} = S + \frac{\partial}{\partial x_i}(K_i \frac{\partial C}{\partial x_i}) \tag{7.45}$$

where S is a source term and $K_i = (K_x, K_y, K_z)$ the eddy diffusivity. In principle, it is possible to predict the spatial and time variability of concentration C given appropriate boundary and initial conditions, plus a knowledge of the time and space fields of U_i and K_i. But there are two fundamental difficulties with this approach: (1) the time and space fields of the meteorological variables are not as well-known as we would like; and (2) when this equation is solved by computer, the introduction of finite differences can lead to erroneous solutions. The last problem can be eliminated through the use of analytical solutions, but such solutions are possible only for very simple forms of the meteorological variables.

A fundamental assumption in (7.44) and (7.45) is that the size of the cloud being diffused is larger than the characteristic size of the turbulent eddies. This technique is therefore not appropriate for calculating diffusion of a plume at short distances downwind of an elevated source. It is appropriate for calculating vertical diffusion from ground level sources, from large area or volume sources, and for calculating horizontal and vertical diffusion from any type of source on regional or global scales. A simple analytical solution to Equation (7.45) for uniform conditions yields $\sigma_y = (2K_y t)^{\frac{1}{2}}$, which is clearly not valid for travel times less than a few days (see Figure 7.4).

Analytical Solutions to the Diffusion Equation

Roberts (1923) gave a solution to the equation

$$U \frac{\partial C}{\partial x} = \frac{\partial}{\partial z}(K_z \frac{\partial C}{\partial z}) \tag{7.46}$$

for a line source, Q_ℓ, and U and K_z varying with height raised to some power

$$U = u_1(z/z_1)^m, \quad K_z = K_{z_1}(z/z_1)^n. \tag{7.47}$$

The predicted concentration distribution is given by the analytical solution

$$C(x,z) = \frac{Q_\ell z_1^m r}{u_1 \Gamma(s)} \left[\frac{u_1 z_1^{n-m}}{r^2 K_{z_1} x}\right]^s \exp\left[-\frac{u_1 z_1^{n-m} z^r}{r^2 K_{z_1} x}\right] \tag{7.48}$$

where $r = m - n + 2$, $s = (m + 1)/r$, and Γ is the Gamma function. In neutral conditions, $m \simeq 0.2$ and $n \simeq 0.8$. Berlyand (1975) provided solutions to the diffusion equations for K_z as a function of bulk Richardson number. The USSR regulatory models are based on the assumption

$$K_z = K_{z_1}\, z/z_1 \quad (z < h), \qquad K_z = K_{z_1}\, h/z_1 \quad (z > h) \tag{7.49}$$

where the height h is usually chosen to be 100 m and the diffusivity K_{z_1} averages 0.2 $m^2 s^{-1}$.

More recently, Nieuwstadt (1980) suggested an analytical solution to the time-dependent, one-dimensional diffusion equation

$$\frac{\partial C}{\partial t} = \frac{\partial}{\partial z}\left(K_z \frac{\partial C}{\partial z}\right) \tag{7.50}$$

with boundary conditions $K_z \partial C/\partial z = 0$, for $z = 0$ and $z = h$, where h is the top of the boundary layer. An instantaneous area source, Q_i, at time $t = 0$ is assumed. The following formula for K_z is used

$$K_z = c u_* z(1-z/h) \tag{7.51}$$

This equation is similar to the one (Equations (7.49)) used by Berlyand (1975) in the lower half of the boundary layer. The main difference is that Equation (7.51) lets K_z drop to zero at the top of the boundary layer. The solution is

$$C = \frac{Q_i}{h}\left[1 + \sum_{n=1}^{\infty} (2n + 1)P_n\left(2\frac{H}{h} - 1\right)P_n\left(\frac{z}{h} - 1\right)\exp\left\{-n(n + 1)\frac{c u_* t}{h}\right\}\right] \tag{7.52}$$

where P_n are Legendre polynomials and H is source height. For the case $H/h = 0$, $t \to 0$, and $h \to \infty$, there is a simple analytical solution (Monin and Yaglom (1971) and Equation (6.60).)

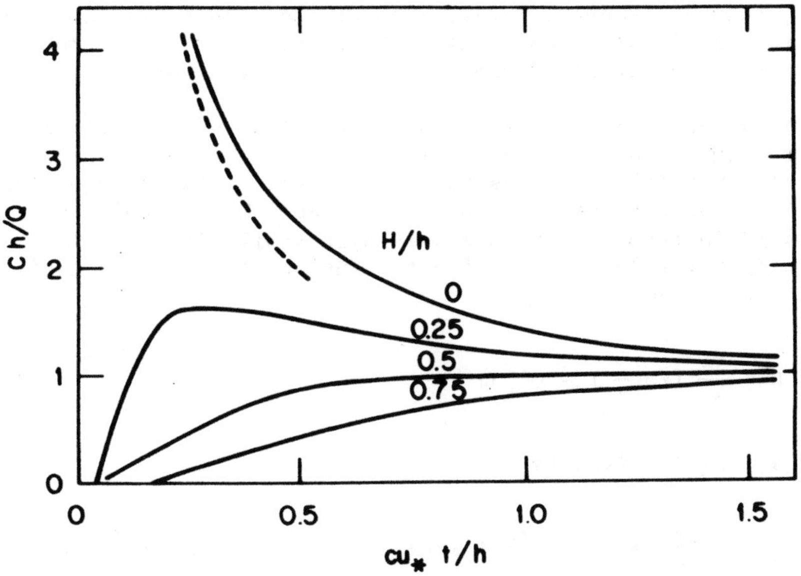

Figure 7.5. Dimensionless concentration Ch/Q at ground level as a function of cu_*t/h for several source heights according to the analytical solution in Equation (7.52). The dashed line is the solution according to Equation (7.53) (Nieuwstadt, 1980).

$$C = \frac{Q_i}{cu_* t} \exp(- \frac{z}{cu_* t}) \; . \tag{7.53}$$

In Figure 7.5, the analytical solutions (7.52) for several source heights are plotted and compared with the simple form (7.53).

We can conclude this section by observing that analytical solutions that employ reasonable U and K profiles give fairly good results. The K_z profile simplifications that must be made to get a solution are not severe, and the results are applicable to many types of problems. However, when using these results, one must always remember that they are valid only for clouds or plumes with scales larger than the turbulence.

Numerical Solutions to the Diffusion Equation

Equation (7.45) can be written in finite difference form, thus opening the door to a countless variety of numerical (computer) solutions. The problem is that several numerical difficulties, such as instabilities and pseudo-diffusion, come through the door while it is open. This has led to the growth of a thriving research area devoted to solving numerical difficulties (e.g., Long and Pepper, 1976). Much of the literature on numerical solutions deals with numerical methods and input/output routines, rather than scientific advances in specifying K_z or U.

As an example of a study that emphasizes atmospheric physics rather than computer science, Nieuwstadt and Van Ulden (1978) derived numerical solutions to the vertical diffusion equation (7.46) in the surface layer using U(z) and $K_z(z)$ profiles expressed in terms of the surface-layer similarity relations obtained in the Kansas experiments. Businger's (1971) velocity profile was used

$$U = \frac{u_*}{k}\{\ell n(\frac{z}{z_o}) - \psi(\frac{z}{L})\} \tag{7.54}$$

where $\psi(z/L)$ is given by

$$\text{for } L \leqq 0 \quad \psi(\frac{z}{L}) = 2\ell n(\frac{1+x}{2}) + \ell n(\frac{1+x^2}{2}) - 2\arctan(x) + \frac{\pi}{2} \tag{7.55}$$

where $x = (1-15z/L)^{1/4}$

$$\text{for } L > 0 \quad \psi(\frac{z}{L}) = -4.7\frac{z}{L}\ . \tag{7.56}$$

Equation (7.55) was derived by Paulson (1970). Two assumptions for K_z were tried. In the first, $K_z = K_h$, where the eddy conductivity K_h is defined by the formula

$$K_h = \frac{ku_* z}{\phi_h(z/L)}\ . \tag{7.57}$$

In the second method, $K_z = \alpha K_m$, where $\alpha = K_h/K_m = 1.35$ at $z/L = 0$, and the eddy viscosity K_m is defined by the formula

$$K_m = \frac{ku_* z}{\phi_m(z/L)}\ . \tag{7.58}$$

The dimensionless functions ϕ_h and ϕ_m are given by the formulas

$$\text{for } L \leqq 0 \quad \phi_h = 0.74(1 - 9\frac{z}{L})^{-1/2} \text{ and } \phi_m = (1 - 15\frac{z}{L})^{-1/4}, \tag{7.59}$$

$$\text{for } L > 0 \quad \phi_h = 0.74 + 4.7\,\frac{z}{L} \quad \text{and} \quad \phi_m = 1 + 4.7\,\frac{z}{L}\,. \tag{7.60}$$

Nieuwstadt and Van Ulden (1978) tried to fit functions of the form

$$C/C_0 = \exp\{-\,b(z/\overline{Z}P)^s\} \tag{7.61}$$

to the results of their numerical calculation, where the mean dispersion height $\overline{Z}P$ (see Equation (6.58)) is defined by the equation

$$\overline{Z}^P = \int_0^\infty zC\,dz \Big/ \int_0^\infty C dz\,. \tag{7.62}$$

The authors tested their numerical model by employing parameters measured during the Prairie Grass diffusion experiment. The numerical profiles $C(z)/C_0$ were found to agree well with the observations, but Equation (7.61) could also be forced to agree well with the observations for appropriate choice of the shape parameter s.

Another method of estimating K_z is being used by F.B. Smith (1972, 1977) to solve Equation (7.46). He calculates the vertical distribution of C due to emissions from a continuous line source at the surface and then extracts $\sigma_z(x)$ from the data, intending to use these results to revise the Pasquill-Gifford σ_z curves. The eddy diffusivity K_z is estimated from a relationship suggested by Hanna (1968)

$$K_z = 0.15\,\sigma_w \lambda_{mw}\,. \tag{7.63}$$

The parameters σ_w and λ_{mw} can be estimated from the results of the Minnesota experiment, given as Equations (7.15), (7.17), (7.19), (7.21), (7.26) and (7.27). All scales in that set of equations are given as $T_L{}^w$. From (7.5) and (7.16) it follows that the wavelength λ_{mw} is related to the Lagrangian time scale $T_L{}^w$ through the following identity

$$\lambda_{mw} = 10\,T_L^w\,\sigma_w\,. \tag{7.64}$$

In practice, Equation (7.63) has much the same behavior during neutral and unstable conditions as Equations (7.49) or (7.51). It is desirable during these stability conditions to have K_z proportional to height near the surface, constant near the middle of the PBL, and decrease near the top of the PBL. These features are illustrated by the graph of the variation of K_z profiles from 8 am to 6 pm in Figure 7.6, based on F.B. Smith's (1977) results.

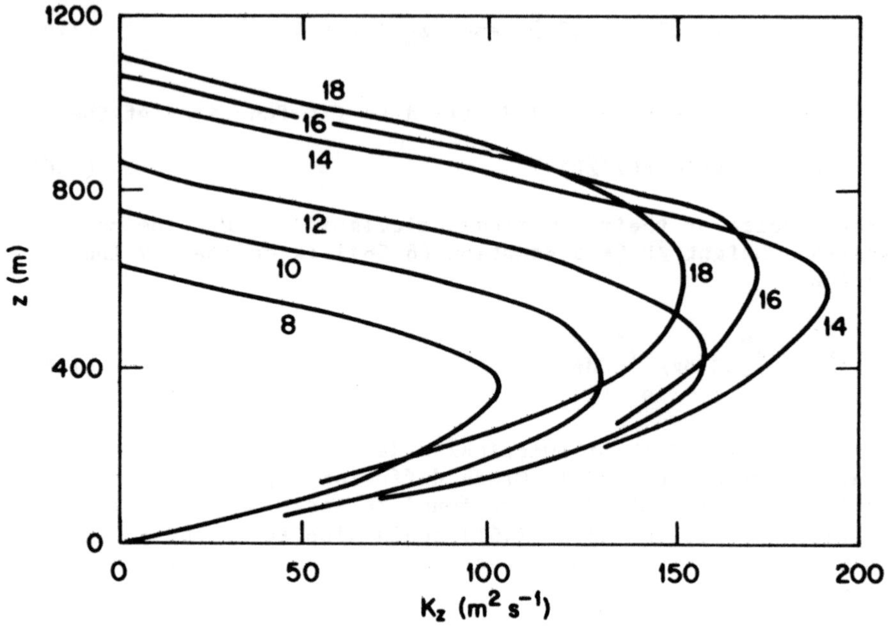

Figure 7.6. An example of the evolution of K_z with time for a typical spring day over flat farmland. (Smith, 1977)

7.5. PROGRESS IN THE SIMILARITY THEORY OF DIFFUSION

Useful models of diffusion in the PBL can often be derived solely from principles of similarity theory or dimensional analysis. Physical insight is required in order to choose the proper variables and parameters. Furthermore, care must be taken in interpreting the results. Hicks (1978) has studied problems associated with the dimensional analysis of the situation in which two variables x and y are uncorrelated. A false correlation between x and y can occur if both variables are divided by a third variable, z. The dimensionless ratios x/z and y/z may be correlated even if x, y and z are independent random variables.

Dimensional analysis can be used to arrive at a prediction for σ_z in the middle of the convective PBL. We know that the convective velocity scale, w_*, is the dominant velocity scale in the convective PBL and thus can make the prediction

$$\sigma_z = c\, w_* t = c\, w_* x/U \ . \tag{7.65}$$

The constant, c, must be determined by observations. Results of field studies and applications of other types of models lead to the conclusion that $c = 0.6$ (see Equation 7.42). Equation (7.65) will apply only as long as the edge of the plume is not close to the ground surface or mixing height, h. In 'mixed-layer' or 'convective' similarity, all diffusion and turbulence statistics can be scaled by the characteristic parameters w_* and h (see Chapters 2, 4 and 5).

Much more work has been done with similarity theory in the surface layer where Monin-Obukhov similarity laws apply to vertical turbulence. Here all diffusion and turbulence statistics in the vertical can be scaled by the characteristic parameters u_* and L (see Chapters 2 and 5). Gifford (1962), Chaudhry and Meroney (1973), Pasquill (1974), and Hunt (1979) are among those who have analyzed this problem. They all begin with the equation

$$\frac{d\overline{Z}^p}{dt} = c\ u_* f\left(\frac{\overline{Z}^p}{L}\right) \tag{7.66}$$

where $\overline{Z}^p$ is the mean height of the diffusing material, which is assumed to be continuously released from ground level. (See Equation (6.58).) The time derivative can be converted to a space derivative by use of the relation

$$d\overline{X} = U(a\overline{Z}^p/z_o,\ a\overline{Z}^p/L)\ dt \tag{7.67}$$

where $\overline{X}$ is the mean downwind position of particles after time t, and $a\overline{Z}^p$ is the effective height of plume transport.

Chaudhry and Meroney (1973) assume that the function $f(\overline{Z}^p/L)$ is equal to $\phi_h^{-1}(\overline{Z}^p/L)$, which follows from Businger et al. (1971) (see Equations (7.59) and (7.60)). The formulas suggested by Businger et al. (1971) for the wind speed are also used (see Equations (7.54)-(7.56)). Pasquill (1974) recommends on theoretical grounds that $c = 0.4$ in Equation (7.66) and $a = 0.56$ in Equation (7.67). Horst (1979) suggests that $a = 0.63$, 0.55, and 0.85 for neutral, unstable, and stable conditions, respectively. Equations (7.66) and (7.67) can be integrated for neutral conditions, with $\overline{Z}^p = z_0$ at $x = 0$, and $a = 0.56$, to give

$$\frac{\overline{X}}{z_o} = 6.25\left[\frac{\overline{Z}^p}{z_o}\ \ell n\left(\frac{\overline{Z}^p}{z_o}\right) - 1.58\left(\frac{\overline{Z}^p}{z_o}\right) + 1.58\right] . \tag{7.68}$$

It is not very useful to know the function $\overline{Z}^p(\overline{X})$, since we are mainly interested in surface concentrations. To obtain the cross-wind integrated surface concentration, $C^y(\overline{X},0)$, Horst (1979) makes use of experiments which give the vertical distribution of C^y in the surface layer due to continuous ground level releases

$$C^y(\overline{X},z)/C^y(\overline{X},0) = \exp\{-(z/b\overline{Z}^p)^s\} \, . \tag{7.69}$$

The factor $b = \Gamma(1/s)/\Gamma(2/s)$, where Γ is the Gamma function. The Gaussian value, $s = 2$, is found to agree with observations only during stable conditions. During neutral conditions, $s \simeq 1.5$, and during unstable conditions, s is less than 1.5.

The mass-continuity equation

$$\int_0^\infty U(z)\, C^y(\overline{X},z)\, dz = Q \tag{7.70}$$

must be used to obtain a final estimate of $C^y(\overline{X},0)$:

$$C^y(\overline{X},0) = (\frac{Q}{z_o})/[\int_1^\infty U(\frac{z}{z_0},\frac{z}{L}) \exp\{-(z/b\overline{Z}^p)^r\}\, d(\frac{z}{z_0})] . \tag{7.71}$$

This equation cannot be solved analytically, but Horst (1979) has solved it by computer for 15 values of z_0/L. He assumed that $s = 1.5$ for all stabilities. His results are plotted in Figure 7.7, showing the expected increased dilution with decreasing stability. It is possible to fit power laws to any of the lines on the figure, e.g., the following formulas apply to the most stable and unstable curves

$$\frac{u_* z_0 C^y(\overline{X},0)}{0.4Q} = 0.75(\frac{\overline{X}}{z_0})^{-0.69} \quad \text{at} \quad \frac{z_0}{L} = 10^{-2}, \tag{7.72}$$

$$\frac{u_* z_0 C^y(\overline{X},0)}{0.4Q} = 35(\frac{\overline{X}}{z_0})^{-1.54} \quad \text{at} \quad \frac{z_0}{L} = -10^{-2}. \tag{7.73}$$

The predictions in Figure 7.7 were compared with observations of cross-wind-integrated concentrations from the Prairie Grass diffusion experiment, with good agreement for all stabilities at distances x/z_0 out to $2\ 10^5$.

Nieuwstadt (1980) has further studied the most unstable cases from the Prairie Grass experiment. He points out that surface layer similarity (scaling with u_*, z_0, and L) is valid for $z < L$ and mixed layer similarity (scaling with w_* and h) is valid for $z > 0.1h$. The region from $L < z < 0.1h$ is known as the convective matching layer. In this layer, the following similarity law provides a satisfactory fit to data

$$\frac{C^y(x,0)Uh}{Q} = 0.9\, (\frac{w_* x}{Uh})^{-3/2} \, . \tag{7.74}$$

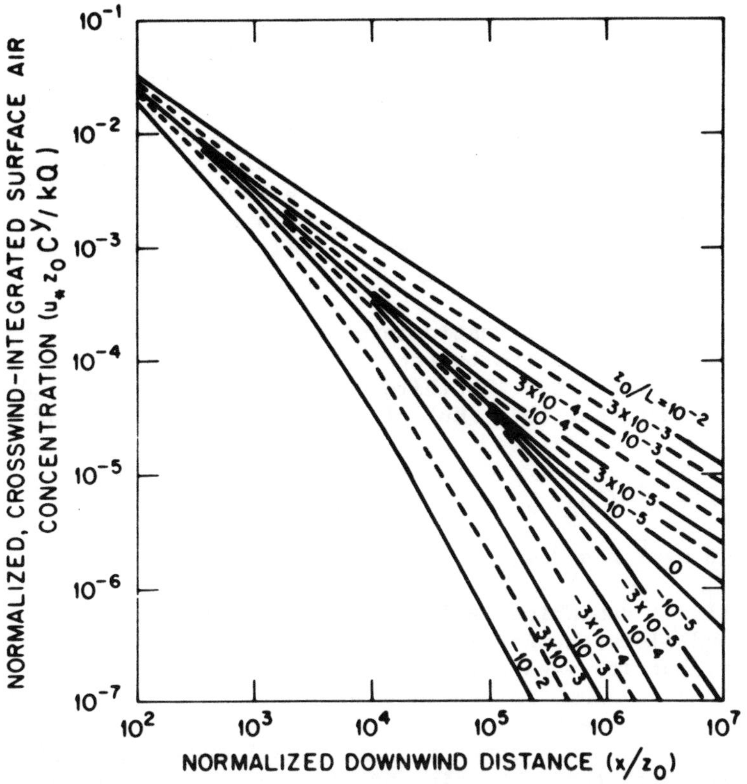

Figure 7.7. Predicted cross-wind integrated concentration at groundlevel as a function of down-wind distance for various stabilities (Horst, 1977).

Note that the power law predicted by Nieuwstadt (1980) closely agrees with that predicted by Horst (1979) (Equation (7.73) above) for his unstable case. Equation (7.74) is susceptible to the false correlation discussed by Hicks (1978), since Uh appears on both sides. Nieuwstadt (1980) claims that the scatter of the data about the line given by Equation (7.74) is much less than that expected for random data showing correlation.

7.6. RECENT SPECIAL APPLICATIONS

There is much research activity in the area of applications of recent findings in boundary-layer turbulence to problems of diffusion. This section reviews three special examples of such studies.

Skewness of Vertical Turbulent Velocity

Fluid dynamicists recognize that if a gradient of turbulent energy exists, then the turbulent velocity distribution must be skewed. Otherwise fluid would accumulate in regions of low turbulent energy. Thus it is not consistent to assume a Gaussian distribution of vertical turbulent velocity w, in the PBL, where the vertical turbulent energy, σ_w^2, is known to vary with height (Equations (7.15), (7.19), (7.26)). This concept is important for Monte Carlo diffusion models which follow the trajectories of thousands of particles.

There is not much information available on the magnitude of the skewness of vertical turbulent velocity and its variation with height and stability. The third moment (skewness, Sk) and the fourth moment (kurtosis, Ku) of a set of observations of w are defined by

$$Sk = \overline{w^3} / \sigma_w^{\ 3} , \qquad (7.75)$$

$$Ku = \overline{w^4} / \sigma_w^{\ 4} . \qquad (7.76)$$

Chiba (1978) summarizes a few observations of Sk in the surface layer and plots them against z/L. There is a great deal of scatter, but it appears that Sk averages about 0.3 in unstable conditions and -0.2 in stable conditions. Observations and models of the convective PBL indicate that updrafts occupy 40% of the area. Based on these concept, Weil (1981) develops a diffusion model in which the convective PBL is assumed to have only two types of vertical motions: $w = 0.6 \text{ m s}^{-1}$ for 40% of the time, and $w = -0.4 \text{ m s}^{-1}$ for 60% of the time. This dual distribution yields Sk = 0.43 and Ku = 1.27. (Note that for a Gaussian distribution Sk = 0 and Ku = 3.)

Lenschow (1970) has published a probability distribution function for w as measured by an aircraft at the 100 m level. The probability distribution function is plotted in Figure 7.8, showing the typical daytime skewness towards upward turbulent fluctuations. For these data, Sk = 0.63 and Ku = 3.17. The dashed line on the figure is a theoretical probability distribution function, determined by methods suggested by Hahn and Shapiro (1967, pp. 198-218). The predicted curve is a Johnson type B distribution, given by the formula

$$p(w) = \frac{1.26}{\sqrt{2\pi}} \frac{6.5}{(w+2.5)\ (4-w)} \exp[-\tfrac{1}{2}(0.74 + 1.26\,\ell n(\frac{w + 2.5}{4 - w}))^2] . \qquad (7.77)$$

Of course, this curve is valid for only one turbulence experiment

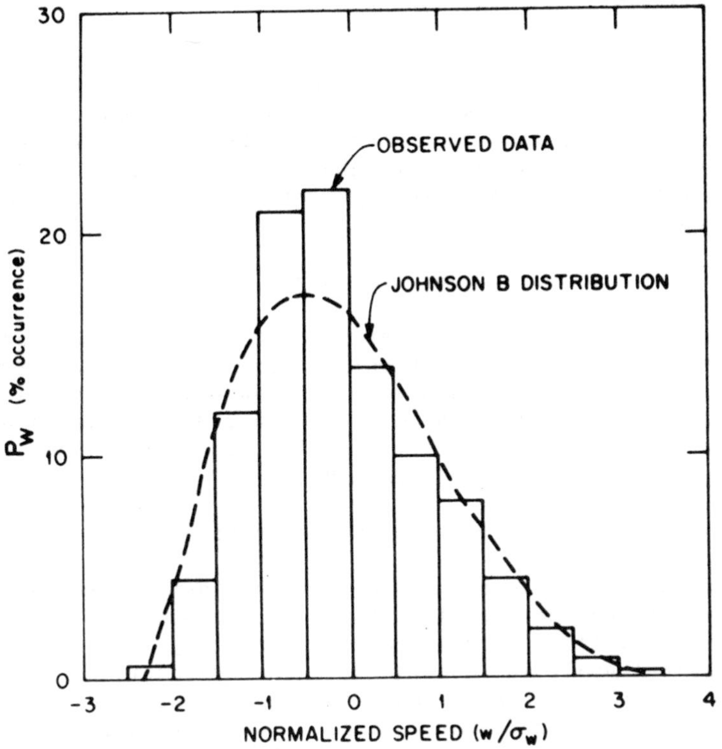

Figure 7.8. Observed and simulated probability density functions for w as measured by an aircraft at the 100m level. The Johnson B-function is given by Equation (7.77).

at 100 m, and it is necessary to develop curves for other stability and heights.

Theoretical estimates of skewness $\overline{w^3}$ are possible from the equations of motion after turbulent fluctuations are introduced and Reynolds averaging criteria applied. The terms $\overline{w^3}$ can be thought of as the vertical flux of vertical turbulent energy ($\overline{w^3} = \overline{w\ w^2}$). However, this problem is used as a classic example of why flux-gradient concept do not apply to all situations (Wyngaard, 1973), as shown in Chapter 3, Equation (3.79). Other terms in the turbulent energy equation must be considered in order to estimate $\overline{w\ w^2}$. Chiba (1978) uses Wyngaard's (1973) analysis of the turbulent energy equation to produce the following formula for the variation of Sk in the surface layer

$$Sk = \frac{-0.77z/L}{(1 - 15z/L)^{-1/4} - 1.8z/L} + 0.1 \ . \qquad (7.78)$$

As stated above, the observations are widely scattered about this curve.

Modelers produce estimate of the flux of total turbulent energy, $e^2 = \frac{1}{2}(u^2 + v^2 + w^2)$, in the convective PBL. For example, André et al. (1978) find that $\overline{w\, e^2}/w_*^3$ increases nearly linearly with height near the surface, reaches a maximum of about 0.08 at z = 0.5h and then decreases towards zero at the mixing height h. If we assume that $u \simeq v \simeq w$, then it can be concluded that

$$\overline{w^3}/w_*^3 = \overline{w\, w^2}/w_*^3 \simeq 0.06 \quad \text{at } z = 0.5h. \qquad (7.79)$$

From the above review it is apparent that skewness is important for diffusion models, but that only a few estimates of skewness are available in the literature. Diffusion modelers should encourage boundary layer modelers and experimentalists to publish estimates of skewness whenever possible, so that the function $Sk(z/z_o, z/L, z/h)$ can be determined.

Natural Variability of Pollutant Concentrations

There is much interest in defining performance measures and performance standards for diffusion models (Fox, 1981). Surprisingly little work has been done in the proper application of statistical analysis techniques to the evaluation of models. Even fewer people recognize that an observed pollutant concentration, averaged over a time, T_a, is a turbulent variable. Venkatram (1979) points out that a model, on the other hand, predicts an ensemble-mean concentration, which is not expected to agree with a single observed concentration even if the model were 'perfect'. For comparison with the model prediction, an ensemble mean of observed concentrations should be calculated. To generate an ensemble mean, observed concentrations should be counted only if they are characterized by similar meteorological (wind speed and direction, stability) conditions and source emissions. The ensemble mean and standard deviation of the members of this set can then be calculated from the set of observations taken under similar conditions. The degree of turbulence or stochastic variability of observed concentrations should be calculated for several data sets in order to determine the natural variability of the data, which determines how well models can be expected to perform.

A long-running field experiment is required for the analysis of natural variability, so that as large a number of observations as possible can be included in each set of data with similar meteorological and source conditions. Hanna (1981b) chose the hourly St. Louis RAPS data for CO and SO_2, since data were available from 25 stations for all hours of 1976. Also, some preliminary evaluations of models had been performed (Ruff, 1980).

To eliminate the effects of variable emissions, the ratio C/Q was used, where Q is total emissions of CO or SO_2 for that particular hour over the entire region. In order to assure that 10 to 100 hours of data would be available for each calculation of natural variability, the data at each station were broken down into the following classes:

18 wind direction classes	0°-20°,20°-40°...320°-340°,340°-360°,
10 wind speed classes	0-1ms^{-1}, 1-2ms^{-1}...8-9ms^{-1}, > 9ms^{-1},
7 stability classes	Turner classes 1 through 7.

Because the hourly concentration observations are nearly log-normally distributed, the variable $\ell n(C/Q)$ was analyzed rather than C/Q. A standard deviation, $\sigma(\ell nC/Q)$, of 0.69 can be interpreted as 'factor of two' variability in C/Q.

The standard deviation, $\sigma(\ell nC/Q)$, was calculated for each joint class containing ten or more hourly concentration observations. It was found that $\sigma(\ell nC/Q)$ for a given pollutant was nearly independent of monitoring station, wind direction, wind speed, and stability. An example of one way of combining the data is given in Figure 7.9. In this figure there is little systematic variation of $\sigma(\ell nC/Q)$. We find that the average variability is given by the equations $\overline{\sigma(\ell nC/Q)} = 0.67$ for CO and $\overline{\sigma(\ell nC/Q)} = 0.96$ for SO_2. These σ's imply factor of 1.9 variability for CO and factor of 2.6 variability for SO_2. It can be concluded that the natural variability of hourly average CO and SO_2 concentration observations in St. Louis for given meteorological and source conditions, is typically a factor of two. This variability contributes to poor correlations between predicted and observed hourly SO_2 concentrations found by Ruff (1980) in St. Louis. Instrument problems and model inaccuracies cause further errors between model predictions and observations.

Representativeness of Wind-Speed Observations

Most diffusion models discussed in this report are based on the idea that meteorological parameters are spatially homogeneous. Furthermore, all models assume that an observation of a meteorological parameter at a certain nearby monitoring station is valid for calculating diffusion over the region. The representativeness of meteorological data is important in regulatory applications. For example, the USEPA requires that five years of on-site meteorological data be used in model applications. Rather than installing and operating an expensive meteorological tower, it is often satisfactory to use data from a

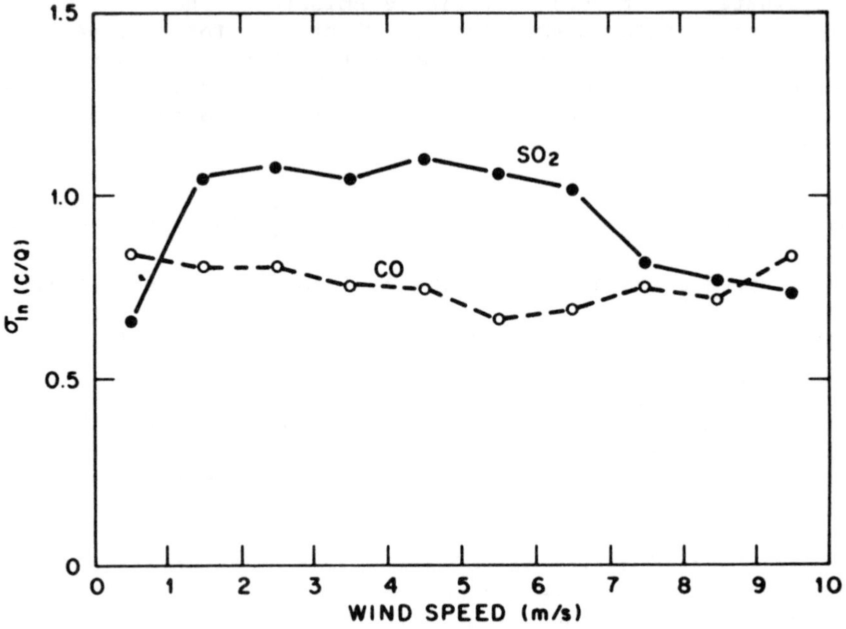

Figure 7.9. Variation of spatial $\sigma(\ell nC/Q)$ for given meteorological conditions for hourly SO_2 and CO data from St. Louis. Values of σ have been averaged over 25 stations and seven stability classes. When $\sigma(\ell nC/Q)=0.69$, there is factor of two variability (Hanna, 1981b)

nearby existing tower, provided that the data from this tower are representative of those that would be obtained from the new site.

A measure of representativeness is the variability of wind speed observations over a regional network of meteorological stations. There are not many networks with high quality data available. The St. Louis RAPS data, studied in the previous section, can be used here also. Lockhart and Irwin (1980) calculate the standard deviation of the difference between hourly observations of wind speed, U, and wind direction, θ, at paired stations 'a' and 'b' for a year's worth of data and plot the resulting $\sigma(\theta_a - \theta_b)$ and $\sigma(U_a - U_b)$ as a function of the distance, x, between the stations. They obtain the following formulas from these data

$$\sigma(\theta_a - \theta_b) = 15 + 5.7 \ \ell n \ x \ , \tag{7.80}$$

$$\sigma(U_a - U_b) = 0.47 + 0.24 \ \ell n \ x \ , \tag{7.81}$$

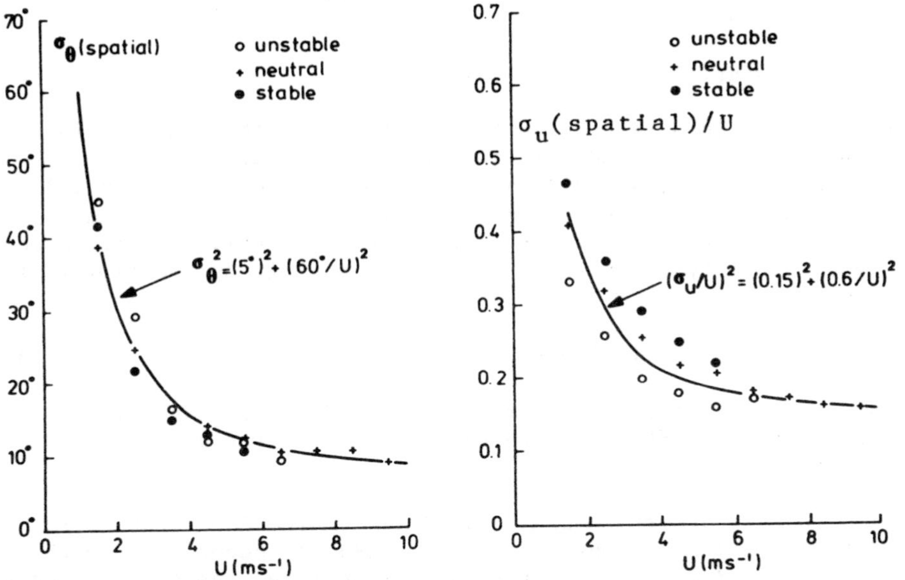

Figure 7.10. Variation of spatial σ_θ and σ_u/U with wind speed and stability for hourly St. Louis data. The spatial standard deviations are calculated from concurrent observations at 25 monitoring stations. Empirical Equations (7.82) and (7.83) are also drawn in the figures.

where θ is in degrees, U is in m s^{-1}, and x is in kilometers. Distances x range from 3 to 80 km for the St. Louis data. The inital 'constants' in Equations (7.80) and (7.81) are partially related to instrument inaccuracies, which are highly site dependent. In St. Louis the instruments were of good quality and calibrated frequently, but in most regional studies the instruments and the quality control are less perfect.

Intuition suggests that $\sigma(\theta_a - \theta_b)$ and $\sigma(U_a - U_b)$ should also depend on wind speed, since higher wind speeds are usually associated with more uniform flow. C.J. Nappo has been looking at this question by calculating the spatial standard deviation for each hour of the 25 wind direction observations, σ_θ (spatial), and 25 wind speed observations σ_u (spatial). The calculated σ_θ and σ_u values for the year's worth of data are then collected into common wind speed classes and stability classes. The resulting σ_θ and σ_u/U values as a function of stability and wind speed are plotted in Figure 7.10. The parameters σ_u/U is plotted because it is the relative wind-speed difference. It is seen that σ_θ (spatial) is not dependent on stability but is a strong function of wind speed, varying from 60° to 10° as wind speeds increase from 0.5 to 10 m s^{-1}. An empirical formula that fits these data is

$$\sigma_\theta^2 \text{ (spatial)} = (5^\circ)^2 + (60^\circ/U)^2 \quad (7.82)$$

where the 5° is a base value that is probably the contribution due to instrument error and U is in m s^{-1}. The relative speed difference σ_u(spatial)/U is a slight function of stability and a strong function of wind speed in Figure 7.10b, varying from 0.5 to 0.15 as wind speeds increase from 1.5 to 10 m s^{-1}. Neglecting the slight stability dependence, these data are fit by the empirical formula

$$(\sigma_u/U)^2 = (0.15)^2 + (0.6/U)^2 \quad (7.83)$$

where the first term on the right hand side is probably the relative speed difference due to instrument error and U is in m s^{-1}.

Other data sets should be used to combine Equations (7.80) and (7.82) and Equations (7.81) and (7.83) to account for both wind speed and station separation. Furthermore, representativeness is also a function of terrain roughness, with greater variability over rough terrain than over smooth terrain.

ACKNOWLEDGEMENT

This work was performed under an agreement among the National Oceanic and Atmospheric Administration, the U.S. Department of Energy and the Nuclear Regulatory Commission.

8. REPORT FROM THE PANEL DISCUSSION

L. Kristensen

Risø National Laboratory, Physics Department

4000 Roskilde, Denmark

The course, from which this book originates, ended with a panel discussion. The purpose was to facilitate a free exchange of ideas and opinions on subjects, relevant within the context of the course, between the participants and the lecturers. The outcome of this discussion may be valuable to a wider audience. Therefore, it was thought to be useful to include in this book an impression of the questions, comments and remarks expressed during this panel discussion.

The panel consisted of J.A. Businger (JAB), S.J. Caughey (SJC), S.R. Hanna (SRH), J.C.R. Hunt (JCRH) and J.C. Wyngaard (JCW). L. Kristensen acted as the chairman. (This report has not been reviewed and approved by the participants of the panel.)

As mentioned in the preface, the first part of the book (Chapters 1-4) is mainly concerned with the description of the atmospheric boundary layer, whereas the second part (Chapters 5-7) contains discussions of current modeling of atmospheric transport and dispersion of pollutants. The understanding of the entire boundary layer has now reached the point where it should be considered to improve the pollution models in current use.

The first general question that must be asked is then: can the adoption of the concepts and the parameters derived from modern boundary-layer scaling improve predictions of air pollution concentrations? The answer is definitely affirmative as illustrated by the following comments.

The 'golden rule' by Pasquill, Gifford and Turner should be replaced.

Firstly, this rule is not quite consistent with surface layer stability classification or, in other words, there is no unique relation between the Pasquill stability classes A - G and the Obukhov length L. Since the latter quantity is so well established experimentally as scaling parameter, it seems more rational to use it rather than the 'golden rule'. Secondly, L can be determined, with reasonable accuracy, from the same observations as those that determine the stability class, i.e. from the cloud cover, solar elevation, and the wind speed. Finally, the Obukhov length describes the stability in a continuous and consequently in a much more specific way, which can easily be used in a climatological sense.

One potential problem in the application of boundary-layer parameters was revealed by the question:

It was mentioned in Dr. Caughey's lecture that it is possible to have convective conditions in the boundary layer if this is capped by a complete stratus cover, because of radiative cooling of the top of the cloud deck. If this is so, how should the relevant boundary-layer parameters be defined and measured?

SCJ reaffirmed that he had encountered cases, on two similar occasions, of turbulent exchange being initiated by a radiative cooling of the cloud top of about 80-90 W/m^2. He was unable to indicate how often this occurred and therefore to assess its importance. JAB commented that it seemed important to know the boundary-layer structure beneath stratus clouds because of their ubiquity. One question that arises is what heat flux should enter L. Measured at the ground, the heat flux may be zero, thus indicating a neutral boundary layer. Yet the conditions must be convective because of the cooling aloft. Could the boundary layer perhaps be considered upside down - so to speak - and the relevant parameters derived from the heat flux at the top, or is the scaling of a completely different nature? JCRH speculated that, according to Dr. Caughey's observations, one should see negative skewness of the vertical velocity distribution, rather than the usual positive skewness. One consequence of that would be that the non-buoyant plumes discussed in Chapter 5 would be strongly rising instead of descending.

Apart from these reservations concerning the availability of the boundary-layer parameters the recommendation to use them seems obvious in the context of the Gaussian plume model, which is the most commonly used model for regulatory purposes; other and more rational methods to derive the σ_y and σ_z are now available. However, it should be noted that the adoption of the concepts of the boundary-layer scaling implies that the Gaussian plume model in the form it is used for elevated sources is basically wrong, since it does not take into account the turning of the wind with height.

The key comment from the audience was:

Anyone trying to compare experimental results with model predictions is going to be faced with the fact that the wind turns with height under all atmospheric conditions - convective and non-convective - and unless they model this turning, they will find that their measurements beyond a few kilometers from the source do not compare at all with the model.

To this JCRH replied that he agreed and that regulatory models do not allow for this directional shear. Perhaps it is not surprising that σ_y seems to be growing linearly with distance from the source out to 100 km or more. The wind turning will make σ_y grow with some larger power of the travel time t than in the case of no directional shear where the horizontal plume growth is proportional to t close to the source and to $t^{\frac{1}{2}}$ at far distances. One rudimentary way to take the turning into account is to let σ_y keep on growing proportional to t. SRH believed that wind turning alone was not responsible for the rapid growth of σ_y, but that also large horizontal eddies took part in the dispersion at large distances. The real problem about wind turning with height is that the plume direction cannot be defined easily and predictions about location and magnitude of maximum concentration are therefore very difficult. The two parameters σ_y and σ_z are not sufficient to characterize the plume geometrically. JAB saw a need for inclusion in the diffusion model descriptions of baroclinic effects as well as stability, so that the wind shear variation with height is properly accounted for. This of course requires more observations, but there is no reason to believe that the wind turning cannot be taken into account in the model.

JCRH felt that it is important to recognize the purpose of the prediction to decide whether to include wind shear in the model. Some people are thinking in terms of research field-experiments, some are thinking of specific calculations for a range of meteorological conditions, and others are thinking about rule books.

Perhaps one may conclude that air pollution models can be improved by use of modern boundary-layer scaling, but that it is difficult to make a general statement about the details that should be included.

Stating that air pollution modelling can benefit from the progress in boundary-layer modelling is not the same as claiming that all problems with the dispersion models in use today can now be solved. Indeed, one of the most fundamental problems, modeling of diffusion in connection with flow over irregular terrain, cannot be approached in terms of present boundary-layer concepts. It must be emphasized that boundary-layer scaling assumes that the atmosphere is horizontally homogeneous, which is in general not the case on scales larger than 100 km and in particular not in mountainous terrain.

In this context a question was asked, slightly facetiously:

How should we use boundary-layer scaling in Norway?

The point was further emphasized by the comment, 'It is important to distinguish between flow and diffusion. If you study dispersion in a system of valleys, the question is where the flow goes. Put up instrumented masts by all means, but if you put up more than one, you are certain to be confused'.

JCRH found that this was a very good point. If the terrain is really complicated there is no need to worry about turbulence because the transport is dominated by the flow. Unfortunately the situation is most often that one has to take into account both the flow and the diffusion, because the terrain features are sufficiently complicated with houses, hills, and so on, and yet not so complicated that turbulent diffusion can be ignored. In strongly convective situations the plume behaviour as a function of t quite rapidly becomes independent of the source height, and boundary-layer scaling will evidently be an excellent frame of work. But in Europe, and particularly in the north, strongly convective conditions are not very common. Further, the terrain is such that the atmospheric motion is not horizontally homogeneous. The rate of mixing of the plume from an elevated source even at rather large distances depends on the turbulence intensity at the height of the plume. The vertical variation of the turbulence is extremely sensitive to local changes of roughness, to topography and to proximity of the sea. Under all these circumstances it can be concluded that boundary-layer scaling will not perform well in predicting maximum ground concentration.

Even if the terrain is uniform there are situations in which boundary-layer scaling is of limited use. JAB agreed with a comment from the audience that often mesoscale phenomena such as helical rolls influence and may even dominate the transport of material. These rolls are frequently observed and fairly well understood. There are often large scale convection systems that occur under light wind conditions. For example, Benard cells have been observed from satellite pictures, and such systems will certainly provide important transport mechanisms. In this context JCW discussed the role of what he called anomalous features of mesoscale atmospheric motion. In the past few years very complex, organized mesoscale convective systems have been observed which, at least over the United States, can persist for a day or two. These systems have horizontal scales of a few hundred kilometers and they have shown up on recent satellite photographs as very deep organized convection covering entire states. Clearly they have a strong impact on regional scale atmospheric motion.

In conclusion, one should expect the boundary-layer scaling to work well in air pollution modelling only under convective conditions and on horizontal scales less than about 100 km. However, in order to apply it to particular studies of for example the pollution from a power plant stack it is necessary to incorporate information about the pollutants, since they are in general extrinsic to the atmosphere. For example, if the plume has a density different from the ambient air at the source, its mean height will change downstream. For a hot plume this is known as plume rise.

Discussions between the audience and the panel about plume rise took place at a couple of occasions. One question was:

Are present methods of calculating effective source heights satisfactory?

SRH believed that they are, in particular under stable conditions. Under neutral and unstable conditions they predict the initial rise well. There are some special, and sometimes very difficult problems which arise if there are strong wind gradients or strange looking temperature gradients, since profile assumptions have to be made in the plume rise formula. JCRH added that there is a real problem with the prediction of plume rise in the vicinity of a hill in otherwise level terrain. The extent to which the plume continues to rise, following the streamlines over the hill, depends on the wind speed.

To the question:

There are on the market now over a hundred plume rise formulas. The United States and Canada use Briggs' while other countries use different formulas. Which set of formulas should one use?

SRH answered that he might be biased, because he had worked so long with Briggs. Nevertheless, Briggs' formulas seemed to provide better - or at least as good - agreement with observations as is the case with other formulas. SRH agreed however that there are very few observations of the final plume rise. Recently there has been some success in determining this quantity by means of an airborne LIDAR.

There is an interesting feature of the Russian plume-rise model in that it predicts plume rises that are more than an order of magnitude less than those predicted by all other models. Briggs believes that the parameters in the Russian formula have been determined from observations which were not far enough downstream. When the Russians use their formulas to estimate maximum concentration, they get nearly the same result as others. This is partly because their formula for maximum concentration is very similar to other formulas, but also because they use compensating, empirical factors. JCRH mentioned that he and Briggs together had looked at some Russian data and found that the rate of rise agreed very well with Briggs' formulas. JCRH continued by pointing out that one of the most sensitive elements in final plume rise formulas is the connection between the external turbulence and the mixing within the plume, and that there are many hypotheses to establish that connection, (e.g. that of Briggs',which equates the rates of dissipation of kinetic energy inside and outside the plume). It should now be possible by means of remote sensing to measure the local turbulence near plumes and look in greater detail at the basis for these final plume rise formulas.

Still in the context of plume rise it was asked:

Is it possible to determine whether a very hot buoyant plume will penetrate through an elevated inversion?

SRH answered that Briggs' handbook covers that problem. Whether or not the plume will make it through the inversion depends on the excess temperature ΔT_p of the plume with respect to the ambient air and the strength of the inversion ΔT_i. If ΔT_p is greater than ΔT_i Briggs' rule is that the plume will penetrate. If ΔT_p is less than ΔT_i the plume will only go partially into the inversion layer. Admittedly, this rule has not been tested on a lot of data. JCRH commented that laboratory experiments show clearly that in the last case some of the plume will go through the inversion and the rest will be left behind, so that there will be a double structure. This point seems not to have been brought out by Briggs.

Even though plume rise is very important, it is not the only extra ingredient, and it is often necessary to include other assumptions about the interactions between the pollutant and the environment. An example is the way in which the different removal processes are accounted for. There are chemical and photochemical processes that change the composition, wet removal processes that can either wash it out or move it to a completely different place by means of the so-called rain-out process. A very important removal process is dry deposition, because it specifies directly the boundary surface conditions for the flux of the pollutant.

SRH agreed and added that this is one of the largest areas of research activity. There are great difficulties in estimating deposition velocities V_d for example of sulphate. It has not been possible to resolve conclusively the question whether V_d is positive or negative over forests during the night. There is also an argument about what happens over water surfaces. It is very important to determine these deposition velocities, in particular in connection with long range transport of sulphates. JCRH added that very careful laboratory experiments are beginning to give some insight into how snow and tiny rain drops actually wash out particles. Eventually, i.e. in a few years, this will lead to a better understanding and a parameterization of these fundamental processes on the micro-hydrodynamical level.

The discussion so far seemed to indicate a lot of uncertainty in the determination of pollution concentration. This was accentuated by a specific remark from the audience:

From a practical point of view the uncertainty of calculations is very important. There is a natural variability in measurements and on top of that the inaccuracies of the model. People who don't know very much about modeling at all make decisions based on model results without having this in mind. It seems that model results should be presented as magnitudes of concentrations plus error specifications or error bands, where the last quantities are just as important as the first.

There was general agreement on that and SRH had some additional comments. In the back of Pasquill's book there are estimates which indicate that models are accurate typically within a factor of two. However, recent evaluations of standard EPA models have revealed that the agreement is rather within a factor of three to four or even more in determinations of the hourly average concentration at a particular point and time. But most of the regulatory decisions are made on the basis of maximum concentrations rather than on the average concentrations hour by hour. In fact, EPA is now taking the stand that the models are good if they are able to predict the maximum reasonably well, but without specifying space or time. For example, the observed maximum might occur in July and the predicted maximum might be in January. The decision makers don't care as long as they are close.

So there is the problem of model accuracy and the problem of translating this to decision making. In SRH's laboratory it is found very difficult to talk to the lawyers and politicians in a scientific way, and when the scientists start talking about including the uncertainties in the regulations, they don't want to hear about it.

The general situation about decision making is also illustrated in the reply to the following question about observational practice:

In Yugoslavia the weather is often in a state which is classified as calm, i.e. the wind speed is so low that the anemometers don't indicate any movement of the air. How should the ground concentration of pollution be estimated in these situations which occur in about 30% of the time?

SRH refered to some investigations made by the Nuclear Regulatory Commission. According to these zero wind speed never occurs in the atmosphere. However, anemometers need a starting speed of about 0.5 m/s. The U.S. Regulatory approach during calms is to take the wind speed as 0.5 m/s and as the direction the one that occurred just before the calm. Then the Gaussian plume model is used with these values as parameters. At the Savannah River Laboratory a different convention is used. The method, which could be called 'the constipated puff model', is to pretend that there is no emission during calms. When the wind starts blowing again the accumulated material is released instantaneously. JAB found that this was an alarming state of affairs and that whenever a calm was reported there should be an alert to stop all air pollution. To this SRH replied that if standard instruments record calm 30% of the time it is not practical to have alarm conditions for that length of time.

Observational problems occupied a significant part of the discussion. As was pointed out, boundary-layer parameters should be included in diffusion modelling, at least in the convective case. The question is whether that is possible from a practical point of view. Certainly, the Obukhov length L and the friction velocity u_* can be estimated by careful interpretation of, for example, routine observations in airports. The depth h of the boundary layer and the complete profiles of wind and temperature are parameters that are not easily accessible. They can only be estimated very crudely from ground-based measurements. On several occasions a great deal of confidence was expressed in the performance of modern remote sensing technique, in particular in acoustic sounding by use of a SODAR. This instrument utilizes the fact that atmospheric velocity- and temperature fluctuations scatter sound waves. The issue was initiated by the question: What are the perceived difficulties of using acoustic Doppler techniques to measure the wind speed, direction and turbulence profiles in the atmosphere from 100 m to 600 m above the ground and are these difficulties preventing widespread use of the method?

SJC answered that the most serious problem is the strong signal reduction with height. In the monostatic case the echo strength is determined by the temperature structure parameter C_T^2 which in convective conditions decreases rapidly with height. This makes it difficult to detect a sensible Doppler shift through the noise in the data, and so measurements of the vertical velocity deteriorate quite rapidly with height. It is possible to avoid this to a large extent by using a bistatic configuration, in which case the velocity structure parameter C_V^2 becomes significant. Using this technique SJC had made good measurements of turbulent dissipation rates in convective conditions at heights up to a few hundred meters. However, it should be realized that rather long base lines are required, and generally it can be said that it is difficult to extend the capabilities of acoustic systems to reach higher than a few hundred meters. SJC commented that very useful information about the surface layer may be obtained with what can be called mini-SODAR'S, i.e. SODAR'S with diameters of about 30 cm and working at higher frequencies, typically 5 kHZ. Recent work at the University College of London shows that one can determine the vertical velocity spectrum extending down to very small scales. Perhaps one should aim to develop these mini-SODAR'S rather than the more powerful units designed for vertical sounding up to several hundred meters.

In apparent contradiction to this last discussion a comment came for the audience that a commercially available bistatic system actually performed very well up to altitudes of more than 500 m. It was reported that such a system had operated virtually unattended for six weeks at a remote site in Canada for two periods, in January and in June of this year. From the recordings it was possible to determine statistically how often different altitudes were reached. In 80% of the time, the instrument reached 500 m for mean wind and direction. It never failed to reach 350 m. The system can also be used to measure the variances of velocity components up to altitudes that are about 60% of those for which the mean velocity components can be measured. To operate the instrument it is necessary to have someone to supervise it. On the other hand, its maintenance does not require a sophisticated machine shop nor skilled electronics people to keep it running. In a few years it will probably be possible to measure temperature profiles by a method developed by E.H. Brown. It is based on the principle called interference fringes and seems to be capable of determining the temperatures with an accuracy of better than 1 °C. From both the panel and the audience there was a general appreciation of the future importance of SODAR's, but a comment was made that SODAR's might work better at remote sites than in industrialized areas and airports, where the back ground noise will often destroy the measurements.

It thus seems that there are future possibilities for making routine remote measurements of the important boundary-layer parameters without establishing a dense network of tall meteorological towers. Nevertheless, it should be kept in mind that even in situations where advanced and detailed measurements are available, as for example in the Minnesota field experiment, it is not always obvious how to define such an important parameter as the depth h of the boundary layer.

The usefulness of the somewhat arbitrary definition of h in the stable cases of the Minnesota experiment was questioned in this way from the audience:

<u>In his evaluation of the Minnesota data Dr. Caughey specified the stable boundary-layer height as the level where the heat flux had decreased to an arbitrary chosen fraction of the surface-level flux. Why has this been done in preference to other specifications (e.g. Deardorff's profile reflection points)?</u>

SJC's answer was that the method seemed to be the most objective, because the heat flux profiles were well defined. It was possible to get values of h lying in a resonable range. The choice was confirmed by the general behavior of the turbulence field and also by the shape of the spectra. Without remembering all the profiles of mean quantities SJC could say that there was no obvious relation between these profiles and the depth of the turbulent layer and so it would be quite uncertain to base the determination of h on them. Accepting this answer the following comment was added to the question:

<u>The method is not a generally applicable approach. Stressing it will lead people to believe that with a diagnostic equation, using only surface data, one can make reliable estimates of h, and this is not true.</u>

In reply to a question from the audience about what kind of experiment should be encouraged in the future JCRH suggested that more detailed tests of the model assumptions should be done. In the modelling of plumes and the environmental turbulence extremely complicated and detailed local assumptions are often made. For example, in second order modeling people make very complex assumptions about fourth order cumulants or about pressure-velocity correlations. After a lot of computation they end up with predictions about mean quantities. If these gross parameters then are in reasonable agreement with measurements they conclude that their assumptions were right. Of course this is a very insensitive test of the assumptions and instead one should use field data to test the original assumptions directly. If it is possible actually to measure, say, Lagrangian correlations and velocity fields by remote sensing technique, then they should be published as such and not, as is customary suppressed in statements about gross parameters such as σ_y and σ_z.

REFERENCES

André, J.C., F. DeMoor, Pl. Lacerure, G. Therry and R. DuVachat, 1978: Modeling the 24-hour evolution of the mean and turbulent structures of the planetary boundary layer. J. Atmos. Sci., 35, 1861-1883.

Angell, J.K., 1974: Lagrangian - Eulerian time-scale relationship estimated from constant volume balloon flights past a tall tower. Advances in Geophysics, vol. 18A, Academic Press, N.Y., 419-432.

Arya, S.P.S., 1975: Comments on similarity theory for the planetary boundary layer of time-dependence height. J. Atmos. Sci., 32, 839-840.

Arya, S.P.S. and J.C. Wyngaard, 1975: Effect of baroclinicity on wind profiles and the geostrophic drag law for the convective planetary boundary layer. J. Atmos. Sci., 32, 767-778.

Asimakopoulos, D.N., R.S. Cole, S.J. Caughey, S.H. Moss and C.J. Readings, 1975: A comparison between acoustic radar returns and the direct measurement of the temperature structure of the atmosphere. Atmos. Environ., 9, 775-776.

Axford, D.N., 1971: Spectral analysis of an aircraft observation of gravity waves. Quart. J. Roy. Meteor. Soc., 97, 313-321.

Ball, F.K., 1960: Control of inversion height by surface heating. Quart. J. Roy. Meteor. Soc., 86, 483-494.

Barad, M.L. and D.A. Haugen, 1959. A preliminary evaluation of Sutton's hypothesis for diffusion from a continuous point source. J. Appl. Meteor., 16, 12-20.

Batchelor, G.K., 1950: Application of the similarity theory of turbulence to atmospheric diffusion. Quart. J. Roy. Meteor. Soc., 76, 133-146.

Batchelor, G.K., 1952: The relative motion of particles. Proc. Cambr. Phil. Soc., 48, 345-362.

Batchelor, G.K., 1960: The Theory of Homogeneous Turbulence. Cambridge University Press, London, 195 pp.

Batchelor, G.K., 1967: An Introduction to Fluid Mechanics. Cambridge University Press, 615 pp.

Berkowicz,R. and L.P. Prahm, 1980: On the spectral turbulent diffusivity theory for homogeneous turbulence. J. Fluid Mech., 100, 433-448.

Berlyand, M.Y., 1975: Contemporary problems of Atmospheric Diffusion and Pollution of the Atmosphere. Gidrometeozdat, Leningrad, 444 pp., translated into English by NERC, US EPA, Raleigh, N.C., U.S.A.

Blackadar, A.K. and H. Tennekes, 1968: Asymptotic similarity in neutral barotropic planetary boundary layers. J. Atmos. Sci., 25, 1015-1020.

Bradley, E.F., 1980: An experimental study of the profiles of wind speed, shearing stress and turbulence at the crest of a large hill. Quart. J. Roy. Meteor. Soc., 106, 101-124.

Bradshaw, P., 1969: Comments on "On the relation between the shear stress and the velocity profile after change in surface roughness". J. Atmos. Sci., 26, 1353-1354.

Briggs, G.A., 1973: Diffusion estimation for small emissions, in environmental research laboratories, air resources atmosphere turbulence and diffusion laboratory. 1973 Annual Report of the USAEC, Report ATDL-106, National Oceanic and Atmospheric Administration.

Briggs, G.A., 1975: Plume rise predictions, in: Lectures on Air Pollution and Environmental Impact Analyses, Workshop Proceedings, Boston, Mass., 45 Beacon Street, Boston, M.A. 02108, 59-111.

Briggs, G.A., 1979: Plume rise and buoyancy effects, Chapter for Atmospheric Science and Power Production (D. Randerson, ed.), to be published by U.S. Department of Energy.

Briggs, G.A., 1980: What convective scaling does for diffusion data. Paper presented at a workshop held in Toronto, Canada.

Briggs, G.A., 1982: Similarity forms for ground source surface layer diffusion. Boundary-Layer Meteorol., 23, 489-502.

Briggs, G.A., and K.R. McDonald, 1978: Prairie Grass revisited: Optimum indicators of vertical spread. Proc. 9th-NATO-CCMS Int. Tech. Symp. on Air Pollution Modeling and its Applications, Toronto, NATO Report No. 103, 209-220.

Brook, Robert R., 1978: The influence of water vapor fluctuations on turbulent fluxes. Boundary-Layer Meteorol., 15, 481-487.

Brost, R. and J.C. Wyngaard, 1978: A model study of the stably-stratified planetary boundary layer. J. Atmos. Sci., 35, 1427-1440.

Boussinesq, J., 1877: Essai sur la théorie des eaux courantes. Mem Savants Etrange, Paris, 23, 46.

Busch, N.E., 1969: Waves and turbulence. Radio Sci., 4, 1377-1379.

Busch, N.E., 1973: The surface boundary-layer. Boundary-Layer Meteorol., 4, 213-240.

Busch, N.E., 1973: On the mechanics of atmospheric turbulence, Workshop on Micrometeorology (D.A. Haugen, ed.), Am. Met. Soc., Boston, Mass., 1-28.

Businger, J.A., 1973: Turbulent transfer in the atmospheric surface layer. Workshop in Micro-meteorology, Am. Meteorol. Soc., 45 Beacon Street, Boston, Mass., 67-100.
Businger, J.A. and J.W. Deardorff, 1968: On the distinction between "total" heat flux and eddy heat flux. J. Atmos. Sci., 25, 521-522.
Businger, J.A., J.C. Wyngaard, Y. Izumi and E.F. Bradley, 1971: Fluxprofile relationships in the atmospheric surface layer. J. Atmos. Sci., 28, 181-189.
Businger, J.A. and S.P.S. Arya, 1974: Height of the Mixed Layer in the Stably Stratified Planetary Boundary Layer. Advances in Geophysics, 18A, Academic Press, N.Y., 73-92.
Carson, D.J., 1973: The development of a dry inversion capped convectively unstable boundary layer. Quart. J. Roy. Meteor. Soc. 99, 450-467.
Carson, D.J. and F.B. Smith, 1974: Thermodynamic model for the development of a convectively unstable boundary layer. Advances in Geophysics, 18A, Academic Press, N.Y., 111-124.
Caughey, S.J., 1977: Boundary layer turbulence spectra in stable conditions. Boundary-Layer Meteorol., 11, 3-14.
Caughey, S.J. and R. Rayment, 1974: High frequency temperature fluctuations in the atmospheric boundary layer. Boundary-Layer Meteorol., 5, 489-503.
Caughey, S.J. and C.J. Readings, 1975: An observation of waves and turbulence in the earth's boundary layer. Boundary-Layer Meteorol., 9, 279-296.
Caughey, S.J. and J.C. Kaimal, 1977: Vertical heat flux in the convective boundary layer. Quart. J. Roy. Meteor. Soc., 103, 811-815.
Caughey, S.J. and S.G. Palmer, 1979: Some aspects of turbulence structure though the depth of the convective boundary layer. Quart. J. Roy. Meteor. Soc., 105, 811-827.
Caughey, S.J. and J.C. Wyngaard, 1979: The turbulence kinetic energy budget in convective conditions. Quart. J. Roy. Meteor. Soc., 105,231-239.
Caughey, S.J., J.C. Wyngaard and J.C. Kaimal, 1979: Turbulence in the evolving stable boundary layer. J. Atmos. Sci., 6, 1041-1052.
Caughey S.J., B.A. Crease, R.S. Cole, D.N. Asimakopoulos and T.J. Moulsley, 1980: Quatitative interpretation of acoustic echoes from the planetary boundary layer. Radio and Electronic Engineer, 5, 585-597.
Caughey, S.J. B.A. Crease and W.T. Roach, 1982: A field study of nocturnal stratocumulus: II Turbulence structure and entrainment. Quart. J. Roy. Meteor. Soc., 108, 125-144.
Chandrasekhar, S., 1961: Hydrodynamic and Hydromagnetic Stability. Clarendon Press, Oxford.
Chatwin, P.C., 1968: The dispersion of a puff of passive contaminant in the constant stress region. Quart. J. Roy. Meteor. Soc., 94, 350-360.

Chatwin, P.C. and P.J. Sullivan, 1979: The relative diffusion of a cloud of passive contaminant in incompressible turbulent flow. J. Fluid Mech., 91, 337-356.
Chaudhry, F.H. and R.N. Meroney, 1973: Similarity theory of diffusion and the observed vertical spread in the diabatic surface layer. Boundary-Layer Meteorol., 3, 405-415.
Chiba, O., 1978: Stability dependence of the vertical velocity skewness in the atmospheric surface layer. J. Meteorol. Soc. Japan, 56, 140-142.
Chorley, L.G., S.J. Caughey and C.J. Readings, 1975: The development of the atmospheric boundary layer: three case studies. Met. Mag., 104, 349-360.
Chou, P.Y., 1945: On velocity correlations and the solutions of the equations of turbulent fluctuation. Quart. J. Appl. Math., 3, 38-54.
Clarke, R.H. and G.D. Hess, 1974: Geostrophic departure and the functions A and B of Rossby-number similarity theory. Boundary-Layer Meteorol., 7, 267-287.
Coantic, M.F., 1975: An introduction to turbulence in geophysics and air-sea interactions. Univ. of Cal., San Diego, Dept. of Applied Mech. and Eng. Sci.
Cole, J.D., 1968: Perturbation Methods in Applied Mathematics. Blaisdell Publishing Co., Waltham, Mass., 260 pp.
Cole R.S., D.N. Asimakopoulos, T.J. Moulsley, S.J. Caughey and B.A. Crease, 1980: Some aspects of the construction and use of atmospheric acoustic sounders. Radio and Electronic Engineer, 5, 585-597.
Corrsin, S., 1952: Heat transfer in isotropic turbulence. J. Applied Phys., 23, 113-118.
Corrsin, S., 1963: Estimates of the relations between Eulerian and Lagrangian scales in large Reynolds number turbulence. J. Atmos. Sci., 20, 115-119.
Corrsin, S., 1974: Limitations of gradient transport models in random walks and in turbulence. Advances in Geophysics, Vol. 18A, Academic Press, N.Y., 25-60.
Coulman, C.E., 1978a: Boundary-layer evolution and nocturnal inversion dispersal. Boundary-Layer Meteorol., 14, 471-513.
Coulman, C.E., 1978b: Convection in stratiform cloud. J. Rech. Atmos., 12, 21-33.
Cramer, H.E., 1957: A practical method for estimating the dispersal of atmospheric contaminants, in Proceedings of the First National Conference on Applied Meteorology, Section C., American Meteorological Society, Hartford, Conn., 33-35.
Crawford, T.L. and J.H. Coleman, 1979: Plume rise study at Gallatin and Allen steam plants. TVA/ONR-79/07. Tennessee Vallay Authority, Muscle Shoals, Alabama 35660.
Crease, B.A., S.J. Caughey and D.T. Tribble, 1977: Information on the thermal structure of the atmospheric boundary layer from acoustic sounding. Met. Mag., 106, 42-52.

Csanady, G.T., 1964: Turbulent diffusion in a stratified flow. J. Atmos. Sci., 7, 439-447.

Csanady, G.T., 1973: Turbulent Diffusion in the Environment. D. Reidel, Dordrecht.

Deardorff, J.W., 1966: The counter-gradient heat flux in the lower atmosphere and in the laboratory, J. Atmos. Sci., 23, 503-506.

Deardorff, J.W., 1970a: A three-dimensional numerical study of turbulent channel flow at large Reynolds numbers. J. Fluid Mech., 41, part 2, 453-480.

Deardorff, J.W., 1970b: Preliminary results from numerical integrations of the unstable boundary layer. J. Atmos. Sci., 27, 1209-1211.

Deardorff, J.W., 1970c: Convective velocity and temperature scales for the unstable planetary boundary layer and for Rayleigh convection. J. Atmos. Sci., 27, 1211-1213.

Deardorff, J.W., 1972a: Numerical investigation of neutral and unstable planetary boundary layers. J. Atmos. Sci., 29, 91-115.

Deardorff, J.W., 1972b: Theoretical expression for the countergradient vertical heat flux. J. Geophys. Res., 77, 5900-5904.

Deardorff, J.W., 1973a: Three dimensional numerical study of turbulence in an entraining mixed layer. Boundary-Layer Meteorol., 1, 169-196.

Deardorff, J.W., 1973b: An explanation of anomalously large Reynolds stresses within the convective planetary boundary layer. J. Atmos. Sci., 30, 1070-1076.

Deardorff, J.W., 1973c: Three-dimensional numerical modeling of the planetary boundary layer. Ch. 7, Workshop on Micrometeorology, Amer. Meteor. Soc., Boston, Mass., D.A. Haugen (ed).

Deardorff, J.W., 1974a: Three-dimensional numerical study of the height and mean structure of a heated planetary boundary-layer. Boundary-Layer Meteorol., 7, 81-106.

Deardorff, J.W., 1974b: Three-dimensional numerical study of turbulence in an entraining mixed layer. Boundary-Layer Meteorol., 7, 199-226.

Deardorff, J.W., 1980: Stratocumulus - capped mixed layers derived from a three-dimensional model. Boundary-Layer Meteorol., 18, 495-527.

Deardorff, J.W., G.E. Willis and D.K. Lilly, 1969: Laboratory investigation of nonsteady penetrative convection. J. Fluid Mech., 35, 7-31.

Deardorff, J.W. and G.E. Willis, 1974: Computer and laboratory modeling of nonbuoyant particles in the mixed layer. In Advances in Geophysics, vol. 18. Academic Press, New York, N.Y.

Deardorff, J.W. and G.E. Willis, 1975: A parameterization of diffusion into the mixed layer. J. Appl. Meteor., 14, 1451-1458.

Deardorff, J.W. and G.E. Willis, 1982: Ground-level concentrations due to fumigation into an entraining mixed layer. Atmos. Environ., 16, 1159-1170.

Donaldson, C. duP, 1973: Construction of a dynamic model of the production of atmospheric turbulence and the dispersal of atmospheric pollutants. In Workshop on Micrometeorology, D.A. Haugen (ed)., American Meteorological Society publication, 392 pp.

Doran, J.C., T. Horst and P. Nickola, 1978a: Variations in measured values of lateral diffusion parameters. J. Appl. Meteor., 17, 825-831.

Doran, J.C., T.W. Horst and P.W. Nickola, 1978b: Experimental observation of the dependence of lateral and vertical characteristics on source height. Atms. Environ., 12, 2259.

Draxler, R.R., 1976: Determination of atmospheric diffusion parameters. Atmos. Environ., 10, 99-105.

Durbin, P.A. and J.C.R. Hunt, 1980: Dispersion from elevated sources in turbulent boundary layers. J. Mécanique, 19, 679.

Dutton, J.A. and G.H. Fichtl, 1969: Approximate equations of motion for gases and liquids. J. Atmos. Sci., 26, 241-254.

El Tahry, S., A.D. Gosman and B.E. Launder, 1981: The two- and three-dimensional dispersal of a passive scalar in a turbulent boundary layer. Int. J. Heat & Mass Transfer, 24, 35-46.

Ekman, V.W., 1905: On the influence of the earth's rotation on ocean currents. Arkiv. Math. Astron. O. Fysik, 2, 11.

Eymard, L., 1978: Ondes de gravite dans la couche limite planetaire, etude experimentale par sondage acoustique. Technical Note No. 54, Centre National d'etudes des Telecommunications France.

Fairall, C.W., R. Markson, G.E. Sacher and K.L. Davison, 1980: An aircraft study of turbulence dissipation rate and temperature structure function in the unstable marine boundary layer. Boundary-Layer Meteorol., 19, 453-469.

Favre, A., 1976: Equations fondamentales des fluides â mass voluminique variable en ecoulements turbulent. Part 2 in Mecanique Statistique des Fluides Turbulents, Dunod, Paris.

Ferziger, J.H., U.B. Mehta and W.C. Reynolds, 1977: Large eddy simulation of homogeneous isotropic turbulence. Proceedings, Symposium on Turbulent Shear Flows, April 1977, University Park, Pa.

Fleagle, R.G. and J.A. Businger, 1980: An Introduction to Atmospheric Physics, 2nd Edn., Academic Press, New York, 432 pp.

Fox, D.G.. 1981: Judging air quality performance. Bull. Am. Met. Soc., 62, 599-610.

Frangi, J.P., 1979: L'etude de la structure verticale des principals caracteristiques turbulent de la couch limite planetaire. Ph.D. Thesis, University of Paris.

Frank, W.M. and G.D. Emmit, 1981: Computation of vertical total energy fluxes in a moist atmosphere. Boundary-Layer Meteorol., 21, 223-230.

Frisch, A.S., R.B. Chadwick, W.R. Moninger and J.M. Young, 1975: Observation of boundary-layer convection cells measured by dual-Doppler radar and echosounder and by microbarograph array. Boundary-Layer Meteorol., 3, 199-226.

Frisch, A.S. and S.F. Clifford, 1974: A study of convection capped by a stable layer using Doppler radar and acoustic echo sounders. J. Appl. Meteor., 31, 1622-1628.

Garratt, J.R., 1978: Flux profile relationships above tall vegetation. Quart. J. Roy. Meteor. Soc., 104, 199-212.

Gibson, M.M. and B.E. Launder, 1978: Ground effects on pressure fluctuations in the atmospheric boundary layer. J. Fluid Mech., 86, 491-511.

Gifford, F.A., 1955: A simultaneous Lagrangian-Eulerian turbulence experiment, Mon. Wea. Rev., 83, 293-301.

Gifford, F.A., 1961: Use of routine meteorological observations for estimating atmospheric dispersion. Nuc. Safety, 2, 47-51.

Gifford, F.A., 1962: Diffusion in a diabatic surface layer. J. Geophys. Res., 67, 3207-3212.

Gifford, F.A., 1981: Horizontal diffusion in the atmosphere. A Lagrangian dynamical theory. LA-8667-MS, Los Alamos Scient. Lab., P.O. Box 1663, Los Alamos, N.M. 87545, 19 pp.

Golder. D., 1972: Relations among stability parameters in the surface layer. Boundary-Layer Meteorol., 3, 47-58.

Guillemet, B., S. Jouveneaux, P. Mascart and H. Isaka, 1978: Normalisation des monents du deuxieme ordre de la vitesse verticale de l'air et de la temperature dans la conche limite convective, J. Rech. Atmos., 12, 229-243.

Hahn, G.J. and S.S. Shapiro, 1967: Statistical Models in Engineering. John Wiley and Sons, New York, 355 pp.

Hanna, S.R., 1968: A method of estimating vertical eddy transport in the planetary boundary layer using characteristics of the vertical velocity spectrum. J. Atmos. Sci., 25, 1026-1032.

Hanna, S.R., 1978: Some statistics of Lagrangian and Eulerian wind fluctuations. J. Appl. Meteor., 18, 518-525.

Hanna, S.R., 1980: Effects of release height on σ_y and σ_z in daytime conditions. Proceedings, Eleventh NATO-CCMS Internat. Tech. Meet. on Air Poll. Modeling and its Applications, Science Policy Programming, Prime Minister's Office, Wetenschapsstraat 8, Rue de la Science, 1040 Brussels, Belgium, 198-215.

Hanna, S.R., 1981a: Lagrangian and Eulerian time-scale relations in the day-time boundary layer. J. Appl. Meteor., 20, 242-249.

Hanna, S.R., 1981b: Natural variability of observed hourly SO_2 and CO concentrations in St. Louis. Sumbitted to Atmos. Environ., ATDL Rept No. 81/6, ATDL, P.O. Box E, Oak Ridge, Tenn., 37830, 21 pp.

Hanna, S.R., G.A. Briggs, J. Deardorff, B.A. Egan, F.A. Gifford and F. Pasquill, 1977: AMS Workshop on Stability Classification Schemes and Sigma Curves - Summary of Recommendations. Bull. Amer. Meteor. Soc., 58, 1305-1309.

Hardy, K.R. and H. Ottersten, 1969: Radar investigation of convective patterns in the clear atmosphere. J. Atmos. Sci., 26, 666-672.

Haugen, D.A., J.C. Kaimal, C.J. Readings and R. Rayment, 1975: A comparison of balloon - borne and tower-mounted instrumentation for probing the atmospheric boundary-layer. J. Appl. Meteor., 14, 540-545.

Hay, J.S. and F. Pasquill, 1959: Diffusion from a continuous source in relation to the spectrum and scale of turbulence. Advances in Geophysics, Vol. 6, Academic Press, N.Y., 345-365.

Heffter, J.L., A.D. Taylor and G.J. Ferber, 1975: A regional-continental scale transport, diffusion and deposition model. NOAA Tech. Memo, ERL-ARL-50, 28 pp.

Herring, J.R., 1979: Subgrid scale modeling -- An introduction and overview. Turb. Shear Flows I. Springer-Verlag, Berlin, 347-352.

Hicks, B.B., 1978: Some limitations of dimensional analysis and power laws. Boundary-Layer Meteorol., 14, 567-569.

Hilst, G.R. and C.I. Simpson, 1958: Observations of vertical diffusion rates in stable atmosphere. J. Met, 15, 125-126.

Hinze, J.O., 1959: Turbulence. McGraw-Hill Book Co., New York, 586 pp.

Horst, T.W., 1979: Lagrangian similarity modeling of vertical diffusion from a ground level source. J. Appl. Meteor., 18, 733-740.

Hunt, J.C.R., 1980: Wind over hills. Survey paper for AMS Workshop, Boulder Colorado, August 1978. Am. Met Soc. (1980), 107-157.

Hunt, J.C.R., 1981: Turbulent stratified flow over low hills. Symposium on "Designing with the wind" Centre Scient. Technique du Batiment, Nantes, France, June 1981.

Hunt, J.C.R. and P.J. Mulhearn, 1973: Turbulent dispersion from sources near two-dimensional obtacles. J. Fluid Mech., 61, 245.

Hunt, J.C.R., J.S. Puttock and W.H. Snyder, 1979: Turbulent diffusion from a point source in stratified and neutral flows aroun a three-dimension hill. Part 1. Diffusion equation analysis. Atmos. Env., 13, 1227-1239.

Hunt, J.C.R. and A.H. Weber, 1979: A Lagrangian statistical analysis of diffusion from a groundlevel source in a turbulent boundary layer. Quart. J. Roy. Meteor. Soc., 105, 423-443.

Hunt, J.C.R., R.E. Britter, and J.S. Puttock, 1979: Mathematical models of dispersion of air pollution around building and hills. Proc. I.M.A. Symp. Maths. Models of Turbulent Diffusion in the Environment. Academic Press, N.Y.

Hunt, J.C.R., J.C. Kaimal, J.E. Gaynor and A. Korell, 1983: Observations of stable layers at the Boulder Atmospheric Observatory. Studies of Nocturnal stable layers at BAO (J.C. Kaimal, ed.), NOAA/ERL, Boulder, U.S.A.

Iribarne, J.V. and W.L. Godson, 1973: Atmospheric Thermodynamics. D. Reidel Publ. Co., Boston, 222 pp.

Irwin, J.S., 1979a: Estimating plume dispersion - a recommended generalized scheme. Preprints Fourth Synmposium on Turbulence, Diffusion and Air Pollution. Am. Meteorol. Soc., 45 Beacon Street, Boston Mass. 02108, 62-69.

Irwin, J.S., 1979b: A theoretical variation of the wind profile power law exponent as a function of surface roughness and stability. Atmos. Environ., 13, 191-194.

Irwin, J.S., 1983: Estimating plume dispersion - A comparioson of several sigma schemes, J Clim. Appl. Meteor., 22, 92-114.

Ivanov, V.N., A.Ye. Ordanovich and L.I. Petrova, 1973: Certain properties of wind velocity and air temperature in the low-frequency range in the presence of convection. Iz. Atmos. Ocean. Phys., 9, 445-452.

Izumi, Y. and S.J. Caughey, 1976: Minnesota 1973 atmospheric boundary layer experiment data report. AFGL Environmental Research Paper no. 547.

Jackson, P.S. and J.C.R. Hunt, 1975: Turbulent wind flow over a hill. Quart. J. Roy. Meteor. Soc., 101, 929-956.

Kaimal, J.C., 1973: Turbulence spectra, length scales and structure parameters in the stable surface layer. Boundary-Layer Meteorol., 4, 289-309.

Kaimal, J.C., D.A. Haugen and J.T. Newman, 1966: A computer controlled mobile micrometeorological observation system. J. Appl. Meteor., 5, 411-421.

Kaimal, J.C. and D.A. Haugen, 1967: Characteristics of vertical velocity fluctuations observed on a 430 m tower. Quart. J. Roy. Meteor. Soc., 93, 305-317.

Kaimal, J.C. and J.A. Businger, 1970: Case studies of a convective plume and a dust devil. J. Appl. Meteor., 9, 612-620.

Kaimal, J.C., J.C. Wyngaard, Y. Izumi and O.R. Coté, 1972: Spectral characteristics of surface layer turbulence. Quart. J. Roy. Meteor. Soc., 98, 563-589.

Kaimal, J.C., J.C. Wyngaard, D.A. Haugen, O.R. Coté, Y. Izumi, S.J. Caughey and C.J. Readings, 1976: Turbulence structure in the convective boundary layer. J. Atmos. Sci., 33, 2152-2169.

Kofoed-Hansen, O., 1962: On the interpretation of smoke diffusion and wind analysis data at Risø. J. Geophys. Res., 67, 3217.

Kolmogorov, A.N., 1941: The local structure of turbulence in compressible turbulence for very large Reynolds numbers. Compt. Rend. Akad. Nauk SSSR, 30, 301-305.

Konrad, T.G., 1970: The dynamics of a convective process in the clear air as seen by radar. J. Atmos. Sci., 27, 1138-1147.

Krishna, K., 1980: The planetary-boundary layer model of Ellison (1956) - A retrospect. Boundary-Layer Meteorol., 19, 293-301.

Kukharets, V.P., 1974: Spectra of the vertical wind velocity components in the atmospheric boundary layer. Izo. Atmos. Ocean Phys., 10, 375-378.

Kukharets, V.P. and L.R. Tsvang, 1979: Maximum in the vertical profile of the temperature field structural characteristics. Akad. Nauk. Doklady, 248, 832-835.

Lamb, R.G., 1975: The calculation of long-term atmospheric pollutant concentration statistics using the concept of a macroturbulence. Seminar on Air Pollution Modeling held in Venice, Italy. IBM Italy Scientific Center, Report No. 48.

Lamb, R.G., 1978: A numerical simulation of dispersion from an elevated point source in the convective planetary boundary layer. Atmos. Environ. 12, 1297-1304.

Lamb, R.G., 1979: The effects of release height on material dispersion in the convective planetary boundary layer. Preprint vol. AMS Fourth Symposium on Turbulence, Diffusion and Air Pollution, Reno, N.V.

Lamb, R.G., 1980: Mathematical principles of turbulent diffusion modeling. In Atmospheric Planetary Boundary Layer Physics, A. Longhetto, (ed.), Elsevier Scientific Publishing Co.

Lamb, R.G., 1981a: A scheme for simulating particles pair motions in turbulent fluid. J. Comp. Physics, 39, 329-346.

Lamb, R.G., 1981b: A numerical investigation of tetroon versus fluid particle dispersion in the convective planetary boundary layer. J. Appl. Meteor., 20, 391-403.

Lamb, R.G. and D.R. Durran, 1978: Eddy diffusivity derived from a numerical model of the convective planetary boundary layer. Il Nuovo Cimento, Cl, 1-17.

Landau, L.D. and E.M. Lifshitz, 1959: Fluid Mechanics. Pergamon Press, London, 536 pp.

Launder, B.E., 1976: Heat and Mass Transport. Ch. 6, Turbulence, P. Bradshaw, (ed.), Springer-Verlag, Berlin.

Launder, B.E., G.J. Reece and W. Rodi, 1975: Progress in the development of a Reynolds-stress turbulence closure. J. Fluid Mech., 68, 537-566.

Lauwerier, H.A., 1954: Diffusion from a source in a skew velocity field. Appl. Sci. Res., 4, 153.

Lemone, M.A., 1973: The structure and dynamics of horizontal roll vortices in the planetary boundary layer. J. Atmos. Sci., 30, 1077-1091.

Lenschow, D.H., 1970: Airplane measurements of planetary boundary layer structure, J. Appl. Meteor., 9, 874-884.

Lenschow, D.H., 1974: Model of the height variation of the turbulene kinetic energy budget in the unstable planetary boundary layer. J. Atmos. Sci., 31, 99-102.

Lenschow, D.H., J.C. Wyngaard and W.T. Pennell, 1980: Mean-field and second-moment budgets in a baroclinic convective boundary layer. J. Atmos. Sci., 37, 1313-1326.

Lenschow, D.H. and P.L. Stephens, 1980: The role of thermals in the convective boundary layer. Boundary-Layer Meteorol., 19, 509-532.

Lenschow, D.H. and P.L. Stephens, 1981: Mean vertical velocity and turbulence intensity inside and outside thermals. Atmos. Environ., 16, 761-764.

Leonard, A., 1974: Energy cascade in large-eddy simulations of turbulent flows. Advances in Geophysics, Vol. 18A, Academic Press, N.Y., 237-248.

Lewellen, W.S. 1977: Use of invariant modeling. Ch. 9, Handbook of Turbulence, W. Frost and T.H. Moulden (eds.), Plenum Publ. Corp., N.Y.,

Lewellen, W.S. and M. Teske, 1975: Turbulence modeling and its application to atmospheric diffusions. Part I: Recent program development, verification and application; Part II: Critical review of the use of invariant modeling. EPA-600/4-75-16a,b. Part I, 79 pages; Part II, 50 pages.

Lilly, D.K., 1967: The representation of small-scale turbulence in numerical simulation experiments. Proceedings, IBM Sci. Comput. Symp. Environmental Sci., IBM Form No. 320-1951, 195-210.

Lin, C.C., 1945: Theory of Hydrodynamic Stability. Cambridge University Press.

Lockhart, T.J. and J.S. Irwin, 1980: Methods for calculating the representativeness of data. Proceedings, Symp. on Intermed. Range Atm. Transport Proc. and Tech. Asses., U.S. DOE, Oak Ridge, TN.

Long, P.E. and D.W. Pepper, 1976: A comparison of six numerical schemes for calculating the advection of atmospheric pollution. Proc. Third Symp. on Atm. Turb., Diffusion and Air Quality, 19-22 Oct. 1976, Raleigh, N.C., 181-186.

Lumley, J.L., 1970: Toward a turbulent constitutive relation. J. Fluid Mech., 41, 413-434.

Lumley, J.L., 1975a: Modeling turbulent flux of passive scalar quantities in inhomogeneous flows. Phys. Fluids, 18, 619.

Lumley, J.L., 1975b: Pressure-strain correlation. Phys. Fluids, 18, 750.

Lumley, J.L., 1979: Computational modeling of turbulent flows. Adv. Appl. Mech., 18, 123-176.

Lumley, J.L. and H.A. Panofsky, 1964: The structure of Atmospheric Turbulence. Interscience, N.Y., 239 pp.

Lumley, J.L., O. Zeman and J. Siess, 1978: The influence of buoyancy on turbulent transport. J. Fluid Mech., 84, 581-597.

Mahrt, L. and D.H. Lenschow, 1976: Growth dynamics of the convectively mixed layer. J. Atmos. Sci., 33, 41-51.

Mahrt, L., 1979: Penetrative convection at the top of a growing boundary layer. Quart. J. Roy. Meteor. Soc., 105, 469-486.

Mahrt, L., R.C. Heald, D.H. Lenschow and B.B. Stankov, 1979: An observational study of the nocturnal boundary layer. Boundary-Layer Meteorol., 17, 247-264.

Mansour, N.N., P. Moin, W.C. Reynolds and J.H. Ferziger, 1977: Improved methods for large-eddy simulations of turbulence. Proceedings Symposium on Turbulent Shear Flows, April, 1977, University Park, PA.

Mason, P.J. and R.I. Sykes, 1979: Flow over an isolated hill of moderate slope. Quart. J. Roy. Meteor. Soc., 105, 383-396.

Maxworthy, T., 1972: The structure and stability of vortex rings. J. Fluid Mech., Vol. 51, part 1, 15-32.

McBean, G.A., (ed.), 1979: The planetary boundary layer. WMO Technical note No. 165, 201 pp.

McBean, G.A. and J.A. Elliott, 1975: The vertical transports of kinetic energy by turbulence and pressure in the boundary layer. J. Atmos. Sci., 32, 753-766.

McDougall, T.J., 1979: Measurements of turbulence in a zero mean shear mixed layer. J. Fluid Mech., 94, 409-431.

McNider, R.T., S.R. Hanna and R.A. Pielke, 1980: Sub-grid scale plume dispersion in coarse resolution mesoscale models. Proceedings, Second Joint Conference on Applications of Air Pollution Meteorology, Am. Meteorol. Soc., 45 Beacon Street, Boston, MA. 02108, 424-429.

Melgarejo, J.W. and J.W. Deardorff, 1974: Stability functions for the boundary-layer resistance laws based upon observed boundary-layer heights. J. Atmos. Sci., 31, 1324-1333.

Monin, A.S. and A.M. Obukhov, 1954: Basic laws of turbulent mixing in the atmosphere near the ground. Tr. Akad. Nauk. SSSR Geofiz. Inst., No. 24 (151), 163-187.

Monin, A.S. and A.M. Yaglom, 1971: Statistical Fluid Mechnanics. Vol. I, M.I.T. Press, Cambridge, Mass.

Monin, A.S. and A.M. Yaglom, 1975: Statistical Fluid Mechanics. Volume II, M.I.T. Press, Cambridge, Mass., 847 pp.

Monin, A.S. and S.S. Zilitinkevich, 1974: Similarity theory and resistance laws for the planetary boundary layer. Boundary-Layer Meteorol., 7, 391-397.

Montgomery, R.B., 1948: Vertical eddy flux of heat in the atmosphere. J. Meteorol., 5, 265-274.

Moore, D.J., 1974. A comparison of the trajectories of rising buoyant plumes with theoretical/empirical models. Atmos. Environ., 8, 441-457.

Moulsley, T.J., D.N. Asimakopoulos, R.S. Cole, B.A. Crease and S.J. Caughey, 1981: Measurement of boundary-layer structure parameter profiles by acoustic sounding and comparison with direct measurements. Quart. J. Roy. Meteor. Soc., 107, 203-230.

Neff, W.D., 1975: Quatitative evaluation of acoustic echoes from the planetary boundary-layer. NOAA Tech. Rep. ERL 322-WPL-38. Boulder, Colorado.

Neumann, J., 1978: Some observations on the simple exponential function as a Lagrangian correlation function in turbulent diffusion. Atmos. Environ., 12, 1965-1968.

Nicholls, S., 1978: Measurements of turbulence by an instrumented aircraft in a convective atmospheric boundary layer over the sea. Quart. J. Roy. Meteor. Soc., 104, 653-676.

Nicholls, S. and C.J. Readings, 1979: Aircraft observations of the structure of the lower boundary layer over the sea. Quart. J. Roy. Meteor. Soc., 105, 785-802.

Nickerson, E.C. and V.E. Smiley, 1975: Surface layer and energy budget parameterizations for meso-scale models. Journ. Appl. Met., 14, 297-300.

Nieuwstadt, F.T.M., 1980a: An analytic solution of the time-dependent one-dimensional diffusion equation in the atmospheric boundary layer. Atmos. Environ., 14, 1361-1364.

Nieuwstadt, F.T.M., 1980b: Application of mixed-layer similarity to the observed dispersion from a ground level source. J. Appl. Meteor., 19, 157-162.

Nieuwstadt, F.T.M., 1984: The structure of the stable boundary layer. Submitted to J. Atmos. Sci.

Nieuwstadt, F.T.M. and A.P. van Ulden, 1978: A numerical study on the vertical dispersion of passive contaminants from a continuous source in the atmospheric surface layer. Atmos. Environ., 12, 2119-2124.

Normand, C. and Y. Pomeau, 1977: Convective instability. A physicist's approach. Reviews of Modern Physics, 49, 581-624.

Obukhov, A.M., 1946: Turbulence in an atmosphere with a non-uniform temperature. Trudy Adad. Nauk. USSR, Inst. Teortet. Geofys., No. 1. Translated and published in Boundary-Layer Meteorol., 2, 7-29, 1971.

Ohmstede, W. and E. Stenmark, 1980: A model for characterizing transport and diffusion of air pollution in the battlefield environment. Second Joint Conf. on Applic. of Air Poll. Meteorol., Am. Meteorol. Soc., 45 Beacon Street, Boston Mass., 416-423.

Okamoto, M. and E.K. Webb, 1970: The temperature fluctuations in stable stratification. Quart. J. Roy. Meteor. Soc., 46, 591-600.

Palmer, S.G., S.J. Caughey and K.W. Whyte, 1979: An observational study of entraining convection using balloon-borne turbulence probes and high power Doppler radar. Boundary-Layer Meteorol., 16, 261-278.

Panofsky, H.A., H. Tennekes, D.H. Lenschow and J.C. Wyngaard, 1977: The characteristics of turbulent velocity components in the surface layer under convective conditions. Boundary-Layer Meteorol., 11, 355-361.

Pasquill, F., 1961: The estimation of the dispersion of windborne material. Meteorol. Mag., 90, 33-49.

Pasquill, F., 1974: Atmospheric Diffusion. 2nd Ed. John Wiley and Sons, New York.

Pasquill, F., 1976: Atmospheric dispersion parameters in Gaussian plume modelling. Part II. Possible requirements for change in the Turner Workbook Values. EPA-600/4-76-030b. 44 pp.

Pasquill, F., 1978: Atmospheric dispersion parameters in plume modeling. EPA-600/4-78-021, 58 pp.
Paulson, C.A., 1970: The mathematical representation of wind speed and temperature profiles in the unstable atmospheric surface layer. J. Appl. Meteor., 9, 857-861.
Pearson, H.J. and R.E. Britter, 1980: A statistical model for vertical turbulent diffusion in stably stratified flows. Proc. 2nd Int. Symp. on Stratified Flows. Tapir Press, p. 269.
Pearson, H.J., J.S. Puttock and J.C.R. Hunt, 1983: A statistical model of fluid element motions and vertical diffusion in a homogeneous stratified turbulent flow. Submitted to J. Fluid Mech., 129, 219-249.
Pennel, W.T. and M.A. Le Mone, 1974: An experimental study of turbulence structure in the fair weather, trade wind boundary layer. J. Atmos. Sci., 31, 1308-1323.
Peterson, E.W., 1969: On the relation between the shear stress and the velocity profile after a change in surface roughness. J. Atmos. Sci., 26, 773-774.
Pielke, R.A., 1974: A three-dimensional numerical model of the sea breezes over south Florida. Mon. Wea. Rev., 102, 115-139.
Pielke, R.A. and Y. Mahrer, 1975: Representation of the heated planetary boundary layer in mesoscale models with coarse vertical resolution. J. Atmos. Sci., 32, 2288-2308.
Prandtl, L., 1932: Meteorologische Anwendungen der Strömungslehre, Beitr. Phys. Atmos., 19, 188-202.
Prandtl, L. and O. Tietjens, 1957: Applied Hydro- and Aeromechanics. Dover, New York, 311 pp.
Rao, K.S. and H.F. Snodgrass, 1979: Some parameterizations of the nocturnal boundary layer. Boundary-Layer Meteorol., 17, 15-28.
Rayment, R., 1973: An observational study of the vertical profile of the high frequency fluctuations of the wind in the atmospheric boundary layer. Boundary-Layer Meteorol., 3, 284-300.
Rayment, R. and C.J. Readings, 1974: A case study of the structure and energetics of an inversion. Quart. J. Roy. Meteor. Soc., 100, 221-233.
Rayment, R. and S.J. Caughey, 1977: An investigation of the turbulence balance equations in the atmospheric boundary layer. Boundary-Layer Meteorol., 11, 15-26.
Readings, C.J. and H.E. Butler, 1972: The measurement of atmospheric turbulence from a captive balloon. Met. Mag.,101, 286-298.
Reid, J.D., 1979: Markov chain simulations of vertical dispersion in the neutral surface layer for surface and elevated releases. Boundary-Layer Meteorol., 16, 3-22.
Reynolds, O., 1895: On the dynamical theory of incompressible viscous fluids and the determination of the criterion. Phil. Trans. Roy. Soc., London, 186, 123-164.
Reynolds, W.C., 1976: Computation of turbulent flows. Ann. Rev. Fluid Mech., 8, 183-208.

Reynolds, W.C. and T. Cebeci, 1976: Calculation of turbulent flows. Ch. 5, Turbulence, P. Bradshaw (ed.) Springer-Verlag, Berlin,

Richards, J.M., 1963: Experiments on the motions of isolated cylindrical thermals through unstratified surroundings. Int. J. Air Water Poll., 7, 17-34.

Richardson, L.F., 1920: The supply of energy from and to atmospheric eddies. Proc. Roy. Soc. London, A97, 354-373.

Richter, J.H., D.R. Jensen, V.R. Noonkester, T.G. Konrad, A. Arnold and J.R. Rowland, 1974: Clear air convection: A close look at its evolution and structure. Geophys. Res. Letts., 1, 173-176.

Rivlin, R.S. 1957: The relation between the flow of non-Newtonian fluids and turbulent Newtonian fluids. Quart. Appl. Math., 15, 212-215.

Roberts, O.F.T., 1923: The theoretical scattering of smoke in a turbulent atmosphere. Proc. Roy. Soc. London, A104, 640-654.

Robins, A.G. and J.E. Fackrell, 1979: Continuum plumes, their structure and prediction. Proc. Conf. on Math. Modelling of Turbulent Diffusion in the Environment. Academic Press.

Rodi, W., 1980: Turbulence models for enviromental problems. Prediction Methods for Turbulent Flows. W. Kollmann (ed.), Hemisphere Publ Corp. (McGraw-Hill International, New York), 259-350.

Rotta, J.C., 1951: Statistische theorie nichthomogener turbulenz. Z. Phys., 129, 547.

Rowland, J.R. and A. Arnold, 1975: Vertical velocity structure and geometry of clear air convection elements. Preprints 16th Radar Meteorology Conf., 296-303.

Ruff, R.E., 1980: Evaluation of the RAM using the RAPS Data Base. Final Rept. Cont. No. 68-02-2770, Atmos. Science Center, SRI Int., Menlo Park, CA. 94025, prepared for ORD, US EPA, Research Triangle Park, NC, 83 pp.

Sacré, C., 1979: An experimental study of the airflow over a hill in the atmospheric boundary layer. Boundary-Layer Meteorol., 17, 381-401.

Sedefian, L. and E. Bennett, 1980: A comparison of turbulence classification schemes. Atmos. Environ., 14, 741-750.

Saffman, P.G., 1960: On the effect of the molecular diffusivity in turbulent diffusion. J. Fluid Mech., 8, 273.

Sagendorf, J., 1979: Diffusion under low wind speed and inversion conditions. NOAA Environmental Research Laboratories, Air Resources Laboratory, Technical Memorandum 52.

Schumann, U., 1975: Subgrid scale model for finite difference simulations of turbulent flows in plane channels and annuli. J. Comp. Phys., 18, 376-404.

Schumann, U., 1977: Realizability of Reynolds stress turbulence models. Phys. Fluids, 20, 721-725.

Schumann, U., G. Grotzbach and L. Kleisen, 1980: Direct numerical simulation of turbulence. pp. 123-258 in Precition Methods for Turbulent Flows. W. Kollman (ed.), Hemisphere Pub. Corp. (McGraw-Hill International, New York), W. Kollman, Ed.

Scorer, R.S., 1959: The behavior of chimney plumes. Int. J. Air Poll., 1, 198-220.

Scriven, R.A., 1969: Variability and upper bounds for maximum ground-level concentration. Phil. Trans., 265, 209.

Sherman, C.A., 1978: A mass consistent model for wind fields over complex terrain. J. Appl. Meteor., 17, 312-319.

Singer, I.A. and M.E. Smith, 1966: Atmospheric dispersion at Brookhave National Laboratory. Int. J. of Air and Water Pollution, 10, 125-135.

Smith, F.B., 1972: A scheme for estimating the vertical dispersion of a plume from a source near ground level. In Proceedings of the Third Meeting of the Expert Panel on Air Pollution Modeling, NATO-CCMS Report 14, North Atlantic Treaty Organization, Brussels.

Smith, F.B., 1977: Application of data from field programs to estimation of K profiles and vertical dispersion. TDN No. 86, Meteorol. Office, Boundary Layer Res. Branch, Bracknell, Berkshire, U.K.

Smith, F.B. and R.M. Blackall, 1979: The application of field-experiment data to the parameterisation of the disperion of plumes from ground level and elevated sources. Proc. I.M.A. Symp. Math. Models of Turbulent Diffusion in the Environment. Academic Press.

Smith, M.E., 1951: The forecasting of micrometeorological variables. Meteorol. Monogr., No. 4, 50-55.

Smith R.B., 1980: Linear theory of stratified hydrostatic flow past on isolated mountain. Tellus, 32, 348-364.

Snyder, W.H., R.E. Britter and J.C.R. Hunt, 1980: A fluid modeling study of the flow structure and plume impingement on a three-dimensional hill in a stably stratified flow. Proc. 5th Int. Conf. on Wind. Engineering, Fort Collins, July 1979. Pergamon Press.

Snyder, W.H. and J.L. Lumley, 1971: Some measurements of particle velocity autocorrelation functions in a turbulent flow. J. Fluid Mech., 48, 41-71.

Sreenivasan, K.R., S. Tavoularis and S. Corrsin, 1981: Turbulent transport in passively heated homogeneous and inhomogeneous flows. Proc. 3rd Symp. Turbulent Shear Flows, Davis, California.

Steward, R.W., 1969: Turbulence and waves, in stratified atmosphere. Radio Sci., 4, 1269-1278.

Taylor, G.I., 1921: Diffusion by continuous movements. Proc. London Math. Soc., 20, 196-202.

Taylor, G.I., 1931: Effect of variations in density on the stability of superposed streams of fluids. Proc. Roy. Soc. London, A132, 499-523.

Tennekes, H., 1968: Outline of a second-order theory for turbulent pipe flow. AIAA Journ., 6, 1735-1740.
Tennekes, H., 1970: Free convection in the turbulent Ekman layer of the atmosphere. J. Atmos. Sci., 27, 1027-1034.
Tennekes, H., 1975: Reply to comments on "A model for the dynamics of the inversions above a convective boundary layer". J. Atmos. Sci., 32, 992-995.
Tennekes, H, 1977: Turbulence: Diffusion, Statistics, Spectral Dynamics. Handbook of Turbulence, Vol. 1, Walter Frost and Trevor H. Moulden (eds.), Plenum publishing Corporation.
Tennekes, H., 1979: The exponential Lagrangian correlation function and turbulent diffusion in the inertial subrange. Atmos. Environ., 13, 1565-1567.
Tennekes, H. and J.L. Lumley, 1972: A First Course in Turbulence. M.I.T. Press, Cambridge, Mass. 300 pp.
Thorpe, A.J. and T.H. Guymer, 1977: The nocturnal jet. Quart. J. Roy. Meteor. Soc., 103, 633-654.
Townsend, A.A., 1956: The Structure of Turbulent Shear Flow. Cambridge University Press, Cambridge, Mass.
Townsend, A.A., 1958: The effects of radiative transfer on turbulent flow of a stratified fluid. J. Fluid Mech., 4, 361-375.
Turner, D.B., 1969: Workbook of atmospheric dispersion estimates. U.S.E.P.A., Office of Air Programs, Pub. No. AP-26.
Ulden, A.P. van, 1978: Simple estimated for vertical diffusion from sources near the ground. Atmos. Environ., 12, 2125-2129.
Ulden, A.P. van and A.A.M. Holtslag, 1980: The wind at heights between 10 and 200 m in comparison with the geostrophic wind. Proceedings of the Seminar on Radioactive releases and their dispersion in the Atmosphere, Risø, 22-25 April 1980. Commission of the European Communities, Luxembourg, Vol. I, 83-92.
Van Dyke, M.D., 1964: Perturbation Methods in Fluid Mechanics. Academic Press, New York, 229 pp.
Venkatram, A., 1979: The expected deviation of observed concentrations from predicted ensemble means. Atmos. Environ., 13, 1547-1550.
Venkatram, A., 1980a: Estimating the Monin-Obukhov length in the stable boundary layer for dispersion calculations. Boundary-Layer Meteorol., 19, 481-485.
Venkatram, A., 1980b: Dispersion from an elevated source in a convective boundary layer. Atmos. Environ., 14, 1-10.
Volkovitskaya, Z.I. and V.N. Ivanov, 1970: Turbulent energy dissipation in the atmospheric boundary layer. Izv. Atmos. Ocean. Phys., 5, 249-253.
Wamser, H. and C. Muller, 1977: On the spectral scale of wind fluctuations within and above the surface layer. Quart. J. Roy. Meteor. Soc., 103, 721-730.
Weber, A.H.,1976: Atmospheric dispersion parameters in Gaussian plume modeling. Part I. Review of current system and possible future developments. EPA-600/4-76-030a. 59 pp.

Weil, J.C., 1979: Assessment of plume rise and dispersion models using lidar data. Maryland Power Plant Siting Prog. Rep. PPSP-MP-24. Martin Marietta Environmental Center, Baltimore, MD.

Weil, J.C., 1980: Performance of simple models for stack plume dispersion during convective conditions. Maryland Power Plant Siting Program. Report PPSP-MP-30. Martin Marietta Environmental Center. Baltimore, MD.

Weil, J.C. and W.F. Furth, 1981: A simplified numerical model of dispersion from elevated sources in the convective boundary layer. Preprint volume, Fifth AMS Symposium on Turbulence, Diffusion and Air Pollution, Atlanta, GA.

Wengle, H., 1979: Numerical solution of advection-diffusion problems by collocation methods. Paper presented at the Third GAMM-Conference on Numerical Methods in Fluid Mechanics. Koln-Porz.

Wieringa, J., 1980: A revaluation of the Kansas mast influence on measurements of stress and cup anemometer overspeeding. Boundary-Layer Meteorol., 18, 411-430.

Willis, G.E. and J.W. Deardorff, 1974: A laboratory model of the unstable planetary boundary layer. J. Atmos. Sci., 31, 1297-1307.

Willis, G.E. and J.W. Deardorff, 1976: A laboratory model of diffusion into the convective planetary boundary layer. Quart. J. Roy. Meteor. Soc.,102, 427-445.

Willis, G.E. and J.W. Deardorff, 1978: A laboratory study of dispersion from an elevated source within a modeled convective planetary boundary layer. Atmos. Environ., 12, 1305-1311.

Willis, G.E. and J.W. Deardorff, 1979: Laboratory observations of turbulent penetrative-convection platforms. J. Geo. Res., 84, 295-302.

Willis, G.E. and J.W. Deardorff, 1981: A laboratory study of dispersion from a source in the middle of the convective mixed layer. Atmos. Environ., 15, 109-117.

Wilson, D.J., 1981: Along wind diffusion of source transients. Atmos. Environ., 15, 489.

Wippermann, F., 1973: The planetary boundary layer of the atmosphere. Deutschen Wetterdienst, 346 pp.

Wyngaard, J.C. 1973: On surface layer turbulence. Ch. 3 Workshop on Micrometeorology, D.A. Haugen, (ed.), Amer Meteor. Soc., Boston, Mass.

Wyngaard, J.C., 1975: Modeling the planetary boundary layer -- extension to the stable case. Boundary-Layer Meteorol., 9, 441-460.

Wyngaard, J.C., 1980: The atmospheric boundary layer -- modeling and measurements. pp. 352-365 in Turbulent Shear Flows 2, Springer-Verlag, Berlin.

Wyngaard, J.C., O.R. Coté and Y. Izumi,1971: Local free convection, similarity and the budgets of shear stress and heat flux. J. Atmos. Sci., 28, 1171-1182.

Wyngaard, J.C. and O.R. Coté, 1971: The budgets of turbulent kinetic energy and temperature variance in the atmospheric surface layer. Boundary-Layer Meteorol, 9, 441-460.

Wyngaard, J.C. and O.R. Coté, 1974: The evolution of a convective planetary layer - a higher-order-closure model study. Boundary-Layer Meteorol., 7, 284-308.

Wyngaard, J.C., O.R. Coté and K.S. Rao, 1974: Modelling the atmospheric boundary layer. Advances in Geophysics, 18A, 193-211.

Wyngaard, J.C. and M.A. Lemone, 1980: Behaviour of the refractive index structure parameter in the entraining convective boundary layer. J. Atmos. Sci., 37, 1573-1585.

Yaglom, A.M., 1977: Comments on wind- and temperature flux-profile relationships. Boundary-Layer Meteorol., 11, 89-102.

Yanskey, G.R., E.H. Markee and A.P. Richter, 1966: Climatology of the National Reactor Testing Station. USAEC, 1.

Zeman, O., 1981: Progress in the modeling of planetary boundary layers. Ann. Rev. Fluid Mech., 13, 253-272.

Zeman, O. and J.L. Lumley, 1976: Modeling buoyancy-driven mixed layers. J. Atmos. Sci., 33, 1974-1988.

Zeman, O. and J.L. Lumley, 1979: Buoyancy effects in entraining turbulent boundary layers: A second-order closure study, pp. 295-306 in Turbulent Shear Flows I, Springer-Verlag, Berlin.

Zilitinkevich, S.S. 1972: On the determination of the height of the Ekman boundary layer. Boundary-Layer Meteorol., 3, 141-145.

Zilitinkevich, S.S., 1975: Comments on "A model for the dynamics of the inversion above a convective boundary layer". J. Atmos. Sci., 32, 991-992.

Zilitinkevich, S.S. and D.V. Chalikov, 1968: The laws of resistance and of heat moisture exchange in the interaction between the atmosphere and the underlying surface. Izv. Akad, Nauk. SSR, Fiz. Atmos. Okeana, 4, 765-772.

Zilitinkevich, S.S., and J.W. Deardorff, 1974: Similarity theory for the planetary boundary layer of time-dependent height. J. Atmos. Sci., 31, 1449-1452.

Zubkovskiy, S.L. and B.M. Koprov, 1970: On the turbulent energy balance in the boundary layer of the atmosphere. Izo. Atmos. and Oceanic Phys., 6, 989-995.

[illegible] J.C. and [illegible], 197[illegible]: The budgets of turbulent kinetic energy and temperature variance in the atmospheric surface layer. Boundary-Layer Meteorol., 7, 441-460.

Wyngaard, J.C. and [illegible]: The evolution of a convective [illegible] boundary-layer [illegible] Boundary-Layer Meteorol., [illegible], 284-308.

[illegible] modelling the atmospheric boundary layer. Advances in Geophysics, [illegible]

[illegible] of the refractive index structure parameter in the [illegible] convective [illegible] Atmos. [illegible], [illegible]-1285.

[illegible] temperature flux profile [illegible] Boundary-Layer Meteorol., 11, 89-[illegible]

[illegible] and A.R. [illegible]: The [illegible]

[illegible]

AUTHORS INDEX

André, J.C.	97, 306, 323
Angell, J.K.	108
Arnold, A.	118, 337
Arya, S.P.S.	49, 50, 76, 323, 325
Asimakopoulos, D.N.	142, 323, 325, 326, 334
Axford, D.N.	145, 323
Ball, F.K.	131, 232
Barad, M.L.	186, 323
Batchelor, G.K.	27, 33, 86, 90, 91, 219, 234, 323
Bennett, E.	272, 337
Berkowicz,	244, 323
Berlyand, M.Y.	296, 324
Blackadar, A.K.	45, 324
Blackall, R.M.	262, 271, 338
Boussinesq, J.	17, 324
Bradley, E.F.	108, 324, 325
Bradshaw, P.	84, 324
Briggs, G.A.	185, 186, 206, 207, 209, 222, 223, 285, 289, 290, 292, 293, 324, 329
Britter, R.E.	330, 335, 338
Brook, R.R.	33, 324
Brost, R.	76, 154, 155, 156, 324
Busch, N.E.	20, 113, 114, 193, 253, 255, 324
Businger, J.A.	7, 14, 47, 76, 113, 128, 152, 298, 301, 324, 328, 331
Butler, H.E.	108, 109, 336
Carson, D.J.	124, 125, 130, 325
Caughey, S.J	108, 109, 110, 114, 116, 117, 118, 121, 122, 123, 128, 132, 134, 135, 136, 137, 138, 139, 140, 142, 143, 144, 145, 146, 147, 148, 150, 152, 153, 154, 155, 158, 253, 254, 256, 261, 283, 323, 325, 326, 327, 331, 334, 335, 336
Cebeci, T.	97, 105, 336

Chadwick, R.B. 329
Chalikov, D.V. 50, 341
Chandrasekhar, S. 7, 9, 325
Chatwin, P.C. 235, 264, 267, 268, 325
Chaudhry, F.H. 301, 326
Chiba, O. 286, 304, 305, 326
Chorley, L.G. 109, 110, 125, 326
Chou, P.Y. 77, 326
Clarke, R.H. 50, 326
Clifford, S.F. 114, 130, 329
Coantic, M.F. 9, 326
Cole, J.D. 41, 326, 334
Cole, R.S. 111, 323, 326, 334
Coleman, J.A. 226, 326
Corrsin, S. 104, 243, 246, 326, 338
Coté, O.R. 25, 56, 79, 81, 113, 124, 139, 149, 154, 331, 340
Coulman, C.E. 110, 158, 326
Cramer, H.E. 292, 326
Crawford, T.L. 226, 326
Crease, B.A. 111, 141, 128, 141, 325, 326, 334
Csanady, G.T. 248, 259, 326

Davison, K.L. 328
Deardorff, J.W. 14, 49, 58, 59, 76, 85, 99, 100, 104, 105, 108, 113, 121, 123, 125, 131, 132, 133, 134, 135, 163, 165, 177, 178, 183, 186, 187, 188, 190, 191, 194, 206, 325, 327, 329, 334, 340, 341
DeMoor, F. 323
Donaldson, C. duP 162, 328
Doran, J.C. 266, 267, 270, 279, 280, 291, 328
Draxler, R.R. 291, 328
Durbin, P.A. 262, 271, 328
Durran, D.R. 195, 332
Dutton 29, 328
DuVachat, R. 323

Egan, B.A. 329
Einaudi 256
Ekman, V.W. 33, 75, 328
El Tahry, S. 271, 328
Elliott, J.A. 81, 139, 334
Emmitt, G.D. 33, 328
Eymard, L. 142, 328

Fackrell, J.E. 272, 337
Fairall, C.W. 137, 328
Favre, A. 9, 328
Ferber, G.J. 330

Ferziger, J.H. 105, 328, 333
Fichtl, G.H. 29, 328
Finnegan 256
Fleagle, R.G. 7, 328
Fox 306, 328
Frangi, J.P. 118, 132, 135, 328
Frank, W.M. 33, 328
Frisch, A.S. 114, 329
Furth, W.F. 194, 340

Garrett, J.R. 113, 329
Gibson, M.M. 94, 329
Gifford, F.A. 277, 286, 287, 291, 293, 301, 329
Godson, W.L. 6, 330
Golder, D. 289, 329
Gosman, A.D. 328
Grotzbach, G. 337
Guillemet, B., 132, 135, 136, 329
Guymer, T.H. 139, 339

Hahn, G.J. 304, 329
Hanna, S.R. 136, 278, 280, 284, 292, 299, 306, 308, 329, 334
Hardy, K.R. 118, 119, 330
Haugen, D.A. 114, 133, 186, 323, 330, 331
Hay, J.S. 277, 330
Heald, R.C. 333
Heffter, J.L. 288, 293, 330
Herring, J.R. 105, 330
Hess, G.D. 50, 326
Hicks, B.B. 300, 303, 330
Hilst, G.R. 252, 259, 330
Hinze, J.O. 2, 47, 330
Holtslag, A.A.M. 41, 339
Horst, T.W. 301, 302, 303, 328, 330
Hunt, J.C.R. 108, 232, 244, 245, 262, 265, 267, 271, 272, 301, 328, 330, 331, 336, 338

Iribarne, J.V. 6, 330
Irwin, J.S. 161, 289, 291, 308, 331, 333
Isaka, H. 329
Ivanov, V.N. 114, 118, 331, 339
Izumi, Y. 110, 114, 325, 331, 340

Jackson, P.S. 108, 331
Jensen, D.R. 337
Jouveneaux, S. 329

Kaimal, J.C. 85, 108, 110, 111, 113, 114, 115, 117, 118, 119, 120, 121, 122, 123, 126, 127, 128, 131, 132, 133, 136, 148, 149, 150, 151, 172, 183, 253, 255, 256, 282, 325, 326, 330, 331
Kleisen, L. 337
Kofoed-Hansen, O 256, 259, 331
Kolmogorov, A.N. 69, 331
Konrad, T.G. 119, 331, 337
Koprov, B.M. 138, 341
Krishna, K. 73, 331
Kukharets, V.P. 118, 136, 331

Lacerure, Pl. 323
Lamb, R.G. 166, 174, 175, 176, 179, 181, 182, 187, 189, 194, 195, 197, 200, 201, 205, 220, 280, 332
Landau, L.D. 2, 332
Launder, B.E. 88, 94, 97, 328, 329, 332
Lauwerier, H.A. 266, 332
Le Mone, M.A. 113, 123, 136, 160, 332, 336, 341
Lenschow, D.H. 108, 110, 114, 123, 125, 133, 138, 197, 304, 332, 333, 335
Leonard, A. 98, 100, 101, 102, 333
Lewellen, W.S. 97, 164, 333,
Lifshitz, E.M. 2, 332
Lilly, D.K. 99, 100, 104, 327, 333
Lin, C.C. 9, 333
Lockhart, T.J. 380, 333
Long, P.E. 298, 333

Lumley, J.L. 9, 40, 53, 56, 60, 63, 66, 68, 74, 78, 86, 90, 91, 92, 93, 94, 97, 104, 195, 236, 243, 246, 333, 338, 339, 341

Mahrer, Y. 77, 336
Mahrt, L. 110, 125, 130, 139, 333
Mansour, N.N. 99, 102, 333
Markee, E.H. 341
Markson, R. 328
Mascart, P. 329
Mason, P.J. 108, 333
Maxworthy, T. 218, 334
McBean, G.A. 46, 50, 81, 139, 334
McDonald, K.R. 185, 292, 324
McDougall, T.J. 118, 334
McNider, R.T. 286, 334
Mehta, U.B. 328
Melgarejo, J.W. 43, 334
Meroney, R.N. 301, 326

Moin, P. 333
Monin, A.S. 9, 19, 24, 47, 49, 50, 62, 66, 68, 116, 239, 296, 334
Moninger, W.R. 329
Montgomery, R.B. 334
Moore, D.J. 219, 222, 334
Moss, S.H. 323
Moulsley, T.J. 142, 326, 334
Mulhearn, P.J. 244, 245, 330
Muller, C. 136, 339

Neff, W.D. 110, 125, 334
Neumann, J. 62, 334
Newman, J.T. 331
Nickerson 289, 335
Nicholls, S. 108, 114, 126, 334
Nickola, P.W. 328
Nieuwstadt, F.T.M. 183, 184, 185, 262, 264, 270, 296, 298, 299, 302, 303, 335
Noonkester, V.R. 337
Normand, C. 159, 335

Obukhov, A.M. 24, 47, 334, 335
Ohmstede, W. 286, 335
Okamoto, M. 150, 254, 335
Ordanovich, A.Ye. 331
Ottersten, H 118, 119, 330

Palmer, S.G. 114, 116, 117, 118, 121, 122, 124, 128, 129, 132, 133, 134, 135, 137, 326, 335,
Panofsky, H.A. 9, 56, 133, 145, 178, 282, 333, 335
Pasquill, F. 64, 162, 163, 164, 236, 243, 259, 262, 269, 277, 278, 279, 291, 301, 329, 330, 335
Paulson, C.A. 298, 336
Pearson, H.J. 248, 259, 335, 336
Pennel, W.T. 123, 332, 336
Pepper, D.W. 298, 333
Peterson, E.W. 85, 336
Petrova, L.I. 331
Pielke, R.A. 77, 281, 334, 336
Pomeau, Y. 159, 335
Prahm 244, 323
Prandtl, L. 47, 336
Puttock, J.S. 330, 336

Rao, K.S. 283, 336, 340
Rayment, R. 114, 124, 130, 131, 136, 138, 139, 325, 330, 336

Readings, C.J. 108, 109, 114, 121, 124, 126, 130, 131, 142, 143, 144, 146, 323, 325, 326, 330, 334, 336
Reece, G.J. 332
Reid, J.D. 262, 280, 285, 336
Reynolds, O. 77, 336
Reynolds, W.C. 97, 105, 328, 333, 336
Richards, J.M. 217, 336
Richardson, L.F. 23, 25, 336
Richter, A.P. 341
Richter, J.H. 110, 337
Rivlin, R.S. 2, 337
Roach, W.T. 325
Roberts, O.F.T. 295, 337
Robins, A.G. 272, 337
Rodi, W. 87, 97, 332, 337
Rotta, J.C. 93, 337
Rowland, J.R. 118, 337
Ruff, R.E. 306, 307, 337

Sacher, G.E. 328
Sacré, C. 108, 337
Saffman, P.G. 232, 337
Sagendorf, J. 294, 337
Schumann, U. 94, 102, 105, 337
Scorer, R.S. 213, 337
Scriven, R.A. 271, 337
Sedefian, L. 292, 337
Shapiro, S.S. 304, 329
Sherman, C.A. 281, 337
Siess, J. 333
Smiley 289
Simpson, C.I. 252, 259, 330
Singer, I.A. 161, 338
Smith, M.E. 161, 292, 338
Smith, F.B. 125, 130, 262, 271, 299, 300, 325, 338
Smith, R.B. 272, 338
Snodgrass, H.F. 283, 336
Snyder, W.H. 243, 274, 330, 338
Sreenivasan, K.R. 247, 338
Stankov, B.B. 333
Stenmark, E. 286, 335
Stephens, P.L. 197, 332
Steward, R.W. 143, 338
Sullivan, P.J. 235, 325
Sykes, R.I. 108, 333

Tavoularis, S. 338
Taylor, A.D. 330
Taylor, G.I. 25, 60, 242, 265, 281, 285, 338

Tennekes, H. 9, 40, 45, 47, 53, 58, 60, 62, 63, 66, 68, 74, 78, 86, 104, 125, 243, 324, 335, 338, 339
Teske, M. 164, 333
Therry, G. 323
Thorpe, A.J. 139, 339
Tietjens, O. 47, 336
Townsend, A.A. 77, 104, 339
Tribble, D.T. 326
Tsvang, L.R. 136, 331
Turner, D.B. 280, 339

Ulden, A.P. van 41, 206, 262, 264, 270, 298, 299, 335, 339

Van Dyke, M.D. 41, 339
Venkatram, A. 210, 292, 306, 339
Volkovitskaya, Z.I. 114, 339

Wamser, H. 136, 339
Warhaft, Z. 236, 339
Webb, E.K. 150, 254, 335
Weber, A.H. 162, 265, 267, 330, 339
Weil, J.C. 194, 210, 304, 339, 340
Wengle, H. 164, 340
Whyte, K.W. 335
Wieringa, J. 47, 340
Willis, G.E. 59, 121, 123, 125, 131, 132, 133, 134, 163, 178, 183, 186, 187, 190, 191, 206, 229, 327, 340
Wilson, D.J. 268, 340
Wippermann, F. 75, 340
Wyngaard, J.C. 9, 50, 55, 56, 76, 79, 80, 81, 82, 87, 93, 94, 96, 97, 113, 121, 124, 136, 137, 138, 139, 140, 149, 154, 155, 156, 247, 254, 283, 284, 305, 323, 234, 325, 326, 331, 332, 335, 340, 341

Yaglom, A.M. 9, 19, 48, 50, 51, 62, 66, 68, 116, 239, 296, 334, 341
Yanskey, G.R. 256, 341
Young, J.M. 329

Zeman, O. 93, 97, 333, 341
Zilitinkevich, S.S. 49, 50, 125, 156, 334, 341
Zubkovskiy, S.L.l 138, 341

SUBJECT INDEX

Acoustic sounder	110, 111, 142, 318
Adiabatic lapse rate	5
Albedo	2
Area source	166, 216
Ashchurch experiment	114, 117, 121, 124, 127,
130, 131, 134, 136	
Atmospheric boundary layer (ABL)	1, 29, 107, 163
Average	
ensemble	9, 70, 72, 98, 105, 162, 165, 166, 168, 169, 175, 197, 306
time	9, 197, 306
volume	71, 97, 102, 105, 163
Baroclinicity	125, 313
Boundary-layer height	34, 48, 56, 284
Boundary-layer scaling	312
Boussinesq approximation	13
Brownian motion	195
Brunt-Vaisala frequency	146, 247
Buoyancy	73
destruction	255
flux	207, 221, 290
production	22
Burst phenomena	127, 150
Circulation	215
Closure	17, 72, 78, 87, 162, 166
first-order	16,18, 72, 161, 295
second-order	19, 77, 85, 164, 195, 271
Cloud	
diffusion	263, 276, 297, 287
mean height	239, 263, 266
relation to plume dispersion of	286
shape	268
width	234, 264, 267
Clouds	108, 158, 312

Coherence 145
Complex atmospheric conditions 272, 286
Concentration 8, 16, 166, 235, 271
cross-wind integrated 173, 180, 184, 186, 190, 197, 203, 225, 302
fluctuation 235
instantaneous 233
maximum ground-level 182, 183, 187, 188, 190, 203, 209, 225, 226, 271, 273, 294
mean square 232, 235
moments 233, 263
variability 306, 317
Convective boundary layer (CBL) 58, 110, 163, 169, 176, 191, 206, 282
around sunset 125
height 2, 94,110, 111, 169, 170, 172, 186, 301
cospectrum 121
diffusion 59, 175, 260, 294, 300, 304
energy spectrum 114, 156
heat-flux profile 123
horizontal velocity variance 133, 178, 282
large eddy scale 173
peak wave length 117, 123, 157, 282
plume centerline descent 180, 186, 194, 197, 203, 227
plume centerline rise 164, 178, 180
stress profile 125
temperature scale 58, 113
temperature spectrum 119, 157
third moment of vertical velocity fluctuations 96
turbulent kinetic energy 137
turbulent structure 191
velocity scale 58, 113, 170
vertical velocity variance 131, 178, 192, 196, 282
Convective plumes 111, 118, 156
Coriolis force 5
Coriolis parameter 12, 14
Correlation function 60, 174
Counter-gradient flux 85, 96, 191

Diffusion
in calm conditions 318
coefficient 7, 8, 197, 321
experiment 161, 320
in free convection 161, 184, 302
inertial subrange 64, 219

Diffusion
   mesoscale 315
   near a hill 272, 315
   non-Gaussian 240, 246
   non-homogeneous 240, 247, 262
   spectral approach of 63, 278
   Taylor's statistical theory of 60, 277, 285
   in straining flow 243
   stream wise 186, 188, 190
   unidirectional shear flow 238
   wind shear 192, 243, 266, 268
Diffusion equation 244, 262, 266, 271
   analytical solution 295
   numerical solution 297
Dissipation (rate of) 7, 22, 27, 52, 65, 69, 79, 85, 114, 136, 153, 208, 220, 225, 255, 289, 319
Downdraft 178, 180, 183, 194, 196, 209, 225, 228, 294
Dynamic viscosity 5

Eddy 74, 77
   diffusivity 17, 60, 72, 75, 83, 195, 250, 286, 295, 299
   viscosity 17, 18, 34, 73, 84, 85, 103
Ekman spiral 12, 33, 75, 157
Elevated source(diffusion from) 177, 191, 198, 258, 260, 263, 264, 295, 313
Entrainment 81, 108, 111, 113, 119, 122, 123, 124, 125, 127, 133, 135, 136, 156, 196, 206
Entropy 27
Equation
   dissipation 86
   enthalpy 6, 11, 16, 30
   scalar 8, 11, 16, 164, 236
   continuity 2, 10, 15
   energy 6, 211
   hydrostatic 4
   kinetic energy (turbulent) 22
   of motion 5, 10, 16
   Reynolds stress 19, 77
   scalar covariance 21
   of state 3, 10
   temperature flux 20, 77
   temperature variance 21
   vorticity 211
Ergodic hypothesis 9, 170

Eulerian
   correlation function 67
   diffusion 165
   integral scale 172, 255
   Lagrangian relationship 243, 255, 276, 277, 278
   spectrum 67
   time scale 243, 277, 282
Exponential correlation function 62, 281

Filter 63, 65, 98, 276
Finite differences 165
Fluid element (see particle)
Flux Richardson number 23
   critical 25
FM-CW Radar 110, 118
Forced convection 25, 160
Fourier transform 62, 89, 98
Free convection 25, 55, 113, 159, 183
Free convection scaling 157
   of diffusion 302
   of temperature 113
   of temperature variance 56, 135
   of velocity 113
   of vertical velocity variance 55, 131
Friction velocity 38, 163, 318
Froude number 29
Fumigation 206

Gaussian plume equation 161, 182, 197, 198, 204, 288, 313
   effective source height 205, 288, 315
   sigma parameters 161, 280, 290, 292, 299
   stability class 289, 290, 291, 312
   virtual source height 205
Geostrophic drag coefficient 46, 48
Geostrophic wind 12, 33, 38, 87
Gradient transfer 165, 243, 262, 305
Gradient diffusion approximation 95
Gravity waves 108, 133, 134, 140, 143, 256

Heat flux 31, 33, 83, 252
Horizontal homogeneity 11, 72, 169

Ideal gas 3
Inertial sublayer 45
Inertial subrange 45, 66,97, 114, 116, 119, 120, 121, 144, 148, 150, 157, 174,
Instability (hydrodynamic) 209, 129

Instrumentation	108
Inversion	178
Isotropic tensor modeling	91
Isotropy (local)	23, 79, 86, 89, 90, 116, 174
Jet	222
K-Profiles	75, 76, 85, 263
K-theory (see Closure)	
Kansas experiment	47, 113, 120, 122, 131, 135, 148, 152, 154, 157, 298
Kelvin-Helmholtz instability	130, 142,157, 196
Kinematic viscosity	6, 211
Kinetic energy (turbulent)	22, 53, 80, 87, 138, 196, 255, 305
large eddy	101
subgrid	106
Kolmogorov scale	69, 232
Kurtosis	304
Lagrangian	163, 170
correlation function	61, 243, 277, 281
diffusion	60,61,166
integral time scale	61, 175, 188, 191, 193, 242, 282, 284, 286, 299
Eulerian relationship	243, 255, 276, 277, 278
spectrum	62
Laminar flow	9, 12, 106
Langevin model	186
Large eddy model	97
Large eddy structure	104
Latent heat flux	31, 33
Line source (diffusion from)	163, 186, 262, 266, 271, 295, 299
Line thermal	213, 217
Logarithmic wind profile	44, 46, 54
Mach number	28
Marine boundary layer	108, 113, 126, 133, 137, 139
Matching	43
Meandering	234, 294
Minnesota experiment	110, 112, 113, 117, 121, 124, 126, 127, 131, 134, 136, 137, 139, 150, 156, 157, 282, 299
Mixed-layer (see CBL)	75, 110, 113
Mixed-layer scaling	59, 113, 125, 131, 135, 136, 156, 163, 186, 301

Mixing height (see CBL)
Mixing length 17, 75, 83
Mixing ratio 8, 16
Mode 192, 194, 196, 261
Molecular
   destruction 79, 85, 86
   diffusion 78, 232, 236 247, 250, 252
   term 78
Momentum flux 223
Monin-Obukhov Similarity 43, 50, 55, 94, 112, 118, 301
Monte Carlo model 194, 262, 271, 280, 285, 304

Navier-Stokes equations 5, 99, 168
Neutral atmospheric boundary layer 74, 284
   height 41
   Lagrangian time scale 284
   peak wave length 284
   velocity variance 284

Obukhov length 24, 48, 50, 54, 163, 254, 289, 312, 318

Particle (or fluid element) 60, 166, 232, 234, 260
   buoyant
   displacement 233, 246
   drift 286
   equation of motion 174, 248, 281
   horizontal spread 176, 189
   probability distribution 193
   trajectory 174, 210
   vertical spread 176, 198
Phase angle 145
Penetration (of inversion) 316
Plume
   centerline 180, 186, 288, 290
   convective boundary layer 207, 239
   diffusion 276, 287
   lateral spread of 186, 263
   mean height 239, 261, 263, 266, 299, 301
   relation to cloud diffusion 268
   stable boundary layer 259
   width 234, 240, 259,
Plume rise 164, 207, 209, 289, 290, 315
   1/3 law of 223
   2/3 law. of 209, 221, 222
   effective 209, 290

Plume rise
  in stably-stratified fluid 223
Point source 166, 175, 186, 271
  continuous 180, 191, 235, 238, 261, 277, 287, 288
  instantaneous 176, 233, 238, 287
Potential temperature 4
Prairie Grass experiment 161, 183, 292, 299, 302
Prandtl number 52
Pressure
  covariance 79, 82, 85, 88, 92
  wall effect 94
  fluctuations 30
  transport 81, 94, 139
Probability distribution 170, 208, 232, 238, 244, 261
  Gaussian 192, 281
  lognormal 307
  vertical velocity fluctuations 192, 238, 304

Radiation 7, 77, 312
Relative diffusion (see Cloud diffusion)
Release time 276
Removal processes 317
Reynolds
  number 18, 70, 106
  stress 18, 40, 71, 100
Richardson number 23, 25, 52, 149, 292, 296
Roll vortices 126, 160, 315
Rossby
  number 38, 284
  similarity 39, 41, 47, 57
Rotta parameterization 93
Roughness length 38, 39, 42, 293
  change in 84

Sampling time 276, 278, 280, 288
Scale height 29
Shallow convection 29
Shear production 22
Similarity theory (of diffusion) 161, 206, 300
Skewness 193, 195, 196, 247, 261, 286, 304, 305, 313
Slope effects 157
Small-scale eddy structure 104
Source
  height 195, 197, 209
  strength 166
Spectral gap 9, 146, 148, 149
Spectral window 278

Spectrum 9, 60, 63, 91, 114, 119, 150, 156, 157, 175, 291
St. Louis experiment 306
Stability parameter 49, 57, 292
Stable boundary layer (SBL) 139, 253, 254, 283
   height 139, 156, 283, 320
   cospectra 150
   diffusion 236, 258
   exchange coefficient 258
   horizontal velocity variance 255, 256, 283
   integral scale 255
   low-level jet 139
   peak wave length 151, 152, 157, 283
   structure parameter 154
   surface inversion 140
   temperature profile 254
   temperature variance 258
   vertical velocity variance 254, 283
   wind profile 254
   temperature variance destruction 153
Stationarity 9, 11, 169
Streamline frame of reference 237
Structure function 65
Structure parameter
   temperature 121, 136, 154, 319
   velocity 136, 154, 319
Subgrid scale parameterization 103, 174, 188
Subsidence 160, 196, 228
Surface heat flux 48, 110, 186
Surface inversion 125
Surface layer 26, 38, 42, 46, 50, 79, 82 107
   cospectra 148
   diffusion 260, 298
   exchange coefficient for momentum 51, 299
   exchange coefficient for heat 51, 256, 299
   horizontal velocity variance 133
   K-profiles
   kinetic energy budget 24
   neutral 95
   peak wave length 118
   reduced frequency scale 148
   spectra 148
   stable 52, 56
   temperature scale 50, 254
   temperature variance 55
   unstable 52, 82, 95
   vertical velocity variance 54
   wind profile 298

Surface source (diffusion from) 177, 183, 263, 268, 269, 295, 301
Surface stress 38, 39

Taylor's hypothesis 67, 206
Temperature variance 22, 53, 252
Temperature variance destruction 26, 134, 136
Thermal 111, 118, 127, 128, 129
conductivity 7, 27
diffusivity 11, 211
street 119
wind 48
Time scale 194,243, 277, 282
Topography 108, 232, 272, 286, 314
Townsend number 77
Trajectory 61
Travel distance 206, 240, 268, 278, 286
Travel time 163, 166, 173, 176, 177, 183, 193, 206, 240, 246, 268, 276, 313
Turbulent transport 81, 85, 94, 139

Updraft 180. 183, 191, 194, 196, 209, 225, 294

Virtual temperature 4
Von Karman constant 24, 44, 47, 54, 254
Vortex
ring 218
stretching 214
tube 123
Vorticity 206
variance 86

Wind
direction fluctuations 279, 291, 292
turning 313
speed 307
profile power law 289
tunnel 161